NANOTECHNOLOGY FOR SUSTAINABLE AGRICULTURE

An Innovative and Eco-Friendly Approach

NANOTECHNOLOGY FOR SUSTAINABLE AGRICULTURE

An Innovative and Eco-Friendly Approach

Edited by

Vishnu D. Rajput

Abhishek Singh

Tatiana M. Minkina

Krishan K. Verma

Awani Kumar Singh

AAP | APPLE ACADEMIC PRESS

First edition published 2024

Apple Academic Press Inc.
1265 Goldenrod Circle, NE,
Palm Bay, FL 32905 USA

760 Laurentian Drive, Unit 19,
Burlington, ON L7N 0A4, CANADA

CRC Press
6000 Broken Sound Parkway NW,
Suite 300, Boca Raton, FL 33487-2742 USA

4 Park Square, Milton Park,
Abingdon, Oxon, OX14 4RN UK

Library and Archives Canada Cataloguing in Publication

CIP data on file with Canada Library and Archives

Library of Congress Cataloging-in-Publication Data

CIP data on file with US Library of Congress

ISBN: 978-1-77491-235-5 (hbk)
ISBN: 978-1-77491-236-2 (pbk)
ISBN: 978-1-00333-312-8 (ebk)

About the Editors

Dr. Vishnu D. Rajput

Vishnu D. Rajput, PhD, is Associate Professor (Leading Researcher) at Southern Federal University, Russia. His ongoing research is based on soil contamination, i.e, potentially toxic elements, and metallic nanoparticles, and investigating the bioaccumulation, bio/geo-transformations, uptake, translocation, and toxic effects of metallic nanoparticles on plant physiology, morphology, anatomy, the ultrastructure of cellular and subcellular organelles, cytomorphometric modifications, and DNA damage. He comprehensively detailed the state of research in environmental science in regard to "how nanoparticles/heavy metals interact with plants, soil, microbial community, and the larger environment. He has published over 200 publications, including over 100 peer-reviewed articles, nine books, and many book chapters and conference articles. He is an internationally recognized reviewer and has received outstanding reviewing certificates from Elsevier and Springer. He is an editorial board member of various journals. He received "certificate for appreciation 2019" and "Certificate of Honor 2020" from the Southern Federal University, Russia, for his outstanding contribution in academic, creative research, and publication activities.

Dr. Abhishek Singh

Abhishek Singh is presently a PhD Research Scholar at Sardar Vallabh Bhai Patel University of Agriculture and Technology, Meerut, Uttar Pradesh, India. He has published a book, four Scopus-indexed, and two book chapters, as well as review papers, abstracts, and Hindi English popular articles in different national international level journal and magazines. He presented a poster presentation at the XIV Agricultural Sciences Congress–2019, ICAR, New Delhi, and presented at the 5th Global Outreach Conference in India. He has been a member of the different national and international scientific societies of agriculture. He earned an integrated MS in Plant Biotechnology at Sam Higginbottom University of Agriculture, Technology and Sciences (formerly Allahabad Agricultural Institute) Allahabad, India.

Prof. Dr. Tatiana M. Minkina

Tatiana M. Minkina, PhD, is the Head of the Soil Science and Land Evaluation Department of Southern Federal University, Russia. She is also Head of the International Master's Degree Educational Program "Management and Estimation of Land Resources" (2015–2022), accredited by ACQUIN. Her area of scientific interest is soil science, biogeochemistry of trace elements, environmental soil chemistry, including soil monitoring, assessment, modeling, and remediation using physicochemical treatment methods. Currently, she is handling projects funded by the Russian Scientific Foundation, Ministry of Education and Science of the Russian Federation, and Russian Foundation of Basic Research. She is a member of the Expert Group of Russian Academy of Science, International Committee on Contamination Land, Eurasian Soil Science Societies, International Committee on Protection of the Environment, and International Scientific Committee of the International Conferences on Biogeochemistry of Trace Elements. A prolific author, Dr. Minkina has published numerous scientific publications and has been an invited editor of the MDPI Open Access Journal by Water. She is an editorial board member of several journals as well. Dr. Minkina was awarded a Diploma of the Ministry of Education and Science of the Russian Federation for her many years of long-term work for development and improvement of the educational process, significant contribution to the training of highly qualified specialists.

Dr. Krishan K. Verma

Krishan K. Verma, PhD, is a Visiting Scientist at the Sugarcane Research Institute, Chinese Academy of Agricultural Sciences and Guangxi Academy of Agricultural Sciences, Nanning, Guangxi, China. He has 10 years of research experience in the field of biotic and abiotic stresses, i.e., environmental toxicology, plant physiology, metallic nanoparticles. His research focuses on environmental toxicology, plant physiology and molecular biology, and their impacts on growth and physiological adaptations of plants to environment and how they affect the plant structure. He has published more than 54 scientific research articles, including five books. He has been serving as an editorial board member of various peer-reviewed journals.

Prof. Dr. Awani Kumar Singh

Awani Kumar Singh, PhD, is presently working as Principal Scientist of Horticulture-Vegetable Science at the ICAR-IARI, New Delhi, India. He is a Fellow of CHAI, Noni, Royal Association for Science-led Socio-cultural Advancement (RASSA), and Progressive Horticulture. He has received a Best Scientist Award. To date, he has published many research papers in national and international journals of repute, research articles, research abstracts, book chapters, training manuals, and other publications, as well as three books. He has delivered many TV and radio talks on the topic of protected cultivation in DD1, DD-Kisan, FM-Gold, MGMG etc. He has developed eight high-yielding varieties of different vegetable crops. He has guided several MSc and PhD students in Vegetable Science. He has also contributed responsibility as Chairman and Co-chairman in various society seminars and symposia. He continues to promote protected cultivation cum hi-tech cum smart technology in vegetable crops.

Contents

Contributors

Mainak Barman
Department of Genetics and Plant Breeding, Bidhan Chandra Krishi Viswavidyalaya, West Bengal

Marina Burachevskaya
Academy of Biology and Biotechnology, Southern Federal University, Rostov-on-Don, Russia

Surachita Das
Department of Human Physiology, Vidyasagar University, West Bengal, India

Anuchaya Devi
Institute of Environment and Sustainable Development, Banaras Hindu University, Varanasi, Uttar Pradesh, India

Sadhan Jyoti Dutta
National Institute of Food Technology Entrepreneurship and Management, Kundli, Sonepat, Haryana, India

Fathy Elbehiry
Central Laboratory of Environmental Studies, Kafr El-Sheikh University, Al-GeishStreet, Kafr El-Sheikh, Egypt

Heba Elbasiouny
Department of Environmental and Biological Sciences, Home Economics Faculty, Al-Azhar University, Tanta, Egypt

Aniket K. Gade
Department of Biotechnology, Sant Gadge Baba Amravati University, Amravati, Maharashtra, India

Pooja Garg
ICAR-National Institute for Plant Biotechnology, Delhi 110012, India

Mohammad Reza Ghaffari
Department of Systems Biology, Agricultural Biotechnology Research Institute of Iran (ABRII), Agricultural Research Education and Extension Organization (AREEO), Karaj, Iran

Zahra Ghorbanzadeh
Department of Systems Biology, Agricultural Biotechnology Research Institute of Iran (ABRII), Agricultural Research Education and Extension Organization (AREEO), Karaj, Iran

Pham Thi Thu Ha
Genomic Research Institute and Seed, Ton Duc Thang University, Vietnam

Rasmieh Hamid
Cotton Research Institute of Iran (CRII), Agricultural Research, Education and Extension Organization (AREEO), Gorgan, Iran

Pramod U. Ingle
Department of Biotechnology, Sant Gadge Baba Amravati University, Amravati, Maharashtra, India

Immanuel Suresh J.
Department of Microbiology, The American College, Madurai, Tamil Nadu, India

Anamika Kashyap
ICAR-National Institute for Plant Biotechnology, Delhi, India

Pravin Khaire
Department of Plant Pathology and Agriculture Microbiology, PGI, Mahatma Phule Krishi Vidyapeeth, Rahuri, Maharashtra, India

Sujata Kumari
ICAR-National Institute for Plant Biotechnology, Delhi, India

Monika Mahajan
Institute of Environment and Sustainable Development, Banaras Hindu University, Varanasi, Uttar Pradesh, India

Bahador Maleknia
Department of Entomology, Faculty of Agriculture, Tarbiat Modares University, Tehran, Iran

Saglara Mandzhieva
Academy of Biology and Biotechnology, Southern Federal University, Rostov-on-Don, Russia

Someshree Mane
Department of Plant Pathology and Agriculture Microbiology, PGI, Mahatma Phule Krishi Vidyapeeth, Rahuri, Maharashtra, India

Navodita Maurice
Prophyl Ltd., Dózsa György út 18, Mohács, Hungary

Tatiana M. Minkina
Academy of Biology and Biotechnology, Southern Federal University, Rostov-on-Don, Russia

Shivangi Mishra
Department of Food Science and Technology, National Institute of Food Technology Entrepreneurship and Management, Kundli, Sonepat, Haryana, India

Subhra Mukherjee
Department of Genetics and Plant Breeding, Bidhan Chandra Krishi Viswavidyalaya, West Bengal, India

Narayan Musmade
Department of Plant Pathology and Agriculture Microbiology, PGI, Mahatma Phule Krishi Vidyapeeth, Rahuri, Maharashtra, India

Tanaji Narute
Department of Plant Pathology and Agriculture Microbiology, PGI, Mahatma Phule Krishi Vidyapeeth, Rahuri, Maharashtra, India

Olga Nazarenko
Academy of Biology and Biotechnology, Southern Federal University, Rostov-on-Don, Russia

Vaishnavi S. Parma
Department of Agriculture and Food Engineering, Indian Institute of Technology, Kharagpur, West Bengal, India

Priyadarshani Rajput
Academy of Biology and Biotechnology, Southern Federal University, Rostov-on-Don, Russia

Vishnu D. Rajput
Academy of Biology and Biotechnology, Southern Federal University, Rostov-on-Don, Russia

Hassan El-Ramady
Soil and Water Department, Faculty of Agriculture, Kafrelsheikh University, Kafr El-Sheikh, Egypt

Mahesh Rao
ICAR-National Institute for Plant Biotechnology, Delhi 110012, India

Sapna Rawat
Department of Botany, University of Delhi, New Delhi, India

Bhavisha Sharma
Institute of Environment and Sustainable Development, Banaras Hindu University, Varanasi, Uttar Pradesh, India

Ragini Sharma
Department of Zoology, Punjab Agriculture University, Ludhiana, India

Victoria Shuvaeva
Academy of Biology and Biotechnology, Southern Federal University, Rostov-on-Don, Russia

Abhishek Singh
Department of Agricultural Biotechnology, College of Agriculture, Sardar Vallabhbhai Patel University of Agriculture and Technology, Meerut, Uttar Pradesh, India

Awani Kumar Singh
Centre for Protected Cultivation, ICAR-Indian Agricultural Research Institute, New Delhi, India
Department of Zoology, Punjab Agriculture University, Ludhiana, India

Satish Kumar Singh
Department of Plant Breeding and Genetics, Dr. Rajendra Prasad Central Agricultural University, Bihar, India

Munna Singh
Department of Botany, University of Lucknow, Lucknow, India

Rajeev Pratap Singh
Institute of Environment and Sustainable Development, Banaras Hindu University, Varanasi, Uttar Pradesh, India

Rakhi Singh
Department of Food Science and Technology, National Institute of Food Technology Entrepreneurship and Management, Kundli, Sonepat, Haryana, India

Sanjay Singh
ICAR-National Institute for Plant Biotechnology, Delhi 110012, India

Sukh Veer Singh
Department of Food Science and Technology, National Institute of Food Technology Entrepreneurship and Management, Kundli, Sonepat, Haryana, India

Svetlana Sushkova
Academy of Biology and Biotechnology, Southern Federal University, Rostov-on-Don, Russia

Kunal Tanwar
ICAR-National Institute for Plant Biotechnology, Delhi, India

Feba Jacob Thoppurathu
Centre for Plant Biotechnology and Molecular Biology, Kerala Agricultural University, Thrissur, India

Rukam S. Tomar
Department of Biotechnology, Junagadh Agricultural University, Junagadh, Gujarat, India

Iswareya Lakshimi V.
Department of Microbiology, The American College, Madurai, Tamil Nadu, India

Chhedi Lal Verma
Irrigation and Drainage Engineering, ICAR-Central Soil Salinity Research Institute,
Regional Research Station, Lucknow, India

Krishan K. Verma
Key Laboratory of Sugarcane Biotechnology and Genetic Improvement (Guangxi),
Ministry of Agriculture and Rural Affairs/Guangxi Key Laboratory of Sugarcane Genetic Improvement/
Sugarcane Research Institute, Guangxi Academy of Agricultural Sciences, Nanning, Guangxi, China

Abbreviations

ABC ATP-binding cassette
ADP adenine triphosphate
AFM atomic force microscopy
Ag^+ silver ions
Ag silver
AgNPs silver nanoparticles
$Al2O_3$ aluminum oxide
ALS acetolactate synthase
AMF arbuscular mycorrhizal fungi
As arsenic
BHU Banaras Hindu University
BIS Bureau of Indian Standard
BN boron nitride
Ca calcium
CGPs cyanophycin grana proteins
CNMs carbon nanomaterials
CNTs carbon nanotubes
COD chemical oxygen demand
CP chlorophyll-binding protein
CTAB cetyl trimethylammonium bromide
Cu copper
CuO copper oxide
CuO-NPs copper oxide nanoparticles
$Cu(OH)_2$ copper (II) hydroxide
$CuSO_4$ copper sulphate
DH dehydrogenase
EC European Commission
eCO_2 elevated CO_2
EILs economic injury levels
ENPs engineered nanoparticles
EPS extracellular polymeric substances
FDAH fluorescent diacetate hydrolase
Fe iron
FETs field-effect transistors

FSSAI	Food Safety and Standard Authority of India
Fv/Fm	photosystem II
GA	gum Arabic
GNR	graphene nanoribbons
HYV	high-yielding varieties
IAA	indole acetic acid
ICM	integrated crop management
IESD	Institute of Environment and Sustainable Development
IISR	Indian Institute of Soybean Research
ILs	introgression lines
IOE	Institute of Excellence
IPM	integrated pest management
IRRI	International Rice Research Institute
IVASSD	in vitro assisted single seed descent
K	potassium
LDHs	layered double hydroxides
LEDs	light-emitting diode
LEEP	lipid exchange envelope penetration
LiP	lignin peroxidase
LPS	lipopolysaccharide
MAS	marker-assisted selection
MBC	microbial biomass
Mn	manganese
MnP	manganese peroxidase
Mo	molybdenum
MOFPI	Ministry of Food Processing Industries
MoS_2	molybdenum disulfide
MSN	mesoporous silica nanoparticle
MTBE	methyl tert-butyl ether
MWCNTs	multi-walled carbon nanotubes
N	nitrogen
Na	sodium
NAIPs	nanoagri-input products
NAM	nested association mapping
NC	nanocomposites
$nCeO_2$	cerium oxide
$nCr2O_3$	chromium trioxide
NFs	nano-fertilizers
NFs	nano-formulations

NIFTEM	National Institute of Food Technology Entrepreneurship and Management
NMs	nanomaterials
NPs	nanoparticles
NSs	nanosensors
NT	nanotechnology
NUE	nutrient use efficiency
NW	nanowire
nZVI	nanoscale zero-valent iron
ONT	Oxford Nanopore Technologies
OP	organophosphorus
OTUs	operational taxonomic units
P	phosphorus
PAH	polycyclic aromatic hydrocarbons
PB	Prussian blue
PCB	polychlorineated biphenyl
PDA	polydopamine
PGPR	promoting rhizobacteria
PGPR	plant growth-promoting rhizobacteria
PM	powdery mildews
PPO	polyphenol oxidase
Pst	pseudomonas syringae pv.tomato
PS II	photosystem II
qPCR	quantitative real-time polymerase chain reaction
RFID	radiofrequency identification detector
RGA	rapid generation advance
RNAi	RNA-interference
ROS	reactive oxygen species
RZM	root zone mass
SB	speed breeding
Se	selenium
SiO_2	silicon dioxide
$Si3N_4$	silicon nitride
S-layer	surface layer
SMRT	single-molecule real-time
SNP	single nucleotide polymorphism
SNV	single nucleotide variation
SOM	soil organic matter
SPR	surface plasma resonance

SWCNTs	single-walled carbon nanotubes
TEM	transmission electron microscope
Th	thorium
TiO_2	titanium dioxide
TNE	turmeric oil nanoemulsions
TTI	temperature indicator
UNDESA	United Nations Department of Economic and Social Affairs
WDP	water dispersible powder
WHO	World Health Organization
WS	water-soluble
Zn	zinc
ZnO	zinc oxide
ZnONPs	zinc oxide nanoparticles
$ZnSO_4$	zinc sulphate
ZrO_2	zirconia

Acknowledgment

We are thankful to the contributors for their effort and interest in this project, in which we have found the collaboration of each of the participating authors to be priceless. It is our great pleasure to offer you our sincere thanks for completing this important project for all of us.

We are also very thankful to Apple Academic Press, for their infinite patience in this project.

Finally, we would like to offer our sincere thanks to the Ministry of Science and Higher Education of the Russian Federation, no. 0852-2020-0029 for providing financial support and facilities for our research.

Preface

Nanotechnology is an emerging tool that influences human civilization through its wide range of applications from Earth to space. Nanobioscience has received a lot of interest in the recent couple of decades, involving research focused on its applications in electronics, power, biomechanics, and biological sciences. In the area of agriculture, the effectiveness of nanoscience is unavoidable, and several types of research are demonstrating their capacity to enhance food security, safety, and agro-systems through various strategies due to increased agricultural economic output and the establishment of innovative food materials. The editors are confident that this book provides recent updates on nanobiotechnology, soil science, plant breeding, food science, tool design, conceptualization, utilization, and governance, and the impacts of such approaches on properties of the soil and plant states, as well as environmental factors. Graduate students, research scholars, as well as other decision makers in the area of agriculture and allied sectors, may benefit from this book.

The book is categorized into four sections.

Part I—Nanotechnology: An Introduction to Nanoparticles—presents chapters on the introduction of nanotechnology. The first chapter focuses on the current application of nanotechnology with an introduction. Part II—Nanoparticles for Crop Production—presents chapters related to nanofertilizers and nanopesticides related to crop production and protection from plant diseases and their management. Part III—Improving the Soil Fertility and Crop Protection—presents chapters on modern nano-approaches for improving the soil fertility and productivity. Part IV—Emerging Nanotechnological Tools and Techniques for Crop Improvement—presents chapters on crop improvement with the help of recent and emerging tools like speed breeding, biotechnology, and genetic engineering combined with nanotechnological approaches.

This new book, Nanotechnology for Sustainable Agriculture, provides a significant addition to the field of agriculture and crop sciences. However, this unique compilation work would not be conceivable without the help of experts and the significant contributors globally.

Editors:
Vishnu D. Rajput, Abhishek Singh,
Tatiana M. Minkina, Krishan K. Verma, and
Awani Kumar Singh

PART I

Nanotechnology: An Introduction to Nanoparticles

CHAPTER 1

Nanotechnology as the Cutting Edge for Sustainable Agriculture

VISHNU D. RAJPUT[1], ABHISHEK SINGH[2], SAPNA RAWAT[3], RAGINI SHARMA[4], TATIANA M. MINKINA[1], AWANI KUMAR SINGH[5], VICTORIA SHUVAEVA[1], and OLGA NAZARENKO[1]

[1]*Academy of Biology and Biotechnology, Southern Federal University, Rostov-on-Don, Russia*

[2]*Department of Agricultural Biotechnology, College of Agriculture, Sardar Vallabhbhai Patel University of Agriculture and Technology, Meerut, UP, India*

[3]*Department of Botany, University of Delhi, New Delhi, India*

[4]*Department of Zoology, Punjab Agriculture University, Ludhiana, India*

[5]*Centre for Protected Cultivation, ICAR-Indian Agricultural Research Institute, New Delhi, India*

ABSTRACT

The upcoming decade will be difficult in many areas of life. One of the main challenges could be unfavorable climatic or environmental conditions that affect crop yields and hence the increasing demand for higher products, the demand for various industrial products with higher quality requirements, smart automotive, and many more. Nanotechnology has gained popularity in recent years as a result of its known potential for use and implementation in various key sectors like medical drugs, medicine, catalysis, energy,

Nanotechnology for Sustainable Agriculture: An Innovative and Eco-Friendly Approach. Vishnu D. Rajput, Abhishek Singh, Tatiana M. Minkina, Krishan K. Verma, and Awani Kumar Singh (Eds.)

material, and plant science research. Nanoparticles (NPs) are small in size with larger surface area (1–100 nm) and have many wonderful applications. Such extraordinary functions of NPs are used in sustainable agriculture, for the development of nano-enabled products, that is, nano-insecticides, nano-pesticides, fertilizers, etc. NPs have recently been suggested as an alternate approach for controlling plant pets such as insects, fungus, and weeds. Antimicrobial properties are there in several NPs and are used in food packaging processes, for example, silver NPs are highly used for such purposes. Apart from their antimicrobial properties, NPs like Si, Ag, Fe, Cu, Al, Zn, ZnO, TiO_2, CeO_2, Al_2O_3, and carbon nanotubes have been also shown to have negative impacts on plant development. This chapter examines the field application of NPs in sustainable agriculture, future perspectives, and growing environmental concerns.

1.1 INTRODUCTION

The agricultural sector is the most important as it provides major raw materials for various industries such as food and fodder. Limited natural resources and raising population claim economical, viable, environmental, and efficient agricultural development. Agricultural development is important to eradicate poverty and hunger among raising population in the present situation. Hence, a bold step is required for agricultural development. Sustainable agriculture can strengthen this applied prospect to get rid of hunger and poverty among the growing population. As a result, new and inventive approaches such as nanotechnology are required for sustainable agriculture. This new technique, which focuses on increasing agricultural production, should be used (Yunlong and Smit, 1994). Professor Norio Tanaguchi (1974) coined this concept (Bulovic et al., 2004). Later, nanotechnology developed a variety of methods for precisely isolating nanoparticles (Bonnell and Huey, 2001).

Nanotechnology is named after the prefix "nano" derived from Greek word "dwarf" and the term "nano" technically refers to 1 billionth. The virus, on the other hand, is around 100 nm in size. Nanotechnology is a term refers to materials that are between 0.1 and 100 nm (Ghidan et al., 2017). Because of their diminutive size, they are more active and have a larger surface area. It has been discovered that europium and dopant atoms interact and produce magnetic properties. In the bulk of biological process sectors, altered nanomaterials are more reactive (Popkropinvy et al., 2007; Prasad, 2014). Changing the characteristics of nanomaterials allows for the formation of novel electronic associations, plasmonic capabilities, and quantum confinement-related optical properties. (Prasad et al., 2016). Agriculture forms the foundation of the economy of the third world countries. Agricultural research has always been concerned with improving the productivity and production of crops, food manufacturing, food protection, and the environmental effects of food production, storage, and distribution. In addition to packaging, NM is also used in fortifications and food processing (Shafiq et al., 2020; Singh et al., 2017). New molecular and cellular biology methods for separating, identifying, and quantifying individual molecules are required to be especially developed.

Rapid advances in suitable agricultural research, such as reproductive science and technology, the conversion of agricultural and food wastes to energy, and other useful byproducts via enzymatic, nano-bioprocessing, and disease prevention and treatment in plants and animals, are all possible thanks to nanotechnology (Pilarska et al., 2017). It is documented that the nano-enabled products are already in commerce and applying for sustainable agriculture such as nanofertilizers, growth stimulators, soil-improving agents, or nano-sensors so on, and gained popularity as smart and controlled delivery. However, rapid application of NPs also raised concern for soil microbiota, accumulation in plants, and risks to human health through food chain (Rajput et al., 2020). The introductory chapter of this new book is discussing all the above-discussed possibilities for sustainable agriculture and raised concern for field application of NPs.

1.2 NANOTECHNOLOGY: MANIPULATION OF MATTER AT THE ATOMIC LEVEL

The practice of manipulating, engineering, or gathering of different atoms, molecules or molecular dusters to create new devices and materials with different features is known as nanotechnology (Rajput et al., 2020). It can work from (1) bottom up, that is, nanostructures are created by manipulating individual atoms and molecules, mainly involves biology or chemistry. (2) Top down, that is, the size of tiniest nanoscales is shrinked as in photonics applications in nanoelectronics and nanoengineering (Ghidan and Antary, 2020). Nanotechnology is based on nanometer-sized particles' unique features. The nanoscale structure is the first level of organization occurring in biological systems, where all fundamental functions are rigorously specified. (Ghidan et al., 2017). Nanotechnology is capable of revolutionizing science by allowing scientists to use physics, engineering, chemistry, and biology to manipulate nuclear or molecular matter. Nanotechnology is wide and multidisciplinary field of study and development that has exploded in popularity in recent years around the world. It allows researchers to better comprehend the relationship between macroscopic features and molecular structure in plant, biological, and animal origin materials. It already has a considerable influence on the business which will certainly increase in the future (Fig. 1.1).

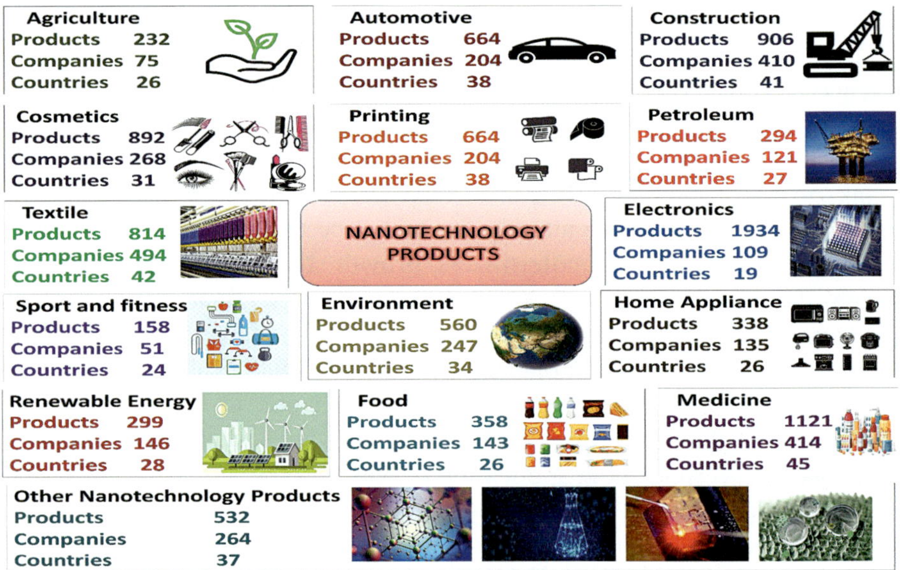

FIGURE 1.1 Diagrammatic representation of nanotechnology product.

1.3 FIELD APPLICATION OF NPs FOR SUSTAINABLE AGRICULTURE

In conventional mode of agriculture, huge amount of fertilizers and pesticides are used that affected adversely our ecosystem. Minimum amount of agro-chemicals should be used to protect ecosystem and environment. Sustainable agriculture is meant to get higher returns at low input. Nanotechnology can help to maintain agriculture in sustainable way with enhanced efficacy. Nano-plant growth promoters such as nanofertilizers, nanopesticides, and nanoherbicides have greater application in agriculture at low cost. The major concern in agriculture about using nanomaterials is the mechanism of their penetration to applied surfaces and risk involved. Although new technologies have been developed for synthesis and application of nanomaterials in agriculture. Hence, this chapter is critically aimed to highlight application of nanoparticles in agriculture in sustainable way to enhance output at low input. In agriculture, some other applications of NTs such as attenuating natural pesticides and biopesticides, the measured and controlled release of fertilizers supported by NM, micronutrients, nanosensors, organic fertilizers (Rajput et al., 2020).

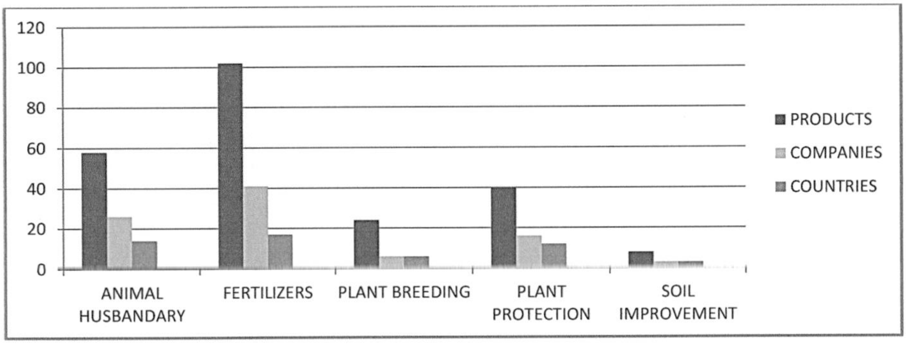

FIGURE 1.2 Number of nanoproducts, companies, and the countries involved in the agriculture related nano products development.

Source: Statnano.

Websites keeping updated records on NPs showed total nano-enabled products: 9245 manufactured by 64 countries including 2658 companies (https://product.statnano.com/; accessed July 7, 2021). In the agriculture sector, 232 products related to plant protection, fertilizers, soil improvement, plant breeding, and animal husbandry are making available. Figures 1.1 and 1.2 illustrate different fields and uses of NPs in agriculture.

1.3.1 NPs AND CROP PROTECTION

Due to limited resources and progressive climate change, food security has become a key concern for the growing population. That means change in climatic baselines like ecological pollution, water scarcity, cold, alkalinity, salinity, and toxicity of metals. Advanced nanomaterial engineering nanofertilizers can enhance the crop production in adverse environment. Salinity can create a severe issue in crop production. Under NaCl stress condition in tomato and squash, nano-SiO_2 can improve proline accumulation, fresh and dry plant weight, seed germination as well as chlorophyll content (Haghighi et al., 2012). Similarly, Foilar spray of $FeSO_4$ showed positive response toward salinity stress, highest photochemical efficiency of photosystem II (Fv/Fm) and iron (Fe) content, reduced substantial amount of sodium (Na) content in sunflower leaves while increased shoot dry weight, leaf area, net CO_2 assimilation rate, sub-stomatal CO_2 concentration (Ci), and chlorophyll content (Torabian et al., 2017). Nanofertilizers application accelerated plant growth and productivity thus they can act as effective tool in agriculture. Nanomaterials can also act as an effective tool in detoxification and remediation of heavy metals. In rice, Cd stress tolerance is improved on applying nano-Si at 2.5 mM (Wang et al., 2015). Crop production is also severely affected by abiotic as well as biotic stress (Oerke et al., 2006). Nanomaterials can reduce the risk of pest and diseases (Rizwan et al., 2019; Wang et al., 2016). *Botrytis cinerea, Alternaria alternate, Colletotrichum gloeospoiodes, Monilinia fructicola, Verticillium Dahliae, Rastonia solanacearum, Fusarium solani, Fusarium oxysporum, and Phytophthora infestans* cause a variety of plant- and soil-borne diseases which can be controlled by nanomaterial oxides such as CuO, Mgo, and ZnO (Malandrakis et al., 2019; Shenashen et al., 2017). Hence, nanofertilizers have great potential toward crop protection.

1.3.1.1 NANO-INSECTICIDES

NPs can contain organic and inorganic ingredients in different forms such as micelles and particles. Nanoformulations can lead to increased solubility of less soluble active ingredient and thus protecting active ingredient from premature degradation. Nanoformulations have significant impact on active ingredient and can introduce of new ingredient into the environment. Gopal et al. prepared nanopesticides and compared their efficacy with conventional ones. Pathogen control was found to be five times as effective with nano-hexaconazole. Similarly, nanosulfur is 10 times more efficient than water dispersible powder (WDP) formulations for mite control. Nicotine carboxylate nanoemulsion was created using a series of fatty acids (C10–C18) and a surfactant, and it demonstrated insecticide formulation bioactivity against *Drosophila melanogaster* adults by determining the lethal time 50 (Casanova et al., 2005). When nanomatrices made of carboxylic acid and metallic cation come into touch with water, they breakdown and produce a bioactive chemical (Beasley and Collins, 1970). Chemical bonding is required to be broken through hydrolysis affecting polymer–insecticide bonds according to Allen et al. (1971). Release of bioactive compounds also depends upon the chemical nature of molecules as well as structure of macromolecule.

1.3.1.2 NANOFUNGICIDES

Plant pathogen can impact a plant severely at any growth stage of plant. *Phythophora, Rhizoctonia solani*, and *Fusarium* spp. attack on soil buried and aerial part of plant, while *B. cinerea* can infect fruit and green tissue of plant. They also cause significant damage to the plant. Worldwide fungicide market is more than €500 million only for *B. cinerea* (Pinmentel et al., 2009). Conventional methods to control plant pathogen are neither cost effective nor environmental friendly. Consequently, it has become the major concern for scientists and public. Estimated fungicide usage is about 2.5 million tons and harm reach 100 billion dollars annually due to pesticides (Pimentel et al., 2009). Harmful effects of pesticides are dual, one is the nonbiodegradable properties and another is harmful effect on human and animal health (Koul et al., 2008). Hence, development of nanomaterials based antifungal is needed as an alternative replacing poisonous elements. Nanofungicides are entities with nanometer size range having unique qualities due to their small size. Wide variety of products have been developed so far (European Commission Joint Research Center, 2010). Nanopesticide formulation includes organic ingredients, silica-based nanoparticles and titanium dioxide, emulsions, and nanoclay (Matthews et al., 2000). Nanotechnology-based product with broad spectrum use, nontargeted effect, and public concern can be abridged.

1.3.1.3 NANOMATERIAL AS NANOHERBICIDES

The nanoherbicide formulation is prepared by exploiting nanotechnological potential of nanosized preparation for their effective delivery into plant. They are based on a particular formulation to improve their efficacy, increased solubility and reduced toxicity as compared to conventional herbicide (Grillo et al., 2014). Nanomaterial-based herbicide system can maintain their prolonged release for the effective herbicide control (Manjunatha et al., 2016). Specific herbicide encapsulated into nanomaterials aims to target receptors present in root cells as they are translocated to inhibit glycolysis once they enter the root system. Thus, create starvation and kills targeted weed. Encapsulated herbicide provides controlled release of herbicide thus weeds can be completely destroyed (Grillo et al., 2012). Adjuvents are also added to the nanomaterial-based formulation for its effectiveness. Nanoparticles can act as nanoformulation in combination with herbicides. Herbicide delivery by nanoparticle system is mostly containing biodegradable polymeric compounds having nontoxic metabolites.

1.3.1.4 NANOMATERIAL AS NANOFERTILIZER

Nanofertilizers are basically conventional fertilizers encapsulated with nanomaterials used for slow entry of nutrients into plants enhances fertility of the soil and productivity (Zulfiquar et al., 2019). Slow release of nanomaterials has minimum impact on environment with high productivity. Due to the use of highly reactive nanomaterials, they interact with fertilizers and make their effective absorption by plants (Prasad et al., 2017). Nanofertilizers with their slow release can enhance nutrient usage, enhance bioavailability with high surface area, prevent leaching, reduce volatilization, and lessen the environmental threats (Solanki et al., 2015). Encapsulation of nutrients can be done either by entrapping with nanomaterials or coating with nanomaterials or by delivery in the form of nanoemulsion (Iqbal et al., 2019). Effectiveness of nanofertilizers depends upon preparation of nanoformulation, soil conditions, and mode of application of nanofertilizers (Zulfiquar et al., 2019). Agricultural growth and their economic distribution are affected by the efficacy of fertilizer distribution. This phenomenon is influenced by numerous factors like uptake efficiency of plants, leaching effect, soil combination, and chemical composition. Chemical fertilizers can severely affect soil fertility and can enhance environmental pollution. Current agricultural research focuses on finding an alternative to chemical fertilizers, that is, environmental friendly and easy to biodegrade.

Hence, nanofertilizers carry a great potential to increase agricultural output while overcoming the imitation of conventional fertilizers.

1.4 PLANT UPTAKE AND TRANSLOCATION OF NANOPARTICLES

A series of transformations are needed to NPs to increase their bioavailability and decreasing their toxicity. Plant roots absorb NPs through their roots, which then accumulate in subcellular and cellular organelles. First step of NPs absorption is bioaccumulation (Nair et al., 2010). Root cells are semipermeable to NPs, restricting large size NPs. After penetrating cell walls, NPs travel apoplastically across extracellular gaps until they reach the vascular cylinder, where they then ascend unidirectionally through the xylem (Fig. 1.3A).

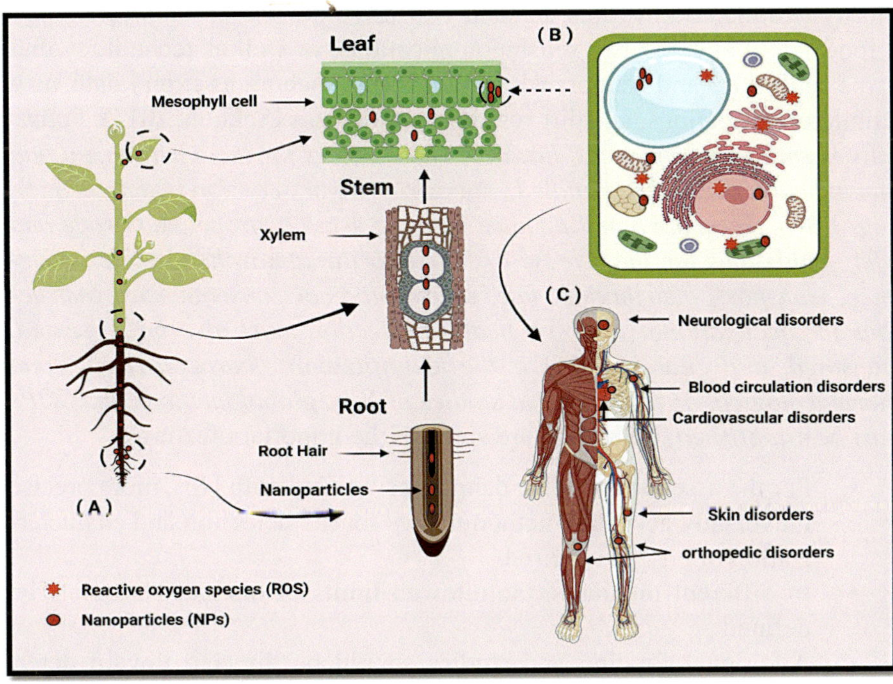

FIGURE 1.3 Diagrammatic representation of translocation of nanoparticle, accumulation, and toxicity.

For symplastic movement, NPs enter the vascular cylinder by crossing the Casparian strip barrier through binding to endodermal carrier proteins. NPs travel through plasmodesmata through cytoplasmic internalization (Tripathi et al., 2017). NPs those are unable to internalize aggregate on casparian strips can modify their absorption (Ruttkay-Nedecky et al., 2017). NPs that are taken up by cell are found in nuclei, cortical cytoplasm, and cell wall of epidermis (Wang et al., 2012). NPs accumulation and translocation are accompanied by physiological structure of cells, the interaction of soil with nanomaterials, and the intrinsic stability of nanoparticles (Fig. 1.3B) (Janmohammadi et al., 2016). Moreover, NP coating and morphology play a substantial role on their action on plants.

1.5 CONCERNS

Because of the significant effect of nanotechnology on food business and commodities which enter the market, safety of food will continue to be a top priority (Kamle, 2020). This demand will drive the use of nanotechnology in food safety and security sensing applications, as well as technology that warns consumers and vendors when food is approaching its expiry date, new antimicrobial coatings, and dirt-resistant plastic bags (Sekhon, 2014). *Potentially harmful consequences, hazards, and dangers for the environment and human health need to be handled with extraordinary caution and dedication (Fig. 1.3C). Accurate handling must be done using appropriate dosage and safety guidelines for the release of NPs into the atmosphere, their persistence, transport, distribution, fate, and ecological consequences. Instructions for occupational workers' health protection must also be presented, sponsored, and guaranteed. For the identification, characterization, and pharmacological or toxicological studies of NPs, global methods and SOPs must be established.* Following are some of the important features:

- For the assessment of the dangers associated with NPs, more precise universally accepted methodologies for the detection and characterization of NP are required.
- In different mediums, the allowed limits of distinct NPs must be defined.
- Additional funding and studies should be directed toward determining the underlining mechanism of interaction of the food chain with NPs, as well as their epigenetic consequences.
- Because a particular attribute of NPs, that is, large volume: mass ratio which makes it more likely to interact with biomolecules, toxicodynamic, and toxicokinetic investigations should be thoroughly understood from the perspective of agroecosystems.
- Other aspects such as distribution of NPs, their fate, transportation, bioavailability, etc. need to be addressed as research areas in agro-nanotechnology.
- To further strengthen the field of research, in particular, the impact on the environment and humans, laboratory, organizational, geopolitical, and other layers of joint research and capacity building are also discussed at different levels.

Depending on the size and surface chemistry of nanoparticles, some types of nanoparticles exhibit large-scale bioretention, and such nanoparticles may accumulate within the body beyond potentially unsafe levels that may affect human health. For example, many cells can penetrate membranes and generate reactive oxygen species (redox oxygen species) in their vicinity, causing the cell to be damaged although any chemical additives and nanoparticles must be collected at the time of food packaging. Keep a certain amount and these certain quantities of nanoparticles are harmful to human health factors were not found. Moreover, within these limits, fertilizers packaged with nanoparticles can be legally sold in the market for the consumer.

1.6 CCONCLUSIONS

Agriculture is the only source of human nourishment that should produce from intermediate and final inputs using cutting-edge technologies. As a result, adopting new contemporary agricultural technology is critical. Despite having relatively modern agricultural methods, emerging nations continue to struggle with a shortage of high-value food crops. Although there is a lot of evidence regarding individual nanoparticles, the toxicity levels of diverse NPs are yet unknown, therefore the use of these nanomaterials is unsatisfac tory owing to a lack of understanding about risk assessments in database and alarm systems, as well as international coordination on directives and regulations.

ACKNOWLEDGMENT

The Ministry of Science and Higher Education of Russian Federation provided financial support for the study (no. 0852-2020-0029).

KEYWORDS

- **fertilizer**
- **nano-insecticide**
- **nanotechnology**
- **pesticides**
- **risk**
- **sustainable agriculture**

REFERENCES

Allan, G. G.; Chopra, C. S.; Neogi, A. N.; Wilkins, R. M. Design and Synthesis of Controlled Release Pesticide-Polymer Combinations. *Nature* **1971,** *234*, 349–351.

Beasley, M. L.; Collins, R. L. Water-Degradable Polymers for Controlled Release of Herbicides and Other Agents. *Science* **1970,** *169,* 769–770.

Bonnell, D. A.; Bulovic, V.; Mandell, A.; Perlman, A. Molecular Memory Device. US 20050116256 A1, 2004.

Bonnell, D. A.; Huey, B. D. Basic Principles of Scanning Probe Microscopy. In *Scanning Probe Microscopy and Spectroscopy: Theory, Techniques, and Applications*, 1st ed.; 2001.

Casanova, H.; Araque, P.; Ortiz, C. Nicotine Carboxylate Insecticide Emulsions: Effect of the Fatty Acid Chain Length. *J. Agric. Food Chem.* **2005,** *53,* 9949–9953.

European Commission Joint Research Center; *Considerations on a definition of nanomaterial for regulatory purposes*; Reference Report, (2010).

Ghidan, A. Y.; Al-Antary, T. M.; Salem, N. M.; Awwad, A. Facile Green Synthetic Route to the Zinc Oxide (ZnONPs) Nanoparticles: Effect on Green Peach Aphid and Antibacterial Activity. *J. Agric. Sci.* **2017,** *9,* 131–138.

Ghidan, A. Y.; Antary, T. M. Applications of Nanotechnology in Agriculture. *Appl. Nanobiotechnol.* **2020.** DOI: 10.5772/intechopen.88390.

Gopal, M.; Kumar, R.; Goswami, A. Nano-pesticides - A Recent Approach for Pest Control. *J. Plant Protec. Sci.* **2012,** *4*(2), 1–7.

Grillo, R.; dos Santos, N. Z. P.; Maruyama, C. R.; Rosa, A. H.; de Lima, R.; Fraceto, L. F. Poly(-capro- lactone)Nanocapsules as Carrier Systems for Herbicides: Physico-chemical Characterization and Genotoxicity Evaluation. *J. Hazard. Mat.* **2012,** 231–232, 1–9.

Grillo, R.; Pereira, A. E.; Nishisaka, C. S.; de Lima, R.; Oehlke, K.; Greiner, R.; Fraceto, L. F. Chitosan/Tripolyphosphate Nanoparticles Loaded With Paraquat Herbicide: An Environmentally Safer Alternative for Weed Control. J. Hazard. Mat. **2014,** *278,* 163–171.

Haghighi, M.; Afifipou, Z.; Mozafariyan, M. The Effect of N-Si on Tomato Seed Germination Under Salinity Levels. *J. Biol. Environ. Sci.* **2012,** *6,* 87–90.

Iqbal, M. A. Nano-Fertilizers for Sustainable Crop Production Under Changing Climate: A Global Perspective. In: *Sustainable Crop Production*; Hasanuzzaman, M., Fujita, M., Filho, M. C. M. T., Nogueira, T. A. R., Eds.; IntechOpen, 2019.

Janmohammadi, M.; Amanzadeh, T.; Sabaghnia, N.; Ion, V. Effect of Nano-silicon Foliar Application on Safflower Growth Under Organic and Inorganic Fertilizer Regimes. *Botanica* **2016,** *22,* 53–64.

Koul, O.; Walia, S.; Dhaliwal, G. S. Essential Oils as Green Pesticides: Potential and Constraints. *Biopestic Int.* **2008,** *4*(1), 63–84.

Manjunatha, S. B.; Biradar, D. P.; Aladakatti, Y. R. Nanotechnology and its Applications in Agriculture: A review. *J. Farm Sci.* **2016,** *29*(1), 1–13.

Matthews, G. A. Pests, Pesticides and Pest Management. In: *Highlights in Environmental Research*; Mason, J., Ed.; Imperial College Press: London, 2000; pp 165–189.

Nair, R.; Varghese, S. H.; Nair, B. G.; Maekawa, T.; Yoshida, Y.; Kumar, D. S. Nanoparticulate Material Delivery to Plants. *Plant Sci.* **2010,** *179*(3), 154–163.

Oerke, E. C. Crop Losses Topests. *J. Agric. Sci.* **2006,** *144,* 31–43. DOI: 10.1017/S0021859605005708.

Pilarska, A.; Wysokowski, M.; Markiewicz, E.; Jesionowski, T. Synthesis of Magnesium Hydroxide and Its Calcinates by a Precipitation Method With the Use of Magnesium Sulfate and Poly (Ethylene Glycols). *Powder Technol.* **2013,** *235,* 148–157.

Pimentel, D. Pesticide and Pest Control. In: *Integrated Pest Management: Innovation-Development Process*; Peshin, P., Dhawan, A. K., Eds.; Springer: Dordrecht, Netherlands, 2009; pp 83–87.

Pokropivny, V.; Lohmus, R.; Hussainova, I.; Pokropivny, A.; Vlassov, S. *Introduction to Nanomaterials and Nanotechnology*; University of Tartu: Tartu, (2007); p 225.

Prasad, R.; Bhattacharyya, A.; Nguyen, Q. D. Nanotechnology in Sustainable Agriculture: Recent Developments, Challenges, and Perspectives. *Front. Microbiol.* **2017,** *8,* 10–14.

Prasad, R.; Kumar, V.; Prasad, K. S. Nanotechnology in Sustainable Agriculture: Present Concerns and Future Aspects. *Afr. J. Biotechnol.* **2014,** *13,* 705–713. DOI: 10.5897/AJBX2013.13554

Prasad, R.; Pandey, R.; Barman, I. Engineering Tailored Nanoparticles With Microbes: Quo vadis. WIREs Nanomed. Nanobiotechnol. **2016,** *8*(2)*,* 316–330. DOI: 10.1002/wnan.1363.

Rajput, V.; Minkina, T.; Mazarji, M.; Shende, S.; Sushkova, S.; Mandzhieva, S.; Burachevskaya, M.; Chaplygin, V.; Singh, A.; Jatav, H. Accumulation of Nanoparticles in the Soil-Plant Systems and Their Effects on Human Health. *Ann. Agric. Sci.* **2020,** *65,* 137–143.

Rizwan, M.; Ali, S.; ur Rehman, M. Z.; Malik, S.; Adrees, M.; Qayyum, M. F.; Alamri, S. A.; Alyemeni, M. N.; Ahmad, P. Effect of Foliar Applications of Silicon and Titanium Dioxide Nanoparticles on Growth, Oxidative Stress, and Cadmium Accumulation by Rice (*Oryza sativa*) *Acta Physiol. Plant.* **2019,** *41,* 35. DOI: 10.1007/s11738-019-2828-7.

Ruttkay-Nedecky, B.; Krystofova, O.; Nejdl, L.; Adam, V. Nanoparticles Based on Essential Metals and Their Phytotoxicity. *J. Nanbiotechnol.* **2017,** *15*(1)*,* 33.

Shafiq, M.; Anjum, S.; Hano, C.; Anjum, I.; Abbasi, B. H. An Overview of the Applications of Nanomaterials and Nanodevices in the Food Industry. *Foods* **2020,** *9,* 148.

Singh, Y.; Meher, J. G.; Raval, K.; Khan, F. A.; Chaurasia, M.; Jain, N. K.; Chourasia, M. K. Nanoemulsion. Concepts, Development and Applications in Drug Delivery. *J. Control. Release* **2017,** *252,* 28–49.

Statnano, 2021 [Online]. https://product.statnano.com/industry/agriculture

Torabian, S.; Zahedi, M.; Khoshgoftar, A. H. Effects of Foliar Spray of Nano-particles of $FeSO_4$ on the Growth and Ion Content of Sunflower Under Saline Condition. *J. Plant Nutr.* **2017,** *40,* 615–623. DOI: 10.1080/01904167.2016.1240187.

Tripathi, D. K.; Singh, S.; Singh, V. P.; Prasad, S. M.; Dubey, N. K.; Chauhan, D. K. Silicon Nanoparticles More Effectively Alleviated UV-B Stress Than Silicon in Wheat (Triticum aestivum) Seedlings. *Plant Physiol. Biochem.* **2017,** *110,* 70–81.

Wang, S. H.; Wang, F. Y.; Gao, S. C. Foliar Application With Nano-silicon Alleviates Cd Toxicity in Rice Seedlings. *Environ. Sci. Pollut. Res.* **2015,** *22,* 2837–2845. DOI: 10.1007/s11356-014-3525-0.

Wang, S. H.; Wang, F. Y.; Gao, S. C.; Wang, X. G. Heavy Metal Accumulation in Different Rice Cultivars as Influenced by Foliar Application of Nano-silicon. *Water Air Soil Pollut.* **2016,** *227,* 228. DOI: 10.1007/s11270-016-2928-6.

Wang, Z.; Xie, X.; Zhao, J.; et al. Xylem-and Phloem-based Transport of CuO Nanoparticles in Maize (Zea mays L.). *Environ. Sci. Technol.* **2012,** *46(8),* 4434–4441.

Yunlong, C.; Smit, B. Sustainability in Agriculture: A General Review. *Agric. Ecosyst. Environ.* **1994,** *49,* 299–307. DOI: 10.1016/0167-8809(94)90059-0

Zulfiqar, F.; Navarro, M.; Ashraf, M.; Akram, N. A.; Munné-Bosch, S. Nanofertilizer Use for Sustainable Agriculture: Advantages and Limitations. *Plant Sci.* **2019,** *289,* 110270.

PART II

Nanoparticles for Crop Production

CHAPTER 2

Nanofertilizers: A Sustainable Alternative to Conventional Means

KRISHAN K. VERMA[1], MUNNA SINGH[2], VISHNU D. RAJPUT[3], CHHEDI LAL VERMA[4], and MARINA BURACHEVSKAYA[3]

[1]Key Laboratory of Sugarcane Biotechnology and Genetic Improvement (Guangxi), Ministry of Agriculture and Rural Affairs/Guangxi Key Laboratory of Sugarcane Genetic Improvement/Sugarcane Research Institute, Guangxi Academy of Agricultural Sciences, Nanning, Guangxi, China

[2]Department of Botany, University of Lucknow, Lucknow, India

[3]Academy of Biology and Biotechnology, Southern Federal University, Rostov-on-Don, Russia

[4]Irrigation and Drainage Engineering, ICAR-Central Soil Salinity Research Institute, Regional Research Station, Lucknow, India

ABSTRACT

Agriculture and agro-industries benefit significantly from nanobioscience. Nano-scale fertilizers allow nutrients to be released slowly and consistently, enhancing soil profile. It improves nutrient use efficiency and reduced fertilizer wastage and the cost of crop cultivation. The increasing availability of nutrients causes plants to upregulate photosynthetic capacity, resulting in maximum productivity and biomass production. Nano-scale fertilizers are advantageous because they reduce synthetic fertilizer application frequency and soil toxicity. Synthetic fertilizers are not only expensive for farmers,

Nanotechnology for Sustainable Agriculture: An Innovative and Eco-Friendly Approach. Vishnu D. Rajput, Abhishek Singh, Tatiana M. Minkina, Krishan K. Verma, and Awani Kumar Singh (Eds.)

but they can also be damaging to humans and the atmospheric environment. Because there are limited cultivated lands and water sources, agricultural system development can only be achieved by increasing resource use efficiency while minimizing loss to the production bed through the practical application of innovative technologies.

2.1 INTRODUCTION

Agriculture with maximum crop yield is required to reduce the risk of hunger and increase food grain security (Singh et al., 2021). Due to a dynamic era of climate variables, an expanding human populace, and limited agri-farms and freshwater resources, food grain production, and delivery are heightened and ongoing stress worldwide (Usman et al., 2020). Combining modern technology capabilities with considerable adjustments to current global food production systems could solve this problem (Dwevedi et al., 2016; Shang et al., 2019; Singh et al., 2021). The application of high rates of agrochemicals is currently significant support for modern agriculture. Agrochemical production will increase in the near future to feed nearly 10 billion population by 2050 (FAO, 2017; Diatta et al., 2020; Seleiman et al., 2020). Synthetic chemical fertilizers are applied to boost crop performance and output. Current farming approaches have not succeeded in improving/upgrading plant mineral accumulation, nutrient usage efficiency (NUE), and crop production (Adnan et al., 2020). Synthetic fertilizers are widely used in agriculture, typically have low NUE values (Guo et al., 2018).

As a result, agricultural sustainability can be achieved by implementing and applying novel strategies (Shang et al., 2019) that increase worldwide food capacity while also protecting resources of nature and the environment (Arora, 2018). Nanobioscience can potentially change the current synthetic framework used in modern agricultural systems (Prasad et al., 2017) by improving the capacity of innovative agrochemicals and delivering solutions to agro-ecological challenges (Usman et al., 2020). The application of nanofertilizers (NFs) in agriculture has garnered a lot of attention in recent years (Kah et al., 2019; Kerry et al., 2017; Seleiman et al., 2020). Nanobioscience has the great potential to create advanced forms of fertilizers that will help enhance worldwide food production capacity and feed the rising global populace (Feregrino-Perez et al., 2018; Diatta et al., 2020; Seleiman et al., 2020).

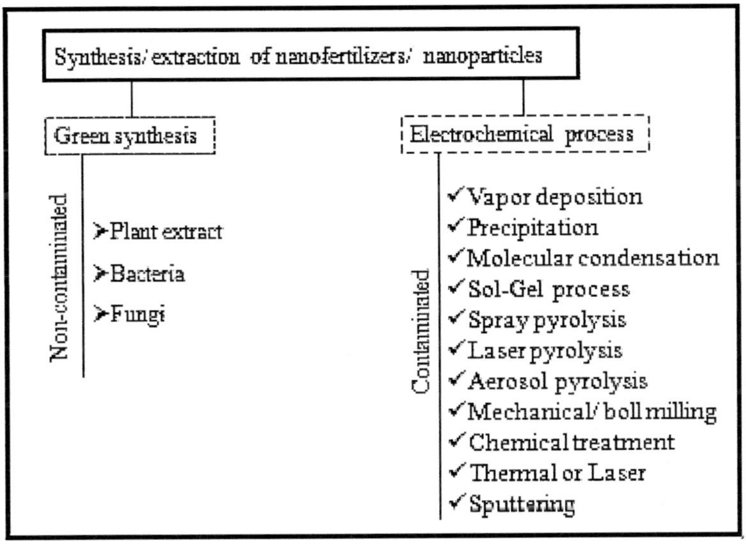

FIGURE 2.1 Bioprocess of nanofertilizers (NFs).

Nanofertilizers are micro or macronutrients that have been encapsulated or coated by nanoparticles (DeRosa et al., 2010; Rajput et al., 2017). NFs can use various synthetic chemicals (e.g., modified synthetic fertilizers) or green produced from multiple plant organs (Singh and Kumar, 2017). Green synthesis is a contamination-free and environmentally beneficial method of producing nanoparticles since it involves bio-organisms, that is, plants, fungus, and bacteria (Lee et al., 2020) (Fig. 2.1). No toxic substances or chemicals are utilized to produce these microorganisms, minimizing, and stabilizing agents (Iravani et al., 2015). NFs are used to enhance the soil fertility profile and quality and the bioavailability of plant nutrients (Chhipa, 2017; Singh and Kumar, 2017) and yield quality (Brunner et al., 2006). NFs are suitable for advanced agriculture because of their vast surface area and a slow and consistent release of nutrients (Prasad et al., 2017; Seleiman et al., 2020; Pitambara and Shukla, 2019).

2.2 ADAPTATION AND UPTAKE MECHANISMS OF NANOFERTILIZER IN PLANTS

The uptake and accumulation of NFs from soil to plants must investigate because this information could help determine which NFs are better for plants. If NFs or NPs prefer to go along the xylem, irrigation is the most convenient approach to apply them. Meanwhile, the exogenous application is advised and acceptable if NFs move by phloem (Pitambara and Shukla, 2019). The content of NFs, particle size of NPs, and plant cell mechanisms influence the nutrients uptake and accumulation processes released by NFs in plants (Rico et al., 2011; Pitambara and Shukla, 2019).

2.3 FOLIAR AND SOIL EXPOSURE OF NFs IN PLANTS

Before entering plant tissues, NPs must pass through the cuticular barrier in foliar applications (Pollard et al., 2008). The cuticular and stomatal pathways make up the cuticle layer of leaves, a waxy coating with two stages. Nonpolar solutes use the lipophilic pathway to reach the plant leaves via diffusion process (Rajput et al.,2021), whereas the stomatal pathway is used by polar solutes (Eichert et al., 2008). The cuticular pathway allows NPs with less than 5.0 nm to enter the cuticle via the cuticular route. NPs greater than 5.0 nm, on the other hand, have been shown to enter plants through foliar application in few studies (Lv et al., 2019). Changes in leaf ultrastructure and the frequency and diameter of stomata between a variety of plants, on the other hand, may affect NP uptake by foliar application (Wiesner et al., 2009). After entering the leaf apoplast, NPs may enter long distances via the plant's circulatory system by following the stomatal pathway (Lough and Lucas, 2006; Lv et al., 2019).

The uptake of NPs by the roots is influenced by plant morphology, growth stage, exposure conditions, particle size, and rhizosphere processes. In the roots of *Arabidopsis thaliana*, silicon (SiO_2) NPs with 50 to 200 nm diameter were discovered (Slomberg and Schoenfisch, 2012). The surface charge impacts NPs absorption and translocation in plants (Lv et al., 2019). Varieties of plants have shown varying NP uptake capacities, likely due to differences in physiological and metabolic mechanisms (Lv et al., 2019).

Nanoparticles are first adsorbed on the roots' surface, then move via a series of barriers to enter the plant's vascular system (Lv et al., 2019). The first barrier is the root cuticle layer analogous to the leaf cuticle layer in a composition. Nanoparticles penetrate the root epidermis after crossing the cuticle on the root surface. NPs can penetrate the root epidermis via apoplastic or symplastic pathways. According to several demonstrations, NPs can penetrate cell wall pores first, then migrate into intercellular gaps in the apoplastic pathway (Lv et al., 2019). The cell wall pore diameter, ranging from 5 to 20 nm, prevents NPs smaller than 20 nm from passing through the apoplastic route. As a result, the Casparian strip around the vascular system is the apoplastic pathway's most significant barrier. Another hypothesis is the symplastic pathway (Rico et al., 2011; Lv et al., 2019), in which NPs travel from one cell to another cell by plasmodesmata. When NPs enter the central cylinder, they flow through the xylem's transpiration stream to the plant's upper parts (LaRue et al., 2012).

2.4 MACRONUTRIENT NFs: THEIR FUNCTION AND IMPACT ON PLANTS

Fertilizers are essential for plant development and increase crop productivity and quality because they provide the necessary elements for plant development. Plants need a sufficient supply of macronutrients as N, P, and K (Czymmek et al., 2020; Kumar et al., 2019; Adeyemi et al., 2020; Batta et al., 2018). Because most of these nutrients are poorly absorbed by plant roots, farmers often apply maximum fertilizer quantity to partly compensate for their minimum NUE, resulting in a well-known adverse effect on soil, water, and the atmospheric environment (Seleiman et al., 2020; Czymmek et al., 2020; Chhipa, 2017; Seleiman et al., 2013). For sustainable agriculture, the use of NFs will enhance fertilizer NUE, improve plant productivity, and reduce the harmful effects of traditional fertilizers (Seleiman et al., 2020; Liu and Lal, 2015; Battaglia et al., 2018). NFs release nutrients precisely in the plant roots, decreasing nutrient losses by avoiding rapid variations in the soil composition. NFs are formed from various materials and carriers, including hydroxyapatite NPs, zeolite, mesoporous silica NPs, N, Cu, Zn, Si, C, and polymeric NPs (Guo et al., 2018; Liscano et al., 2000; Mikhak et al., 2017).

2.4.1 NITROGEN NFs

Nitrogen (N) is a vital mineral element for plants and can be found in various amino acids, proteins, DNA, ATP, chlorophylls, and cellular structural units. The majority of metabolic activities and regulatory mechanisms in plants require an adequate amount of nitrogen. Plants absorb nitrogen as NO^{-3} and NH^{+4} (Seleiman et al., 2020; Preetha and Balakrishnan, 2017). One of the most significant drawbacks of traditional N fertilizer is the excess rate of volatilization and leaching during and shortly after application in the field. Nitrogen-based NFs could deliver a steady supply of N at a moderate frequency to reduce these losses. NPK-coated NFs were recently employed on coffee plants grown in a greenhouse. The authors found that applying

NPK NF to coffee plants increased nutrient uptake and growth by enhancing the leaf numbers/area and leaf gas exchange capacity (Ha et al., 2018). Nitrogen in NFs is advised since it can result in a slower nitrogen release, lower volatilization and leaching rates, increased nutrient absorption, and improved plant performance and yield (Table 2.1).

2.4.2 PHOSPHORUS NFs

After nitrogen, phosphorus (P) is an important element for better plant performance. ATP, ADP (adenine triphosphate), phospholipids, and sugar phosphate are the energy transfer molecules. Photosynthetic capacity, respiration, and DNA biosynthesis are associated (Soliman et al., 2016). The availability of P affects plant growth and production efficiency (Preetha and Balakrishnan, 2017). Because of its lengthy release time and high soil fixation, phosphorus in synthetic fertilizers is not readily available. NFs now supplied P for 45–55 days after treatment, whereas conventional P synthetic fertilizers release all nutrients in 10 days (Liu and Lal, 2015). NFs could boost the NUE of P in some plants (Liu and Lal, 2015). In addition to contributing to high NUE, biosafe NFs that are a P source were found to consider enhancing biomass, production, and crop quality (Patra et al., 2013; Table 2.1).

2.4.3 POTASSIUM NFs

After nitrogen and phosphorus, potassium (K) is a crucial element, and it plays a critical action mechanism in all of a plant's physiochemical functions to maintain proper plant development. Potassium is linked with various processes in plants, including stomatal opening, photosynthetic responses, photosynthate translocation, protein synthesis, ionic balance, water interactions, and enzymatic mechanisms (Preetha and Balakrishnan, 2017). Abiotic stresses, that is, water-deficit and excess light intensity are more resistant in plants with adequate K levels (Sohair et al., 2018; Wang et al., 2013; Taha et al., 2020). According to the scientists, utilizing a nano-K fertilizer could decrease K losses in the soil while simultaneously providing a longer supply of K to crops. *Cucurbita pepo's* growth, biomass, and quality were significantly improved by foliar application of nano-K fertilizer (Gerdini, 2016). By lowering K losses into the soil and leading, K NFs may prevent soil quality, enhance water uptake, and improve physiological and yield parameters (Table 2.1).

2.4.4 *MICRONUTRIENT NANOFERTILIZERS AND THEIR ROLE*

Micronutrients are essential for optimizing plant production and quality and enhance plant resistance capacity to different unfavorable conditions (Seleiman et al., 2012; 2013; 2020). The use of nanoscale structures to synthesize micronutrients could improve their solubility and bioavailability, aid in their uniform distribution of soil dispersion, and reduce micronutrient adsorption and fixation to soil colloids.

2.4.5 *ZINC NFs*

Zn is a structural component co-factor for various proteins and enzymes, and it is essential for proper plant growth. Auxin regulation, protein metabolism, carbohydrate biosynthesis, and plant protection against biotic and abiotic stresses are all zinc-dependent processes (Broadley et al., 2007). Zinc NFs in ZnO are more efficient and cost-effective than synthetic fertilizers (Khanm et al., 2018; Seleiman et al., 2020); they are frequently used in advanced agri-systems (Seleiman et al., 2020). It can be used for soil mixing, seed priming, and foliar application. High concentrations of trace metals like Zn, on the other hand, may harm plant growth by inducing metabolic alterations (Ali et al., 2020). The use of Zn NFs has been shown in studies to improve crop germination, seedling growth, and productivity (Seleiman et al., 2020). Zn NFs in ZnO are the most widely used NFs in modern agriculture, with uses in foliar, soil mixing, and seed priming; they are also less expensive than synthetic Zn fertilizers. They improve crop yield and quality by promoting growth (Table 2.1).

2.4.6 *IRON NFs*

Chlorophyll synthesis, DNA synthesis, chloroplast ultrastructure, respiration, and other metabolic pathways all require iron (Fe). Plants require a small quantity of Fe for growth, but its deficit or excess harms their physiological and metabolic functions (Palmqvist et al., 2017). Iron availability is generally high in well-watered soils. However, Fe in these soils produces insoluble ferric complexes at neutral pH values, leaving it unavailable to plants. As a result, Fe-enriched fertilizers may assist plants in absorbing Fe. In relation to normal and/or traditional Fe sources, Fe NFs have been demonstrated in several studies to improve the germination and growth of various crops. The development of spinach was aided by iron pyrite nanoparticles (Srivastava et al., 2014). Under field conditions, assessed improved root growth in peanut applied with Fe NPs relative to control (Rui et al., 2016). Fe NFs could be an excellent alternative source, especially in soils with low Fe levels (Table 2.1).

2.4.7 MANGANESE NFs

Manganese (Mn) is a micronutrient element important for nitrogen metabolism, photosynthetic capacity, fatty acids, ATP, and protein production (Palmqvist et al., 2017). Despite this, Mn may be hazardous to various plants depending on the chemical characteristics of acidic soil. Manganese also aids plants in coping with a variety of stresses. Mn applications have been shown to increase wheat, maize, sugarcane, soybeans, and common beans growth and productivity (Fageria, 2001; Dimkpa and Bindraban, 2016). Mn NPs binds to the chlorophyll-binding protein (CP_{43}) of photosystem-II at the physiological level, increasing the photosynthetic mechanisms' electron transport chain and overall efficiency (Pradhan et al., 2013). Fertilized plants with Mn NFs have a higher nitrogen uptake and metabolic activities than their bulk counterparts (Pradhan et al., 2013, Table 2.1).

2.4.8 COPPER NFs

Copper (Cu) is a component of regulatory proteins involved in plants' photosynthetic and respiration mechanisms and a cofactor of antioxidative enzymes, that is, SOD and AsO. Copper imbalance causes different problems, that is, chlorosis, necrosis, stunted growth, low seed, grain, fruit yield, and finally, and low crop yield (Rai et al., 2018). Because the amount of organic matter in the soil influences Cu availability, Cu NPs in the soil can be significant due to their broad surface area, maximum solubility, and reactivity (Hong et al., 2015). CuO NPs in the field increased the soybeans and chickpeas germination (%) and root development (Adhikari et al., 2012). Cu NFs can improve biochemical and yield characteristics significantly and positively, but they must be used with caution (Table 2.1).

2.4.9 SILICON NFs

For proper plant fitness, silicon (Si) has been placed between essential and nonessential components. However, it provides some benefits to plants (Rastogi et al., 2019; Seleiman et al., 2019). Although silicon is abundant

TABLE 2.1 Types of NFs and Concentrations With the Application Focus and Effects Found on Crop Plants.

NFs	Plant species	Application	Concentration	Response	Source
nZn	Cotton (*Gossypium barbadense* L.)	Soil irrigation	0–200 mg L^{-1}	Growth and yield traits enhanced except root DW and S/R ratio. Enhance the content of nZn can decrease P uptake and accumulation to leaves and consequently minimize the ratio of P/Zn	Hussein and Abou-Baker, 2018
nZn	Soybean (*Glycine max* L.)	Petridish	0–1 mg L^{-1}	Germination and percentage rate enhanced, but dry mass reduced	Sedghi et al., 2013
nZn	Okra (*Abelmoschus esculentus* L.Moench)	Foliar spray	0–10 mg L^{-}	Increased chlorophyll content, enzyme activities, and decreased proline and sugar level	Alabdallah and Alzahrani, 2020
nZn and nSi	Mango (*Mangifera indica* L.)	Foliar spray	nZn: 0–150 mg L^{-1} and nSi: 0–300 mg L^{-1}	Upgrade plant performance, uptake of minerals, and photosynthetic capacity	Elsheery et al., 2020
nSi	Tomato (*Solanum lycopersicum* L.)	In vitro culture	0–3 mM	Up and downregulated genes help to alleviate the salinity stress	Almutairi, 2016
nSi	Strawberry (*Fragaria × ananassa*)	Nutrient soil	0–100 mg L^{-1}	Improved growth characteristics, enhanced proline, chlorophyll content, improved/balance epicuticular wax layer, LRWC, and canopy temperature	Avestan et al., 2019
nSi	Banana (*Musa acuminata* "Grand Nain")	In vitro	0–600 mg L^{-1}	Increased plant growth, development, and photosynthetic pigments, improved leaf gas exchange, maintain K+ and Na+ balance, reduce cell wall damage	Mahmoud et al., 2020
nSi	Wheat (*Triticum aestivum* L.)	Nutrient solution	0–10 μM	Increased antioxidants enzyme activities to mitigate UV-B generated oxidative stress	Tripathi et al., 2017
nFe	Menthe (*Mentha piperita* L.)	Hoagland solution	0–30 μM	Downregulated proline and MDA content, the suppressed antioxidative activities	Askary et al., 2017

TABLE 2.1 *(Continued)*

NFs	Plant species	Application	Concentration	Response	Source
nFe	Moldavian dragonhead (*Dracocephalum moldavica* L.)	Foliar spray	0–90 mg L^{-1}	Enhanced leaf area-expansion, phenolic, flavonoid, and anthocyanin activities and GPX, APX, CAT, and GR activities	Moradbeygi et al., 2020
nFe	Wheat (*Triticum aestivum* L.)	Soil mixture	0–100 mg kg^{-1} soil	Maintain growth and physiological parameters, Fe content, and decrease cadmium level	Adrees et al., 2020
nZn	Cucumber (*Cucumis sativus* L.)	Soil culture	100–1000 mg kg^{-1} soil	Dose-dependent responses on biomass and antioxidant activities	Moghaddasi et al., 2017
nCu	Maize (*Zea mays* L.)	-	0.04–0.12%	Enhanced growth parameters by enhancing the activities of α-amylase and starch content	Saharan et al.,2016
nMn	Mung bean (*Vigna radiate* L.)	Hoagland solution	0.05 ppm	Increase biomass, stem, and root height. Also, greater photophosphorylation and upregulated oxygen evolution in chloroplasts	Pradhan et al., 2013
nN	Rice (*Oryza sativa* L.)	Soil irrigation	25–100%	Enhanced tillers per plant, height, and dry mass	Rathnayaka et al., 2018
nP	Soybean (*Glycine max* L.)	–	0–100 mg L^{-1}	Enhanced productivity and root length	Liu and Lal, 2015
nP	Lettuce (*Lactuca sativa* L.)	Soil culture	0–100 mg kg^{-1} soil	Enhanced P level and dry mass	Taskin et al., 2018
nK	Peanut (*Arachis hypogaea* L.)	Soil irrigation	1500–2500 mg L^{-1}	Enhanced morphological and biological yield.	Asgari et al., 2018
nB	Pomegranate (*Punica granatum* cv. Ardestani)	Foliar spray	0–6.5 mg L^{-1}	Improve nutrient uptake, increase the number of fruits, quality, and productivity. No changes were noted in antioxidant and total anthocyanins activities.	Davarpanah et al., 2016

in the earth, plants only absorb it in mono-silicic acid from the soil. Due to its varied function in plants' resistance to different stresses, Si has recently received a lot of attention (Seleiman et al., 2019). Si has been shown to enhance plant adaptation strategies to ion toxicity, temperature, drought, chilling, UV, waterlogging, and salinity stresses (Rastogi et al., 2019; Seleiman et al., 2019). Furthermore, using SiO_2 in conjunction with organic fertilizers can boost overall plant performance (Janmohammadi et al., 2016; Seleiman et al., 2020). Si NPs-based nanosensors and nanozeolites have been used successfully in agriculture to monitor soil moisture and increase soil water retention (Rastogi et al., 2019). Si NPs could be used as fertilizers for plants that need a certain amount of silicon to flourish or as nano-carriers to assist agriculture is becoming more sustainable (Table 2.1).

2.4.10 BORON NFs

Boron (B) is a necessary component for forming cellular walls, the movement of photosynthetic organisms from leaves to active sites, and the production of flowers and fruits (Davarpanah et al., 2016). B NFs or NPs have been shown in studies to enhance plant growth and productivity (Ibrahim and Al Farttoosi, 2019). Genaidy et al. (2020) applied as-sprayed nano-boron on olive plants; the plants produced the most fruit with the highest seed oil content. Taherian et al. (2019) used B nanofertilizer on calcareous soil on alfalfa plants. They harvested a high-yielding crop with good forage quality. Finally, NFs of B may help increase crop production (Table 2.1).

2.5 DEVELOPMENT OF NFs FOR AGRO-ECOSYSTEMS

In upcoming years, agriculture will be under enhancing overload to maintain food safety for a fast-growing global populace while also lowering its total environmental effect. Modification of current fertilization methods could be one choice for increasing biomass and grain yields. Plants required sufficient mineral elements for proper development. The minerals are delivered to them in synthetic fertilizers, which have been steadily rising in adoption among growers throughout the globe since the era of the green revolution (De la Luz Mora et al., 2007). Innovative fertilizers, such as NFs have been suggested as a technique to improve fertilizer NUE by allowing for a more regulated and slower nutrient release that better matches crop's long-term nutritional needs (DeRosa et al., 2010; Bley et al., 2017). Semipermeable coatings on the surfaces of or inside fertilizers may be utilized to achieve a coherent and slow release of nutrients for a long time (Naz and Sulaiman, 2016). This would result in a new fertilizer framework that delivers the appropriate quantity of nutrients at the crucial period and significant nutrient loss to the environment.

2.6 CONCLUSIONS AND FUTURE ASPECTS

According to sustainable agriculture, nanobioscience can generate new novel fertilizer methods to boost global food security and feed the rising worldwide populace. Because of their huge surface areas and slow and consistent release of nutrients, NFs have potential as part of smart crop production systems in sustainable agriculture. Because of these promising properties, they are perfect for application in the advanced agricultural system. The usage of NFs in agriculture will benefit both production and resistance to biotic and abiotic challenges. By delivering active ingredients more efficiently, boosting nutrient and NUE uptake, and minimizing fertilizer loss due to volatilization, leaching, runoff, and wasted energy during production, NFs may help minimize fertilizer use. Seed coverings with NFs and nanosensors can also lower agricultural crop production costs and environmental issues. Future research should focus on the safety, bioavailability, and toxicity of various NFs used in agriculture. Bio-synthesized or green synthesized nano-biofertilizers and NFs should be investigated to increase yields in sustainable agriculture.

ACKNOWLEDGMENTS

Dr. Rajput and Dr. Burachevskaya would like to recognize The Ministry of Science and Higher Education of Russian Federation provided financial support for the study (no. 0852-2020-0029).

KEYWORDS

- **agro-industries**
- **fertilizers**
- **innovative technologies**
- **nanoscale**
- **nutrient use efficiency**
- **toxicity**

REFERENCES

Adeyemi, O.; Afshar, R. K.; Jahanzad, E.; Battaglia, M. L.; Luo, Y.; Sadeghpour, A. Effect of Wheat Cover Crop and Split Nitrogen Application on Corn Yield and Nitrogen Use Efficiency. *Agronomy* **2020,** *10*, 1081.

Adhikari, T.; Kundu, S.; Biswas, A. K.; Tarafdar, J. K.; Rao, A. S. Effect of Copper Oxide Nano Particle on Seed Germination of Selected Crops. *J. Agric. Sci. Tech.* **2012,** *2*, 815.

Adnan, M.; Fahad, S.; Zamin, M.; Shah, S.; Mian, I. A.; Danish, S.; Zafar-Ul-Hye, M.; Battaglia, M. L.; Naz, R. M. M.; Saeed, B.; Saud, S.; Ahmad, I.; Yue, Z.; Brtnicky, M.; Holatko, J.; Datta, R. Coupling Phosphate-Solubilizing Bacteria With Phosphorus Supplements Improve Maize Phosphorus Acquisition and Growth Under Lime Induced Salinity Stress. *Plants* **2020,** *9*, 900.

Adrees, M.; Khan, Z. S.; Ali, S.; Hafeez, M.; Khalid, S.; Ur Rehman, M. Z.; Hussain, A.; Hussain, K.; Chatha, S. A. S.; Rizwan, M. Simultaneous Mitigation of Cadmium and Drought Stress in Wheat by Soil Application of Iron Nanoparticles. *Chemosphere* **2020,** *238*, 124681.

Alabdallah, N. M.; Alzahrani, H. S. The Potential Mitigation Effect of ZnO Nanoparticles on (*Abelmoschus esculentus* L. Moench) Metabolism Under Salt Stress Conditions. *Saudi. J. Biol. Sci.* **2020,** *27*, 3132–3137.

Almutairi, Z. M. Effect of Nano-Silicon Application on the Expression of Salt Tolerance Genes in Germinating Tomato (*Solanum lycopersicum* L.) Seedlings Under Salt Stress. *Plant Omics* **2016,** *9*, 106.

Arora, N. K. Agricultural Sustainability and Food Security. *Environ. Sustain.* **2018,** *1*, 217–219.

Asgari, S.; Moradi, H.; Afshari, H. Evaluation of Some Physiological and Morphological Characteristics of Narcissus Tazatta Under BA Treatment and Nano potassium Fertilizer. *J. Chem. Health Risks* **2018,** 4.

Askary, M.; Talebi, S. M.; Amini, F.; Bangan, A. D. Effects of Iron Nanoparticles on Mentha Piperita L. Under Salinity Stress. *Biologija* **2017,** *63*, 65–75.

Avellan, A.; Yun, J.; Zhang, Y.; Spielman-Sun, E.; Unrine, J. M.; Thieme, J.; Li, J.; Lombi, E.; Bland, G.; Lowry, G. V. Nanoparticle Size and Coating Chemistry Control Foliar Uptake Pathways, Translocation, and Leaf-to-Rhizosphere Transport in Wheat. *ACS Nano* **2019,** *13*, 5291–5305.

Battaglia, M. L.; Groover, G.; Thomason, W. E. *Harvesting and Nutrient Replacement Costs Associated With Corn Stover Removal in Virginia*; Virginia Cooperative Extension Publication: Ettrick, VA, 2018; CSES-229NP.

Bley, H.; Gianello, C.; Santos, L. D. S.; Selau, L. P. R. Nutrient Release, Plant Nutrition, and Potassium Leaching From Polymer-Coated Fertilizer. *Rev. Brasil. Ciênc. Solo.* **2017,** *41*, 0160142.

Broadley, M. R.; White, P. J.; Hammond, J. P.; Zelko, I.; Lux, A. Zinc in Plants. *New Phytol.* **2007,** *173*, 677–702.

Brunner, T. J.; Wick, P.; Manser, P.; Spohn, P.; Grass, R. N.; Limbach, L. K.; Bruinink, A.; Stark, W. J. In-Vitro Cytotoxicity of Oxide Nanoparticles, Comparison to Asbestos, Silica, and the Effect of Particle Solubility. *Environ. Sci. Technol.* **2006,** *40*, 4374–4381.

Chhipa, H. Nanofertilizers and Nanopesticides for Agriculture. *Environ. Chem. Lett.* **2017,** *15*, 15–22.

Czymmek, K.; Ketterings, Q.; Ros, M.; Battaglia, M.; Cela, S.; Crittenden, S.; Gates, D.; Walter, T.; Latessa, S.; Klaiber, L.; Albrecht, G. *The New York Phosphorus Index 2.0*;

Agronomy Fact Sheet Series; Fact Sheet #110; Cornell University Cooperative Extension: Ithaca, NY, 2020.

Davarpanah S.; Tehranifar A.; Davarynejad, G.; Abadia, J.; Khorasani, R. Effects of Foliar Applications of Zinc and Boron Nano-Fertilizers on Pomegranate (*Punica granatum* cv. Ardestani) Fruit Yield and Quality. *Sci. Hortic.* **2016,** *210*, 57–64.

De la Luz Mora, M.; Cartes, P.; Núñez, P.; Salazar, M.; Demanet, R. Movement of NO^{3-}-N and NH^{4+}-N in an Andisol and its Influence on Ryegrass Production in a Short-Term Study. *J. Soil Sci. Plant. Nutr.* **2007,** *7*, 46–63.

DeRosa, M. C.; Monreal, C.; Schnitzer, M.; Walsh, R.; Sultan, Y. Nanotechnology in Fertilizers. *Nat. Nanotechnol.* **2010,** *5*, 91.

Diatta, A. A.; Thomason, W. E.; Abaye, O.; Thompson, T. L.; Battaglia, M. L.; Vaughan, L. J.; Lo, M.; Filho, J. F. D. C. L. Assessment of Nitrogen Fixation by Mungbean Genotypes in Different Soil Textures Using 15N Natural Abundance Method. *J. Soil Sci. Plant Nutr.* **2020,** *20*, 2230–2240.

Dimkpa, C. O.; Bindraban, P. S. Fortification of Micronutrients for Efficient Agronomic Production, A Review. *Agron. Sustain. Dev.* **2016,** *36*, 1–26.

Dwivedi, S.; Saquib, Q.; Al-Khedhairy, A. A.; Musarrat, J. Understanding the Role of Nanomaterials in Agriculture. In *Microbial Inoculants in Sustainable Agricultural Productivity;* Springer Science and Business Media LLC: Heidelberg, Germany, 2016; pp 271–288.

Eichert, T.; Kurtz, A.; Steiner, U.; Goldbach, H. E. Size Exclusion Limits and Lateral Heterogeneity of the Stomatal Foliar Uptake Pathway for Aqueous Solutes and Water-Suspended Nanoparticles. *Physiol. Plant.* **2008,** *134*, 151–160.

Elsheery, N. I.; Helaly, M. N.; El-Hoseiny, H. M.; Alam-Eldein, S. M. Zinc Oxide and Silicone Nanoparticles to Improve the Resistance Mechanism and Annual Productivity of Salt-Stressed Mango Trees. *Agronomy* **2020,** *10*, 558. DOI: 10.3390/agronomy10040558.

Fageria, V. D. Nutrient Interactions in Crop Plants. *J. Plant. Nutr.* **2001,** *24*, 1269–1290.

FAO. *The Future of Food and Agriculture-Trends and Challenges*; Annual Report, FAO: Rome, Italy, 2017.

Feregrino-Perez, A. A.; Magaña-López, E.; Guzman, C.; Esquivel, K. A General Overview of the Benefits and Possible Negative Effects of the Nanotechnology in Horticulture. *Sci. Hortic.* **2018,** *238*, 126–137.

Food and Agriculture Organization of the United Nations (FAO). FAO Statistics Division; [Online]. http.//www.fao.org/faostat/en/#data/QC/visualize (accessed 20 Mar 2021).

Genaidy, E. A. E.; Abd-Alhamid, N.; Hassan, H. S. A.; Hassan, A. M.; Hagagg, L. F. Effect of Foliar Application of Boron Trioxide and Zinc Oxide Nanoparticles on Leaves Chemical Composition, Yield and Fruit Quality of *Olea europaea* L. cv. Picual. Bull. *Natl. Res. Cent.* **2020,** *44*, 106.

Gerdini, F. Effect of Nano Potassium Fertilizer on Some Parchment Pumpkin (*Cucurbita pepo*) Morphological and Physiological Characteristics Under Drought Conditions. *Intl. J. Farm Alli Sci.* **2016,** *5*, 367–371.

Guo, H.; White, J. C.; Wang, Z.; Xing, B. Nano-Enabled Fertilizers to Control the Release and Use Efficiency of Nutrients. *Curr. Opin. Environ. Sci. Heal.* **2018,** *6*, 77–83.

Ha, N. M. C.; Nguyen, T. H.; Wang, S. L.; Nguyen, A. D. Preparation of NPK Nanofertilizer Based on Chitosan Nanoparticles and its Effect on Biophysical Characteristics and Growth of Coffee in Green House. *Res. Chem. Intermed.* **2018,** *45*, 51–63.

Hong, J.; Rico, C. M.; Zhao, L.; Adeleye, A. S.; Keller, A. A.; Peralta-Videa, J. R.; Gardea-Torresdey, J. L. Toxic Effects of Copper-Based Nanoparticles or Compounds to Lettuce (*Lactuca sativa*) and Alfalfa (*Medicago sativa*). *Environ. Sci. Process. Impacts* **2015,** *17*, 177–185.

Hussein, M. M.; Abou-Baker, N. H. The Contribution of Nano-Zinc to Alleviate Salinity Stress on Cotton Plants. *R. Soc. Open Sci.* **2018,** *5*, 171809.

Ibrahim, N. K.; Al-Farttoosi, H. A. K. Response of Mung Bean to Boron Nanoparticles and Spraying Stages (*Vigna Radiata* L.). *Plant. Arch.* **2019,** *19*, 712–715.

Iravani, S.; Korbekandi, H.; Mirmohammadi, S.; Zolfaghari, B. Synthesis of Silver Nanoparticles, Chemical, Physical and Biological Methods. *Res. Pharm. Sci.* **2015,** *9*, 385–406.

Janmohammadi, M.; Amanzadeh, T.; Sabaghnia, N.; Ion, V. Effect of Nano-Silicon Foliar Application on Safflower Growth Under Organic and Inorganic Fertilizer Regimes. *Bot. Lith.* **2016,** *22*, 53–64.

Kah, M.; Tufenkji, N.; White, J. C. Nano-Enabled Strategies to Enhance Crop Nutrition and Protection. *Nat. Nanotechnol.* **2019,** *14*, 532–540.

Kerry, R. G.; Gouda, S.; Das, G.; Vishnuprasad, C. N.; Patra, J. K. Agricultural Nanotechnologies, Current Applications and Future Prospects. In *Microbial Biotechnology*; Springer Science and Business Media LLC: Singapore, 2017; pp 3–28.

Kumar, P.; Lai, L.; Battaglia, M. L.; Kumar, S.; Owens, V.; Fike, J.; Galbraith, J.; Hong, C. O.; Farris, R.; Crawford, R.; Crawford, J.; Hansen, J.; Mayton, H.; Viands, D. Impacts of Nitrogen Fertilization Rate and Landscape Position on Select Soil Properties in Switchgrass Field at Four Sites in the USA. *CATENA* **2019,** *180*, 183–193.

LaRue, C.; Laurette, J.; Herlin-Boime, N.; Khodja, H.; Fayard, B.; Flank, A. M.; Brisset, F.; Carriere, M. Accumulation, Translocation and Impact of TiO_2 Nanoparticles in Wheat (*Triticum aestivum* spp.), Influence of Diameter and Crystal Phase. *Sci. Total. Environ.* **2012,** *431*, 197–208.

Lee, K. X.; Shameli, K.; Yew, Y. P.; Teow, S. Y.; Jahangirian, H.; Rafiee-Moghaddam, R.; Webster, T. J. Recent Developments in the Facile Biosynthesis of Gold Nanoparticles (AuNPs) and Their Biomedical Applications. *Int. J. Nanomed.* **2020,** *15*, 275–300.

Liscano, J. F.; Wilson, C. E.; Norman, R. J.; Jr, Slaton, N. A. *Zinc Availability to Rice From Seven Granular Fertilizers*; Arkansas Agricultural Experiment Station: Fayetteville, CA, 2000, pp 963.

Liu, R.; Lal, R. Synthetic Apatite Nanoparticles as a Phosphorus Fertilizer for Soybean (*Glycine max*). *Sci. Rep.* **2015,** *4*, srep05686.

Liu, R.; Lal, R. Potentials of Engineered Nanoparticles as Fertilizers for Increasing Agronomic Productions. *Sci. Total. Environ.* **2015,** *514*, 131–139.

Lough, T. J.; Lucas, W. J. Integrative Plant Biology, Role of Phloem Long-Distance Macromolecular Trafficking. *Annu. Rev. Plant Biol.* **2006,** *57*, 203–232.

Lv, J.; Christie, P.; Zhang, S. Uptake, Translocation, and Transformation of Metal-Based Nanoparticles in Plants, Recent Advances and Methodological Challenges. *Environ. Sci. Nano.* **2019,** *6*, 41–59.

Mahmoud, L. M.; Dutt, M.; Shalan, A. M.; El-Kady, M. E.; El-Boray, M. S.; Shabana, Y. M.; Grosser, J. W. Silicon Nanoparticles Mitigate Oxidative Stress of In Vitro Derived Banana (*Musa acuminata* 'Grand Nain') Under Simulated Water Deficit or Salinity Stress. *South Afr. J. Bot.* **2020,** *132*, 155–163.

Mikhak, A.; Sohrabi, A.; Kassaee, M. Z.; Feizian, M. Synthetic Nanozeolite/nanohydroxyapatite as a Phosphorus Fertilizer for German Chamomile (*Matricaria chamomilla* L.). *Ind. Crop. Prod.* **2017,** *95*, 444–452.

Moghaddasi, S.; Fotovat, A.; Hkoshgoftarmanesh, A. H.; Karimzadeh, F.; Khazaei, H. R.; Khorassani, R. Bioavailability of Coated and Uncoated ZnO Nanoparticles to Cucumber in Soil With or Without Organic Matter. *Ecotoxicol. Environ. Saf.* **2017,** *144*, 543–551.

Moradbeygi, H.; Jamei, R.; Heidari, R.; Darvishzadeh, R. Investigating the Enzymatic and Non-Enzymatic Antioxidant Defense by Applying Iron Oxide Nanoparticles in Dracocephalum Moldavica L. Plant Under Salinity Stress. *Sci. Horticult.* **2020,** *272*, 109537.

Naz, M.; Sulaiman, S. A. Slow-Release Coating Remedy for Nitrogen Loss From Conventional Urea, A Review. *J. Control. Release* **2016,** *225*, 109–120.

Palmqvist, N. M.; Seisenbaeva, G. A.; Svedlindh, P.; Kessler, V. G. Maghemite Nanoparticles Acts as Nanozymes, Improving Growth and Abiotic Stress Tolerance in *Brassica napus*. *Nanoscale Res. Lett.* **2017,** *12*, 1–9.

Patra, P.; Choudhury, S. R.; Mandal, S.; Basu, A.; Goswami, A.; Gogoi, R.; Srivastava, C.; Kumar, R.; Gopal, M. Effect Sulfur and ZnO Nanoparticles on Stress Physiology and Plant (*Vigna radiata*) Nutrition. In *Advanced Nanomaterials and Nanotechnology*; Springer: Guwahati, India, 2013; pp 301–309.

Pitambara, A.; Shukla, Y. M. Nanofertilizers: A Recent Approach in Crop Production. *Nanotechnol. Agric. Crop Prod. Prot.* **2019,** 25–28.

Pollard, M.; Beisson, F.; Li, Y.; Ohlrogge, J. B. Building Lipid Barriers, Biosynthesis of Cutin and Suberin. *Trends Plant Sci.* **2008,** *13*, 236–246.

Pradhan, S.; Patra, P.; Das, S.; Chandra, S.; Mitra, S.; Dey, K. K., Akbar, S.; Palit, P.; Goswami, A. Photochemical Modulation of Biosafe Manganese Nanoparticles on *Vigna radiata*, A Detailed Molecular, Biochemical, and Biophysical Study. *Environ. Sci. Technol.* **2013,** *47*, 13122–13131.

Prasad, R.; Bhattacharyya, A.; Nguyen, Q. D. Nanotechnology in Sustainable Agriculture, Recent Developments, Challenges, and Perspectives. *Front. Microbiol.* **2017,** *8*, 1014.

Preetha, P. S.; Balakrishnan, N. A Review of Nano Fertilizers and Their Use and Functions in Soil. *Int. J. Curr. Microbiol. Appl. Sci.* **2017,** *6*, 3117–3133.

Rai, M.; Ingle, A. P.; Pandit, R.; Paralikar, P.; Shende, S.; Gupta, I.; Biswas, J. K.; Da Silva, S. S. Copper and Copper Nanoparticles, Role in Management of Insect-Pests and Pathogenic Microbes. *Nanotechnol. Rev.* **2018,** *7*, 303–315.

Rajput, V. D.; Minkina, T.; Feizi, M.; Kumari, A.; Khan, M.; Mandzhieva, S.; Sushkova, S.; El-Ramady, H.; Verma, K.; Singh, A.; Hullebusch, E.; Singh, R.; Jatav, H.; Choudhary, R. Effects of Silicon and Silicon-Based Nanoparticles on Rhizosphere Microbiome, Plant Stress and Growth. *Biology* **2021,** 10, 7–9.

Rajput, V. D.; Minkina, T.; Sushkova, S.; Viktoriia, T.; Saglara, M.; Andrey, G.; Dina, N.; Natalya, G. Effect of Nanoparticles on Crops and Soil Microbial Communities. *J. Soils Sediments* **2017,** *18*, 179–187.

Rastogi, A.; Tripathi, D. K.; Yadav, S.; Chauhan, D. K.; Živcák, M.; Ghorbanpour, M.; El-Sheery, N. I.; Brestic, M. Application of Silicon Nanoparticles in Agriculture. *3 Biotech* **2019,** *9*, 1–11.

Rathnayaka, R.; Iqbal, Y.; Rifnas, L. Influence of Urea and Nano-Nitrogen Fertilizers on the Growth and Yield of Rice (*Oryza sativa* L.) Cultivar 'Bg 250'. *Biol. Life Sci.* **2018,** *5*, 7–17.

Rico, C. M.; Majumdar, S.; Duarte-Gardea, M.; Peralta-Videa, J. R.; Gardea-Torresdey, J. L. Interaction of Nanoparticles With Edible Plants and Their Possible Implications in the Food Chain. *J. Agric. Food Chem.* **2011,** *59*, 3485–3498.

Rui, M.; Ma, C.; Hao, Y.; Guo, J.; Rui, Y.; Tang, X.; Zhao, Q.; Fan, X.; Zhang, Z.; Hou, T.; Zhu, S. Iron Oxide Nanoparticles as a Potential Iron Fertilizer for Peanut (*Arachis hypogaea*). *Front. Plant. Sci.* **2016,** *7*, 815.

Saharan, V.; Kumaraswamy, R. V.; Choudhary, R. C.; Kumari, S.; Pal, A.; Raliya, R.; Biswas, P. Cu-chitosan Nanoparticle Mediated Sustainable Approach to Enhance Seedling Growth in Maize by Mobilizing Reserved Food. *J. Agricul. Food. Chem.* **2016,** *64*, 6148–6155.

Sedghi, M.; Hadi, M.; Toluie, S. G. Effect of Nano Zinc Oxide on the Germination Parameters of Soybean Seeds Under Drought Stress. *Annal. West Univ. Timisoara ser Biol.* **2013,** *16*, 73–78.

Seleiman, M. F.; Alotaibi, M.; Alhammad, B. A.; Alharbi, B.; Refay, Y.; Badawy, S. A. Effects of ZnO Nanoparticles and Biochar of Rice Straw and Cow Manure on Characteristics of Contaminated Soil and Sunflower Productivity, Oil Quality, and Heavy Metals Uptake. *Agronomy* **2020,** *10*, 790.

Seleiman, M. F.; Refay, Y.; Al-Suhaibani, N.; Al-Ashkar, I.; El-Hendawy, S.; Hafez, E.; Suhaibani, A.; Ashkar, A.; Hendawy, E. Integrative Effects of Rice-Straw Biochar and Silicon on Oil and Seed Quality, Yield and Physiological Traits of *Helianthus annuus L.* Grown Under Water Deficit Stress. *Agronomy* **2019,** *9*, 637.

Seleiman, M. F.; Santanen, A.; Jaakkola, S.; Ekholm, P.; Hartikainen, H.; Stoddard, F. L.; Makela, P. Biomass Yield and Quality of Bioenergy Crops Grown With Synthetic and Organic Fertilizers. *Biomass Bioenergy* **2013,** *59*, 477–485.

Seleiman, M. F.; Santanen, A.; Stoddard, F. L.; Mäkelä, P. Feedstock Quality and Growth of Bioenergy Crops Fertilized With Sewage Sludge. *Chemosphere* **2012,** *89*, 1211–1217.

Shang, Y.; Hasan, K.; Ahammed, G. J.; Li, M.; Yin, H.; Zhou, J. Applications of Nanotechnology in Plant Growth and Crop Protection, A Review. *Molecules* **2019,** *24*, 2558.

Singh, M. D.; Kumar, B. A. Bio Efficacy of Nano Zinc Sulphide (ZnS) on Growth and Yield of Sunflower (*Helianthus annuus* L.) and Nutrient Status in the Soil. *Int. J. Agric. Sci.* **2017,** *9,* 3795–3798.

Singh, A.; Rajput, V. D., Mehrotra, R.; Pal, N.; Singh, V. K.; Minkina, T.; Chokheli, V. A.; Singh, R. K. In *Sustainable Soil Fertility Management*; Nova. Sci. Publishers, Inc., 2020; vol 1, pp 73–100.

Singh, A.; Rajput, V.; Singh, A. K.; Sengar, R. S.; Singh, R. K.; Minkina, T. Transformation Techniques and Their Role in Crop Improvements: A Global Scenario of GM Crops. In *Policy Issues in Genetically Modified Crops*, 2021; pp 515–542.

Slomberg, D. L.; Schoenfisch, M. H. Silica Nanoparticle Phytotoxicity to *Arabidopsis thaliana*. *Environ. Sci. Technol.* **2012,** *46*, 10247–10254.

Sohair, E. E. D.; Abdall, A. A.; Amany, A. M.; Hossain, M. F.; Houda, R. A. Effect of Nitrogen, Phosphorus and Potassium Nano Fertilizers With Different Application Times, Methods and Rates on Some Growth Parameters of Egyptian Cotton (*Gossypium barbadense* L.). *Biosci. Res.* **2018,** *15*, 549–564.

Soliman, A. S.; Hassan, M.; Abou-Elell, F.; Ahmed, A. H.; El-Feky, S. A. Effect of Nano and Molecular Phosphorus Fertilizers on Growth and Chemical Composition of Baobab (*Adansonia digitata* L.). *J. Plant Sci.* **2016,** *11*, 52–60.

Srivastava, G.; Das, C. K.; Das, A.; Singh, S. K.; Roy, M.; Kim, H.; Sethy, N.; Kumar, A.; Sharma, R. K.; Singh, S. K.; Philip, D.; Das, M. Seed Treatment With Iron Pyrite (FeS_2) Nanoparticles Increases the Production of Spinach. *RSC Adv.* **2014,** *4*, 58495–58504.

Taha, R.; Seleiman, M. F.; Alotaibi, M.; Alhammad, B. A.; Rady, M. M.; Mahdi, A. H. A. Exogenous Potassium Treatments Elevate Salt Tolerance and Performances of *Glycine max* L. by Boosting Antioxidant Defense System Under Actual Saline Field Conditions. *Agronomy* **2020,** *10*, 1741.

Taherian, M.; Bostani, A.; Omidi, H. Boron and Pigment Content in Alfalfa Affected by Nano Fertilization Under Calcareous Conditions. *J. Trace Elements Med. Biol.* **2019,** *53*, 136–143.

Taskın, M. B.; Sahin, O.; Taskin, H.; Atakol, O.; Inal, A.; Gunes, A. Effect of Synthetic Nano-Hydroxyapatite as an Alternative Phosphorus Source on Growth and Phosphorus Nutrition of Lettuce (*Lactuca sativa* L.) Plant. *J. Plant. Nutr.* **2018,** *41*, 1148–1154.

Tripathi, D. K.; Singh, S.; Singh, V. P.; Prasad, S. M.; Dubey, N. K.; Chauhan, D. K. Silicon Nanoparticles More Effectively Alleviated UV-B Stress Than Silicon in Wheat (*Triticum aestivum*) Seedlings. *Plant Physiol. Biochem.* **2017,** *110*, 70–81.

Usman, M.; Farooq, M.; Wakeel, A.; Nawaz, A.; Alam Cheema, S.; Rehman, H. U.; Ashraf, I.; Sanaullah, M. Nanotechnology in Agriculture, Current Status, Challenges and Future Opportunities. *Sci. Total. Environ.* **2020,** *721*, 137778.

Wang, W. N.; Tarafdar, J. C.; Biswas, P. Nanoparticle Synthesis and Delivery by an Aerosol Route for Watermelon Plant Foliar Uptake. *J. Nanoparticle Res.* **2013,** *15*, 1–13.

Wiesner, M. R.; Lowry, G. V.; Jones, K. L.; Hochella, J. M. F.; Di Giulio, R. T.; Casman, E.; Bernhardt, E. S. Decreasing Uncertainties in Assessing Environmental Exposure, Risk, and Ecological Implications of Nanomaterials. *Environ. Sci. Technol.* **2009,** *43*, 6458–6462.

CHAPTER 3

Nano-Pesticides: A Dab Hand at Eliminating Pests

PRAVIN KHAIRE, SOMESHREE MANE, TANAJI NARUTE, and NARAYAN MUSMADE

Department of Plant Pathology and Agriculture Microbiology, PGI, Mahatma Phule Krishi Vidyapeeth, Rahuri, Maharashtra, India

ABSTRACT

The agricultural sector provides, specifically and indirectly, food for human consumption. As the global population grows, new innovations such as biotechnology and nanotechnology (NT) need to be used in agricultural research. NT refers to substances, constructions, and operations that function on a scale of 100 nanometers or less. Every year, plant pests and pathogens damage 20–40% of crops. Synthetic pesticides that are potentially damaging to society and the environment are being used to control plant diseases. Nanomaterials have the potential to improve pesticides by lowering toxicity, lengthening their shelf lives, and enhancing the dispersion of pesticides that are poorly soluble in water, all of which could benefit the environment. In this chapter, we will look at two ways that nanoparticles can be utilized to treat pathogenic microorganisms. As a defensive nanoparticle, or for the molecules of insecticides, fungicides, herbicides, and gene silencing, either independently. Not many nanocomposites products were manufactured for agriculture, along with several potential benefits linked to the utilization of nanoparticles. Several variables could explain the absence of industrial products, such as insufficient field studies and underuse of pest-crop host systems. NT's has made rapid advances in other fields, and the only way to

Nanotechnology for Sustainable Agriculture: An Innovative and Eco-Friendly Approach. Vishnu D. Rajput, Abhishek Singh, Tatiana M. Minkina, Krishan K. Verma, and Awani Kumar Singh (Eds.)

maintain with this advancement for smart agriculture is to consider the basic problems of science and to resolve technical gaps in order to provide rational and promote the production of industrial nano-formulations.

3.1 INTRODUCTION

Taniguchi (1974) initially established the term nanotechnology (NT) in science, which focuses on the manufacture and deployment of nano-particulate matter (1–1-00 nm or 1.0 × 10-9 m). When it is nano compressed, it acts differently and exhibits some new features that in its macro-size type are utterly absent. These days, farmers are continuously dealing with pests and diseases threatening their crops. Fungal pathogens must be regulated if the need for higher quality and diverse food for consumers in developed countries is to be met, whereas elevated cereals, vegetables, and fruits are an indication of development of the economy (Singh et al., 2021b). Crop disease and insects induce loss in crop productivity, with global losses of 20–40% annually (Flood, 2010). The employment of chemical products, such as insects, fungicides, and herbicides, is a significant factor in the current pest control (Stephenson, 2003). The resistance of pathogens and insect pests to agrochemicals (pesticides and fungicides) is increasingly now becoming critical threat (Singh et al., 2021a). These activities are now being reassessed due to concern over potential health and environmental implications. Nanomaterials (NMs) have helped in the invention of modern facilities and agricultural goods with an amazing capability to handle the issues alluded to above (Singh et al., 2020). Application of NT has made considerable advances in medicine and pharmacology but has gained relatively less attention in agriculture (Sinha, 2017; Balaure, 2017). Nowadays in the field of plant hormone, germination of seeds, water treatment, target gene transfer, nano-barccoding, nano-devices, and controlled agricultural chemical produces, the use of NT to agriculture is being researched (Hayles, 2017). NPs, such as structure, pore surface morphology features, can be found by the scientists to serve as protectants or for effective and selective distribution by adsorption, encapsulating of an active component such as a pesticide (Khandelwal, 2016). The pesticides and other active compounds will intricately link the opportunity to suppress crop diseases in a current frontier as agricultural NT is growing. Opportunity to provide a new era with pesticides and other active substances for the suppression of plant diseases will significantly increase as agricultural NT grows. The use of NMs in

PDM is a novel and innovative method that may find very useful in the coming years with the advancement of the NT application component. NT has new opportunities for PDM use in a number of ways. The safest and most successful way is direct soil delivery of NPs to seeds or aerial plant parts to protect crop against pathogenic invasion. Throughout this way, NPs can inhibit the development of pathogens in a manner similar to pesticides. Since NPs should be injected directly into the soil, it would have a significant impact on nontarget species, in particular mineral microorganisms, which are fixed and solubilized. Subsequently, NMs, carbon nanotubes, cups, etc., can be utilized as providers of some trendy compounds like pheromones, chemical substances stimulating the SAR, polyamine synthesis inhibitors, and even pesticide a.i (Khan et al., 2014). NPs may offer a range of benefits, as carriers (Hayles, 2017), like,

a) Improved life span,
b) Enhanced water-soluble (WS) pesticide solubility,
c) Decreased toxicity and increased site-specific absorption of the target pest.

Other potential advantage of nano-carrier is to increase the action power and nanopesticides to stabilize. The number of operations decreases effectively, while reducing toxicity and saving costs under environmental restrictions (UV and rain) (Fig. 3.1).

In this chapter of the book, writers explore recent developments in RNA-interference (RNAi) PDMs that use both NPs themselves as guardians, NPs as providers for pesticides, fungicides, herbicides, and dsRNs.

3.2 PLANT DISEASE MANAGEMENT BY DIFFERENT FORMS OF NPs

3.2.1 NPs AS PROTECTORS

Taniguchi in (1974), the term NT was first used as the main discipline for the composition and deployment of nanoparticles of any substance (1–1-00 nm or 1.0 × 10-9 m). When the substance is reduced to micro-size, it is composed of specific new chemical, physical, and biological qualities that are totally missing in the macro-size. The ability of NPs alone is particular for the protection of plants from pests, flies, bacteria, fungus, and viruses in the seed of leaves and roots. Metal NPs such as silver, copper, zinc oxide, and titanium dioxide are intensively analyzed for their antiviral and antibacterial characteristics (Fig. 3.1; Table 3.1).

TABLE 3.1 AgNPs Antimicrobial Activity Against Different Pathogenic Microorganisms.

Crop	Source/mode of AgNPs synthesis used in the analysis	Pathogens target	References
Pumpkin	The nanosized silica–silver prepared by combining nanosilver with silica molecules and water-soluble polymer and exposing a solution including silver salt, silicate, and water-soluble polymer to radioactive rays	Powdery mildews (PM)	Park et al., (2006)
Oak	AgNPs supplied by BioPlus Co., Ltd.	Wilt pathogen *Raffaelea* sp.	Kim et al., (2007)
Cut gerbera	Nanosilver (Shanghai Huzheng Nano Technology Co. Ltd., China)	Stem-end bacteria	Liu et al., (2009)
Rice	AgNPs provided by Quantum Sphere Inc., Santa Ana, CA	*Bipolaris sorokiniana* and *Magnaporthe grisea*	Jo et al., (2009)
–	AgNPs synthesized using high-voltage arc discharge method	*Fusarium culmorum*	Kasprowicz et al., (2010)
–	Colloidal solution of AgNPs supplied by BioPlus Co. (Pohang, Korea)	*Colletotrichum* sp.	Lamsal et al., (2011a)
Cucumber and pumpkins	AgNPs supplied by BioPlus Co. Ltd. (Pohang, Korea)	PM	Lamsal et al., (2011b)
–	Colloidal solution of AgNPs provided by BioPlus Co. Ltd. (Pohang, Korea)	Different phyto-pathogenic fungi	Kim et al., (2012)
Tomato	Nanosized Ag–silica hybrid complex prepared by γ-irradiation	*Pseudomonas syringae*	Chu et al., (2012)

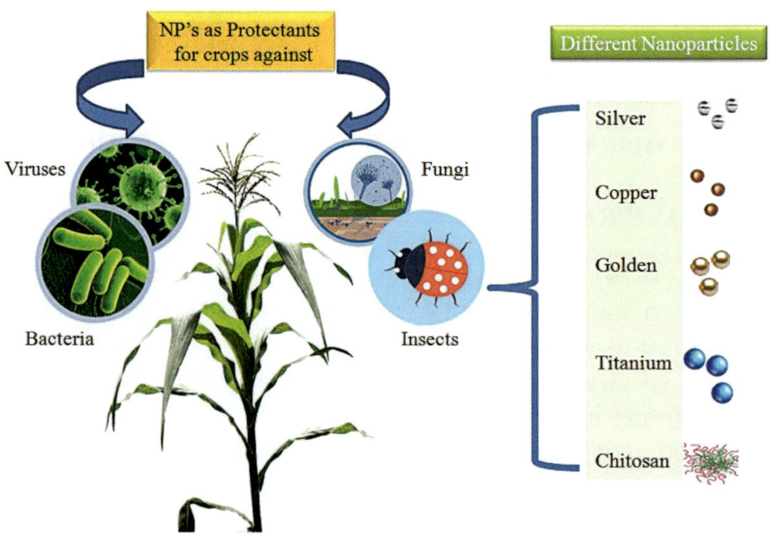

FIGURE 3.1 This schematic shows NPs as protectants combating a broad range of pests and pathogens.

This quantum gives a broad summary and update of the current case studies for the existing NPs. The creation of "green synthesis" in plants, bacteria, fungus, or yeast lately gained in relevance from Silvery NP (Table 3.2).

TABLE 3.2 Number of Living Organisms That Synthesize NPs.

Living organisms		Type of NPs produced	Produced size of NPs	References
Plants/medicinal plants				
Alfalfa	*Medicago sativa*	Gold and silver	20–40 nm	Gardea-Torresdey et al., (1999)
Mustard	*Brassica juncea*	Gold nanotriangles Gold NPs	–	Lamb et al., (2001)
Camphor tree	*Cinnamomum camphora*	Gold and silver	55–80 nm	Huang et al., (2007)
Aloe	*Aloe vera*	Silver NPs	15–15.6 nm	Zhang et al., (2008)
Oat	*Avena sativa*	Gold NPs	25–85 nm	Shanker et al., (2005)
Tamarind leaf extract	*Tamarindus indica*	Gold nanotriangles	20–40 nm	Paciotti et al., (2004)
Chilli	*Capsicum annum*	Silver NPs	–	Chandran et al., (2006)
Cochin grass	*Cymbopogon flexuous*	Gold NPs	–	Hong-Juan et al., (2006)
Neem	*Azadirachta indica*	Silver NPs, gold NPs, and silver and gold bimetallic	50–100 nm	Shanker et al., (2004)
Hops	*Humulus lupulus*	Gold NPs	–	Rai et al., (2006)
Desert-willow	*Chilopsis linearis*	Silver NPs		Ankamwar et al., (2005)
Sweet scented geranium	*Pelargonium graveolens*	Silver NPs	16–40 nm	Shanker et al., (2003)
Indian gooseberry	*Emblica officinalis*	Silver and gold NPs	10–20 nm and 15–25 nm, respectively	Lopez et al., (2005)
Bacterial pathogens				
Klebsiella aerogens		CdS NPs	20–200 nm	Husseiny et al., (2007)
Desulfovibrio desulfuricans		Palladium	–	Nair and Pradeep, (2002)

TABLE 3.2 *(Continued)*

Living organisms	Type of NPs produced	Produced size of NPs	References
Magnetotactic bacteria	Magnetic (Fe_3O_4) and Greigite (Fe_3S_4)	–	Klaus-Joerger, (2001)
Clostridium thermoaceticum	CdS NPs	–	Mandal et al., (2006)
Pseudomonas aeruginosa	Gold NPs	5–15 nm	Yong et al., (2002)
P. stutzeri	Silver NPs	100–200 nm	Beveridge et al., (1980); Joerger et al., (2000)
Shewanella algae	Gold NPs	10–20 nm intracellularly and 50–500 nm extracellularly	Lengke and Southam, (2005)
Rhodopseudomonas capsulate	Gold NPs	–	He et al., (2008)
Acidithiobacillus thiooxidans	Gold NPs	–	Gericke and Pinches, (2006)
Fungal pathogens			
Thermonospora sp.	Gold NPs	8 nm	Ahmad et al., (2003)
Trichothecium sp.	Gold NPs	–	Ahmad et al., (2003)
Phaenerochaete chrysosporium	Silver NPs	–	Basavaraja et al., (2008)
Aspergillus fumigatus	Silver NPs	5–25 nm	Armendariz et al., (2004)
Verticillum sp.	Silver and gold NPs	20–25 nm	Mukherjee et al., (2001)
Fusarium semitectum	Silver NPs	–	Bhainsa et al., (2006)
Fusarium oxysporum	Silver and gold NPs	20–40 nm and 5–15 nm, respectively	Mukherjee et al., (2002); Ahmad et al., (2003)
Colletotrichum sp.	Gold NPs	20–40 nm	Shanker et al., (2003)

3.3 SYNTHESIZING OF THE AGNPs BY MICROORGANISMS

The extraction of NPs through microorganisms is defined as the green method of getting NPs. Production of NPs by organisms is a relatively easy process that can be performed by nutritional media and equipment utilized routinely in a plant pathology laboratory. Enrichment and isolation technologies allow for microorganisms tolerant/resistant to metals and capable of synthesizing NMs ideally through mining sites, sea fields, etc. (Fig. 3.2).

NT research focuses on the creation and application of nanoparticles of various components and combinations. NPs are also employed as antimicrobial agents for a variety of purposes, including the treatment of plant diseases. The generation of NPs can be performed by a number of physiological mechanisms (Rajput et al., 2017; Rajput et al., 2021).

3.3.1 FUNGI AS A SOURCE OF NPs

The use of fungi in NP synthesis is somewhat new. With rapid downstream processing, simple operation (Mandal et al., 2006), and the ability to release a significant share of catalysts, fungi have surpassed bacteria as natural "NF" lines. Nonetheless, because fungi are eukaryotes, they are less susceptible to genetic alteration than prokaryotes. As a result, genetic manipulation of fungus for the generation of additional NPs would be more difficult. To better monitor the form, size, and other required qualities of the synthesized NMs, it is necessary to understand the process of NP synthesis in microbial systems.

3.3.2 BACTERIA AS A SOURCE OF NPs

Prokaryotes have received significantly more attention for nanoparticle biosynthesis than other species (Mandal et al., 2006). Mainly silver, gold, FeS, and magnetite NPs and quantic points of cadmium sulfide (CdS), zinc sulfide (ZnS), and lead sulfide (PbS) are biosynthesized by bacterium (Table 3.2).

3.3.3 VIRUSES AS A SOURCE OF NPs

Plant viruses, in particular spherical/icosahedral plant viruses, are examples of organic compounds NT or NPs. A tobacco necrosis virus satellite measuring only 18 nm in diameter is the smallest known plant virus (Hoglund, 1968). The viruses consist of RNA/DNA single or double-stranded as a protein-coated genome. The protein coat/shell serves as a container for carrying nucleic acid molecules from one host to another, both structurally and functionally. Their ability to penetrate, transmit nucleic acid genomes to specified locations in the host genome, reproduce, package nucleic acid, and escape from the host cell in the ordered manner that nanoscience requires. Young et al. (2008) conducted a systematic investigation of the utilization of plant viruses as bio-templates for NMs and their applications.

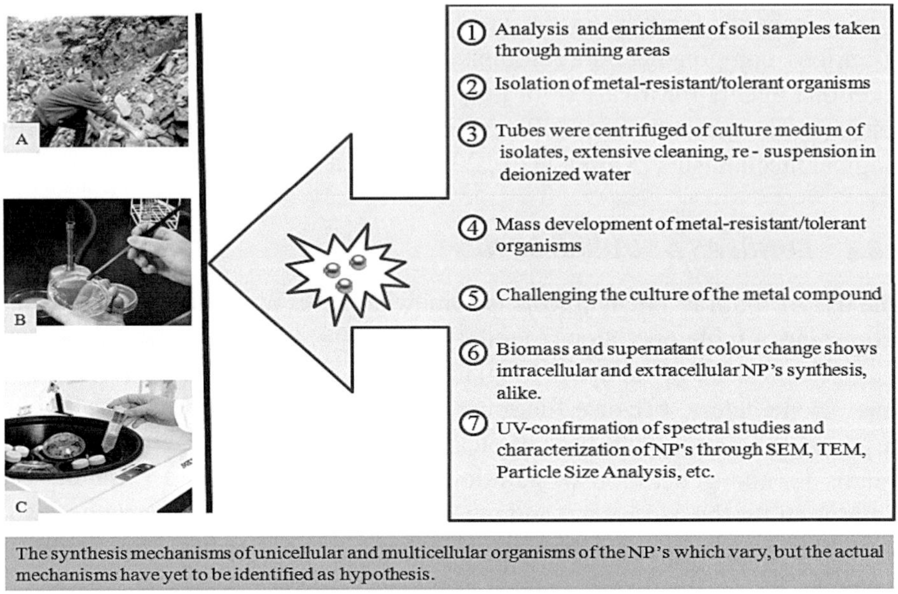

FIGURE 3.2 Typical process for synthesis of microbial NPs.

3.4 THE USE OF NT IN THE TREATMENT OF PLANT DISEASE

In the short to medium term, a variety of NTs can help improve current crop control procedures. The subject of NT field implementations is also being discussed (Kar et al., 2014; Li et al., 2007). Field implementations of nanoscience are likewise a hot topic. By correctly monitoring when and where pesticides, herbicides, and fertilizers are formed, NPs have the ability to strengthen the effectiveness and security of these products (Rai and Ingle, 2012). Metal NPs have been demonstrated to be beneficial against plant diseases, insects, as well as pests in previous studies (Choudhury et al., 2010). A fungicide that utilizes NMs to activate its pathogen-killing capability only when it comes into contact with invading pathogens is being developed. As a result, the use of nanoscience in phytopathology and food contamination is investigated, with effective technological advancements demonstrated (Fig. 3.3).

3.4.1 AGNPs USE AS AN ANTIMICROBIAL AGENT

Researchers have been able to explore the antibacterial properties of metal NPs thanks to the recent development of NMs and NT. To mention a few fields, NPs are used in medicine, pharmaceuticals, pollution monitoring,

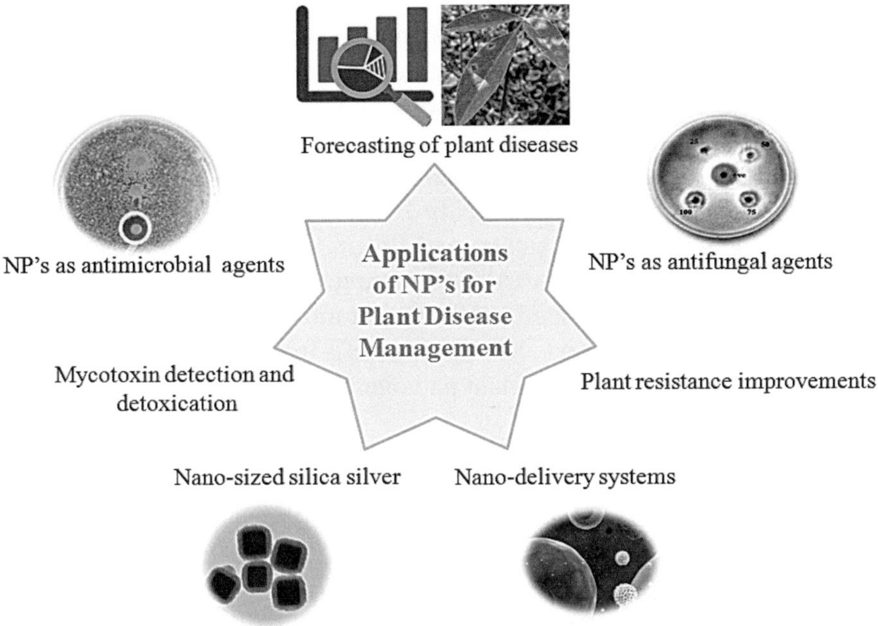

FIGURE 3.3 Applications of NPs for plant disease management.

technology, and agriculture/plant science (Navarro et al., 2008). Chemicals based on silver ions and silver are highly harmful to microbes (Sondi and Salopek-Sondi, 2004). As a result, silver ions have been used in a variety of compositions, and a mixture of Ag nanoparticles and amphiphilic hyperbranched macromolecules has recently been shown to generate an efficient antibacterial surface coating (Retchkiman-Schabes et al., 2006). Copper, zinc, titanium, magnesium, gold, alginate, and silver NPs have all been studied, but Ag NPs have been found to be particularly effective against bacteria, viruses, and fungus (Bhatia, 2016) (Dutta and Kaman, 2017). Soil can influence NPs, but it can also modify other soil characteristics like contaminants and pathogens. Under controlled conditions, the AgNPs activity against 18 phytopathogens has previously been documented (Prabhu and Poulose, 2012).

3.4.2 ANTIMICROBIAL PATHWAYS FOR NANOMETAL TOXICITY

Lemire et al. (2013) proposed five mechanisms of action: (1) release of toxic ions (Cd2C, Zn2C, Ag C) that bind to sulfur-containing proteins; this accumulation prevents proteins from working properly in the membrane and interferes with cell permeability; (2) genotoxic toxic ions that kill DNA, resulting in cell death; (3) genotoxic toxic ions that kill DNA, resulting in cell death. (4) Electron transport is disrupted, protein oxidation occurs, and the membrane potential collapses as a result of contact with CeO_2 or nC60; (5) reactive oxygen species (ROS) are produced (reactive oxygen species) specific types of protein, membrane, or DNA malfunction could be caused by ROS-mediated cellular damage and diverse metal-catalyzed oxidation processes (Zeng et al., 2007). (6) Obstructing nutrition absorption. These processes are not self-contained, meaning that multiple mechanisms may be engaged at the same moment. NPs may benefit from multi-goal operations in their fight against numerous plant pathogens.

3.4.3 ANTIFUNGAL PROPERTIES OF NPs

Disease control in food crops is critical. Falletta's current study focuses on developing nonhazardous management choices that are less hazardous to man and livestock, as well as addressing the scarcity of synthetic fungicides (2008). Only a few articles have looked into the antifungal effects of Ag NPs (Reo, 2008; Kim, 2008). Efficacy against clinical isolates and ATCC strains of *Candida spp.* and *Trichophyton mentagrophytes* has been investigated in numerous previous investigations (Li, 2012; Panacek, 2006; Min, 2009). As technological advancements have made their manufacturing more cost-effective, nanosized silver crystals have become increasingly commonly used as antibacterial agents. Plant disease prevention is one of silver's potential applications. Silver can be used to manage a number of plant microorganisms in a somewhat more practical form than synthetic fungicides since it inhibits plant infections in a variety of ways (Park, 2006; Oh, 2006). *Botrytis cinerea* is resistant to Ag-SiO2 NPs, despite their potent antifungal properties. According to the researchers, Ag_2S NMs on amorphous silica particles show antifungal properties against *Aspergillus niger* (Fateixa et al, 2009). ZnO and $ZnTiO_3$ nanopowders were tested for biocidal efficacy against the fungus *A. niger* (Ruffolo et al., 2010). The ZnTiO3 nanocomposite (NC) was shown to be more effective at inhibiting development than ZnO (Jo et al., 2009). The antifungal activity of silver ions and nanoparticles was investigated using *Bipolaris sorokiniana* and *Magnaporthe grisea*. Both silver ions and NPs have been reported to decrease disease development in phytopathogenic fungi in vitro and in vivo (Woo et al., 2009). The antifungal capabilities of Ag NPs were explored by (Min et al., 2009), who focused on sclerotium-forming phytopathogenic fungi. Panacek et al. (2006) looked explored how Ag NPs affected *Candida albicans* (I and II), *Candida tropicalis*, and *Candida parapsilosis* pathogenic yeasts in terms of fungistatic and fungicidal effects. The antifungal activity of Ag NPs was evaluated against the unknown ambrosia fungus *Raffaelea sp*., which has killed a large number of oak trees in Korea (Woo et al., 2009). Kasprowicz et al. (2010) looked at the effect of Ag NPs on *F. culmorum* plant pathogenic spores. Silver NPs show antifungal efficacy against *F. oxysporum,* according to Musarrat et al. (2010). Silver NPs considerably reduced the quantity of germination pieces and sprout length when compared to the control. The effect of nano-silver liquid on green onion white rot caused by *Sclerotium cepivorum* was investigated (Jung et al., 2010). Synthetic fungicides may provide a greater risk to humans and wildlife than silver nanoparticles. Furthermore, in algae, plants, and fungi, NP toxicity can be accompanied by some beneficial effects (Sondi and Salopek-Sondi, 2004). *Colletotrichum gloeosporioides*, a plant disease that causes anthracnose in a variety of fruits, was used to assess the antifungal effectiveness of Ag NPs. In the presence of Ag NPs, *C. gloeosporioides* development was significantly slowed in a dose-dependent manner (Aguilar-Mendez et al., 2010).

A. niger, a facultative pathogenic food-borne pathogen, was also examined in a comparison of elemental sulfur and nano-sulfur. Sulfur nanoparticles were discovered to be more potent than the natural form (Choudhary et al., 2010). Using different concentrations of Ag NPs, the inhibitory activity of fungal plant pathogenic pathogens *Alternaria alternata, Sclerotinia sclerotiorum, Macrophomina phaseolina, Rhizoctonia solani, B. cinerea,* and *Curvularia lunata* was examined. Surprisingly, Ag NPs at 15 mg had a strong inhibitory effect against a wide range of infections (Krishnaraj et al., 2012). In vitro tests revealed that chitosan and Cu-chitosan NPs were of similar size and consistency, which could explain their superior antifungal efficacy against *A. alternata*, *M. phaseolina*, and *R. solani*. Cu-chitosan NPs also suppressed *A. alternata* fungal growth to the greatest extent. When applied to chitosan and Cu-chitosan NPs, the antifungal effectiveness of chitosansaponin NPs was reported to be modest (Saharan et al., 2013).

3.4.4 NANOSIZED COMPOUNDS

Since it was first used to treat damaged plants, WS silicate solution has proven to be an effective preventative against powdery mildew (PM) and downy mildew (Khaydarov et al., 2011). Plants' physiological function and development were boosted, as well as disease and stress tolerance (Kanto et al., 2004). The inhibitory effect of nanosized silica-silver on Pseudomonas syringae and *Xanthomonas campestris* pv. vesicatoria was investigated at various concentrations, and it was discovered that at 100 ppm, 100% growth prevention of Pseudomonas syringae and *X. campestris* pv. vesicatoria occurred. At 10 ppm of nanosized silica-silver, *M. grisea, B. cinerea, C. gloeosporioides, Pythium ultimum*, and *R. solani* all grew by 100% inhibition (Oh, 2006). Nanosized silica silver at 0.3 ppm effectively controlled PMs in pumpkin in field and greenhouse studies. *Erysiphe cichoracearum* had gone from the leaf surfaces in just 3 days. *Sphaerotheca pannosa var. rosae*, or rose PM, was used to investigate the antifungal effectiveness of colloidal nanosilver (1.5 nm average diameter) (Kim et al., 2008). Nanocopper has been shown to be efficient against bacterial diseases such as rice blight (*Xanthomonas oryzae* pv. *oryzae*) and mung leaf spot (*X. campestris* pv. *phaseoli*) (Gagoi et al., 2009).

3.4.5 DELIVERY NETWORKS FOR NMs

To overcome biological barriers to involvement targeting, smart dosage forms for pesticides can be created using NT with a combination of time-controlled, spatially targeted, self-regulated, remotely regulated, preprogrammed, or multifunctional capabilities (Bouwmeester et al., 2009). Pesticides and herbicides, fungicides, plants, insects, soils, and the ecosystem may all be investigated using smart nano-carriers. There is a lot more potential in using a smarter delivery strategy to increase the performance of fungicides in farming methods. Such advancements in plant protection technologies will allow them to be utilized in crop protection (Boumeester et al., 2009). Quick distribution systems could be installed right away to improve the management of plant diseases with chemicals (fungicides, insecticides, and herbicides). Dispersing pesticides within the crop and approaching the action stage, on the other hand, is a more difficult task. If NPs can be delivered and guided to specified regions via the crop vascular system, they can be used for phytosanitary applications with a small amount of ingredients, reducing the risk of environmental contamination and chemical presence in the crop for future commercialization. Plant pathogens including fungi, viruses, bacteria, and parasitic plants, for example, may be targeted with NPs (Abd-Eslam, 2012).

Food science uses nano-delivery systems, which is an important category of NPs (Letchford and Burt, 2007; Taylor et al., 2005). Commercial products with 100–250 nm NPs are substantially more WS, which means they are more effective. Commercial products containing 100–250 nm NPs are much more WS, resulting in increased efficacy. Nanofungicides, nanopesticides, and nanoherbicides are widely used in agriculture (e.g., CruiserMaxx and Subdue MAXX), and commercial products contain 100–250 nm NPs are much more WS, resulting in increased efficacy (Owalade et al., 2008). Many industries employed nano-emulsions of nanoscale particles, which might be liquid or oil-based, to manufacture consistent pesticide or herbicide NP concentrations ranging from 200 to 400 nanometers in size (Rickman et al., 1999). *M. grisea* was confirmed to be eradicated from contaminated rice plants using "Nano Green," a product made by combining various bio-based chemicals (Gogoi et al., 2009).

3.4.6 PLANT RESISTANCE IS BEING IMPROVED

Plants are vulnerable to a variety of disease-causing agents, such as predatory insects, worms, and a variety of other pathogens, as well as drought, which both result in significant financial loss. The only way to avoid these reductions is to cultivate resistant plant varieties. Crop resistance would aid in the control of the aforementioned agents, allowing for the resolution of the problem of financial harm. Nanobiotechnology uses NPs, nanofibers, and nanocapsules to give a simple set of strategies for copying genes and improving plant resistance (McKnight et al., 2003; Rai et al., 2012). The effective insertion and incorporation of plasmid DNA into the plant genome has been confirmed by gene expression (Filipenko et al., 2007). Nanostructures agricultural changes can improve plant and disease resistance through genetic engineering (McKnight et al., 2003). NT can be used in agriculture to tackle specific plant-pathology issues, including as plant–pathogen interactions, and to bring novel crop disease management strategies (Torney et al., 2007). For example, using nanotechnological ways to transfer resistance genes into plant cells can result in the generation of disease-resistant types that save money on agrochemicals (Boumeester et al., 2009).

3.4.7 NANO-PESTICIDES MADE OF SLIVER NPs

Rapid advances in nano-pesticide technology have encouraged a number of international organizations to explore potential issues related to the need for NT for agricultural production over the last 2 years (Kah and Hofman, 2014). Since Ag has antimicrobial activity towards microbes and is nontoxic to mankind, it has a broad array of applications in metal and compound form (Elchiguerra et al., 2005; Yeo et al., 2003). NT has improved recently the efficacy of Ag NPs (Kim et al., 2008). Ag NPs have a higher surface area-to-volume ratio, which enhances their interaction with microorganisms and permeation potential (Kim et al., 2008). Microbes have wreaked havoc on the atmosphere and ecology. This is the product of new pathogens entering nations, causing disease and the death of tree species (for example, in the United States and Europe) (Boumeester et al., 2009). Insect pests and fungal diseases must also be shielded from invading agricultural products and wildlife. As a result, a disease-control mechanism is required, and nano-pesticide manufacturing could help with plant disease management (Boumeester et al., 2009). In lab conditions, the effect of Ag NPs on the oak wilt-causing fungal phytopathogen *Raffaelea sp* was investigated (Woo et al., 2009). According to the researchers, AgNPs destroy fungal hyphae, interfere with microbial absorption, and inhibit fungal and conidial germination. AgNPs suppress the hyphal growth of *R. solani, S. sclerotiorum*, and *S. minor* in a dose-dependent manner, according to (Min et al., 2009). (Jo et al., 2009) investigated the antifungal activity of various silver ions and nanoparticles against *B. sorokiniana* and *M. grisea*. They discovered that silver ions and Ag NPs affect the growth of spore colonies and the course of disease in phytopathogenic fungus. These data suggest that Ag NPs may have a lot of promise as phytopathogen control nano-pesticides.

3.5 CONCLUSIONS

NT has the ability to revolutionize existing pest control systems and provide solutions for agricultural applications. Nanopesticides are clearly an appealing advancement, as shown by the aforementioned debate, due to their possible benefits for the human health and the environment. Even so, NT in agriculture has yet to be commercialized. The effectiveness and toxicity of nanostructures pesticides on soil and the ecosystem must be investigated further. Only two insecticides loaded on NPs have been tested in the field. Song et al. (2012) loaded chlorfenapyr into silica NPs with extra emulsions, and imidacloprid into sodium alginate NPs (Kumar et al., 2014). Another issue with nanoparticles in plant protection research in its early phases is the scarcity of long-term tests. For example, Mitter et al. (2017) evaluated a topical delivery system called BioClay for the protection of virus plants 20 days after foliar sprays for RNAi/nanoparticle and (Mitter et al., 2017) evaluates a topical delivery system called Bio Clay for the protection of plants from viruses 20 days after foliar sprays for RNAi/nanoparticle (Zhao et al., 2017). Nanopesticides are indeed a new technical development in which regulatory authorities have yet to develop a specific concept of what constitutes a nano-pesticide and what does not (Kookana et al., 2014). The results of nanopesticides, versus traditional pesticides, may be based on the absorption, bioavailability, concentration, and potency of the NPs, and the proportion of the active attached to it, according to (Kookana et al., 2014). In addition, there is a paucity of research on pesticide resistance and how NMs might aid in reducing its frequency. Fungi are becoming more frequent in these applications, and they may be able to develop nanobiofactories for metallic NPs that are both rapid and environmentally sustainable. Pesticide use that has gone unchecked has resulted in a plethora of problems, including affects on human health, pollinating insects, and domestic animals, as well as the entry of this material into the water and soil, and its direct and indirect effects on habitats. Chemicals utilized wisely at the nanoscale could be a feasible solution to the problem.

In order to get greater understanding of the fundamental association processes in an intricate bio-nano environment, material researchers and biologists must cooperate and introduce significant knowledge from other fields. The proper NPs may be chosen from a complete grasp of the structural characteristics of nanoparticles including the shape, the size, the function groups, and the active loading/adsorption capability. The agricultural NT research and innovation landscape is highly interesting, since the possibilities offered by NPs for delivering beneficial products are being thoroughly investigated. The authors state that multiple disciplines and integrated work would provide a solid foundation for bringing NT crop safety applications to fruition.

KEYWORDS

- **agriculture sector**
- **nanoparticles**
- **nano-pesticides**
- **plant diseases management**

REFERENCES

Abd-Elsalam, K. A. Nanoplatforms for Plant Pathogenic Fungi Management. *Fungal Genomics Biol.* **2012,** *2,* 107.

Aguilar-Mendez, M. A.; Martın-Martınez E. S.; Ortega-Arroyo, L.; Cobian-Portillo, G.; Sanchez-Espındola, E. Synthesis and Characterization of Silver Nanoparticles: Effect on Phytopathogen *Colletotrichum gloesporioides*. *J. Nanopart. Res.* **2010,** *13,* 2525–2532.

Ahmad, A.; Senapati, S.; Khan, M. I.; Kumar, R.; Ramani, R.; Srinivas, V.; Sastry, M. Intracellular Synthesis of Gold Nanoparticles by a Novel Alkalotolerant Actinomycete, *Rhodococcus* Species. *Nanotechnology* **2003,** *14,* 824–828. https.//iopscience.iop.org/article/10.1088/0957-4484/14/7/323/pdf

Armendariz, V.; Herrera, I.; Peralta-Videa, J. R. Size Controlled Gold Nanoparticle Formation by *Avena Sativa* Biomass: Use of Plants in Nanobiotechnology. *J. Nanopart. Res.* **2004,** *6,* 377–382. https.//doi.org/10.1007/s11051-004-0741-4

Balaure, P. C.; Gudovan, D.; Gudovan, I. Nanopesticides: A New Paradigm in Crop Protection. *New Pestic. Soil Sens.* **2017,** 129–192.

Basavaraja, S.; Balaji, S. D.; Lagashetty, A. K.; Rajasab, A. H.; Venkataraman, A. Extracellular Biosynthesis of Silver Nanoparticles Using the Fungus *Fusarium semitectum*, *Mater. Res. Bull.* **2008,** *43*(5), 1164–1170.https.//doi.org/10.1016/j.materresbull.2007.06.020.(https.//www.sciencedirect.com/science/article/pii/S0025540807002358)

Bhainsa, K. C.; D'Souza, S. F. Extracellular Biosynthesis of Silver Nanoparticles Using the Fungus *Aspergillus fumigatus*. *Colloids Surf. B Biointerfaces* **2006,** 47(2), 160–164. DOI: 10.1016/j.colsurfb.2005.11.026. https.//pubmed.ncbi.nlm.nih.gov/16420977/

Bhatia, S. Nanoparticles Types, Classification, Characterizatio-n, Fabrication Methods and Drug Delivery Applications. In *Natural Polymer Drug Delivery Systems*; Springer: Cham, 2016. https.//doi.org/10.1007/978-3-319-41129-3_2

Bouwmeester, H.; Dekkers, S.; Noordam, M. Y.; Hagens, W. I.; Bulder, A. S.; Heer, C. de.; Voorde, S. E.C. G. t.; Wijnhoven, S. W. P.; Marvin, H. J. P.; Sips, A. J. A. M. Review of Health Safety Aspects of Nanotechnologies in Food Production. *Regul. Toxicol. Pharmacol.* **2009,** *53*(1), 52–62. https.//pubmed.ncbi.nlm.nih.gov/19027049/

Choudhury, S. R.; Nair, K. K.; Kumar, R.; Gogoi, R.; Srivastava, C.; Gopal, M.; Subhramanyam, B. S.; Devakumar, C.; Goswami, A. Nanosulfur: A Potent Fungicide Against Food Pathogen, *Aspergillus niger*. *AIP Conf. Proc.* **2010,** *1276,* 154–157. DOI: https.//doi.org/10.20546/ijcmas.2017.606.336

Dutta, P.; Kaman, P. K. Nanocentric Plant Health Management With Special Reference to Silver. *Int. J. Curr. Microbiol. App. Sci.* **2017,** *6*(6), 2821–2830.

Elchiguerra, J. L.; Burt, J. L.; Morones, J. R.; Camacho- Bragado, A.; Gao, X.; Lara, H. H.; Yacaman, M. J. Interaction of Silver Nanoparticles With HIV-1. *J. Nanobiotechnol.* **2005,** *3,* 1–10.

Falletta, E.; Bonini, M.; Fratini, E.; Lo Nostro, A.; Pesavento, G.; Becheri, A. Clusters of Poly (acrylates) and Silver Nanoparticles: Structure and Applications for Antimicrobial Fabrics. *J. Phys. Chem. C* **2008,** *112,* 1175811766.

Fateixa, S.; Neves, M. C.; Almeida, A.; Oliveira, J.; Trindade, T. Anti-Fungal Activity of SiO_2/Ag_2S Nanocomposites 234 M.A. Alghuthaymi et al. against *Aspergillus niger*. *Coll. Surf. B* **2009,** *74,* 304308.

Filipenko, E. A.; Filipenko, M. L.; Deineko, E. V.; Shumnyi, V. K. Analysis of Integration Sites of T-DNA Insertions in Transgenic Tobacco Plants. *Cytol. Genet.* **2007,** *41,*199–203.

Flood, J. The Importance of Plant Health to Food Security. *Food Secur.* **2010,** *2,* 215–231. The importance of plant health to food security | SpringerLink.

Gajbhiye, M.; Kesharwani, J.; Ingle, A.; Gade, A.; Rai, M. Fungus Mediated Synthesis of Silver Nanoparticles and Their Activity Against Pathogenic Fungi in Combination With Fluconazole. *Nanomedicine* **2009,** *5,* 382386.

Gardea-Torresdey, J. L.; Tiemann, K. J.; Gamez, G.; Dokken, K. Effects of Chemical Competition for Multi-Metal Binding by *Medicago sativa* (alfalfa). *J. Hazard. Mat.* **1999,** *69*(1), 41–51. DOI: 10.1016/s0304-3894(99)00057-6. https.//europepmc.org/article/med/10502605

Gericke, M.; Pinches, A. Biological Synthesis of Metal Nanoparticles. *Hydrometallurgy* **2006,** *83(*1–4), 132–140. DOI: 10.1016/j.hydromet.2006.03.019.

Ghormade, V.; Deshpande, M. V.; Paknikar, K. M. Perspectives for Nano-Biotechnology Enabled Protection and Nutrition of Plants. *Biotechnol. Adv.* **2011,** *29,* 792–803. Perspectives for nano-biotechnology enabled protection and nutrition of plants - ScienceDirect

Gogoi, R.; Dureja, P.; Singh, P. K. Nanoformulations: A Safer and Effective Option for Agrochemicals. *Indian Farm.* **2009,** *59*(8), 7–12.

Hayles, J.; Johnson, L.; Worthley, C.; Losic, D. Nanopesticides: A Review of Current Research and Perspectives. *New Pestic. Soil Sens.* **2017**, 193–225. Nanopesticides a review of current research and perspectives - ScienceDirect

He. L.; Liu, Y.; Mustapha, A.; Lin, M. Antifungal Activity of Zinc Oxide Nanoparticles Against *Botrytis cinerea* and *Penicillium expansum*. *Microb. Res*. **2010,** *166,* 207–215.

He, S.; Guo, Z.; Zhang, Y.; Zhirui, G.; Ning, G. Biological Synthesis of Gold Manowires Using Extract of *Rhodopseudomonas capsulate*. *Biotechnol. Prog.* **2008,** *24,* 476–480. https.//pubmed.ncbi.nlm.nih.gov/18293997/

Huang, J. W.; Cunnigham, S. D. Lead Phytoextraction: Species Variation in Lead Uptake and Translocation. *New Phytol.* **1996,** *134,* 75–84. https.//nph.onlinelibrary.wiley.com/doi/full/10.1111/j.1469-8137.1996.tb01147.x

Jo, Y. K.; Kim, B. H.; Jung, G. Antifungal Activity of Silver Ions and Nanoparticles on Phytopathogenic Fungi. *Plant Dis*. **2009,** *93,* 1037–1043.

Jung, J. H.; Kim, S. W.; Min, J. S.; Kim, Y. J.; Lamsal, K.; Kim, K. S. The Effect of Nano-Silver Liquid Against the White Rot of the Green Onion Caused by *Sclerotium cepivorum*. *Mycobiology* **2010,** *38*(1), 39–45.

Kah, M.; Hofmann, T. Nanopesticide Research: Current Trends and Future Priorities. *Environ. Int*. **2014,** *63,* 224–235.

Kanto, T.; Miyoshi, A.; Ogawa, T.; Maekawa, K.; Aino, M. Suppressive Effect of Potassium Silicate on Powdery Mildew of Strawberry in Hydroponics. *J. Gen. Plant Pathol.* **2004,** *70,* 207–211.

Kar, P. K.; Murmu, S.; Saha, S.; Tandon, V.; Acharya, K. Anthelmintic Efficacy of Gold Nanoparticles Derived From a Phytopathogenic Fungus, *Nigrospora oryzae*. *PLoS One*, **2014,** *9*(1), 84693.

Kasprowicz, M. J.; Kozio, M.; Gorczyca, A. The Effect of Silver Nanoparticles on Phytopathogenic Spores of *Fusarium culmorum*. *Can. J. Microbiol*. **2010,** *56,* 247–253.

Khandelwal, N.; Barbole, R. S.; Banerjee, S. S.; Chate, G. P.; Biradar, A. V.; Khandare, J. J.; Giri, A. P. Budding Trends in Integrated Pest Management Using Advanced Micro- and Nano-Materials: Challenges and Perspectives. *J. Environ. Manag*. **2016,** *184,* 157–169.

Khaydarov, R. R.; Khaydarov, R. A.; Evgrafova, S.; Estrin, Y. Using Silver Nanoparticles as an Antimicrobial Agent. *NATO Science Peace Security Series A*, 2011; pp 169–177.

Kim, H.; Kang, H.; Chu, G.; Byun, H. Antifungal Effectiveness of Nanosilver Colloid Against Rose Powdery Mildew in Greenhouses. *Solid State Phenom.* **2008,** *135,* 1518.

Kookana, R. S.; Boxall, A. B. A.; Reeves, P. T.; Ashauer, R.; Beulke, S.; Chaudhry, Q.; Cornelis, G.; Fernandes, T. F.; Gan, J.; Kah, M. Nanopesticides. Guiding Principles for Regulatory Evaluation of Environmental Risks. *J. Agric. Food Chem.* **2014,** *62,* 4227–4240. Nanopesticides. guiding principles for regulatory evaluation of environmental risks - PubMed (nih.gov)

Krishnaraj, C.; Ramachandran, R.; Mohan, K.; Kalaichelvan, P. T. Optimization for Rapid Synthesis of Silver Nanoparticles and its Effect on Phytopathogenic Fungi. *Spectrochimica Acta Part A* **2012,** *93,* 95–99.

Kumar, S.; Bhanjana, G.; Sharma, A.; Sidhu, M.; Dilbaghi, N. Synthesis, Characterization and on Field Evaluation of Pesticide Loaded Sodium Alginate Nanoparticles. *Carbohydr. Polym.* **2011,** *101,* 1061–1067.

Lamb, A. E.; Anderson, C. W. N.; Harkamp, R. The Induced Accumulation of Gold in the Plants *Brassica juncea*, *Berkheya coddii,* and Chicory. *Chem. N.Z.* **2001,** *65,* 34–36.

https.//www.researchgate.net/publication/291293247_The_induced_accumulation_of_gold_in_the_plants_Brassica_juncea_Berkheya_coddii_and_chicory

Lamsal, K.; Kim, W. S.; Jung, H. J.; Kim, S. Y.; Kim, S. K.; Lee, S. Y. Inhibition Effects of Silver Nanoparticles Against Powdery Mildews on Cucumber and Pumpkin. *Mycobiology* **2011,** *39*(1), 26–32. https.//www.ncbi.nlm.nih.gov/pmc/articles/PMC3385079/

Letchford, K.; Burt, H. A Review of the Formation and Classification of Amphiphilic Block Copolymer Nanoparticulate Structures: Micelles, Nanospheres, Nanocapsules and Polymersomes. *Eur. J. Pharm. Biopharm.* **2007,** *65,* 259–269.

Li, G.; He, D.; Qian, Y.; Guan, B.; Gao, S.; Cui, Y.; Yokoyama, K.; Wang, L. Fungus-Mediated Green Synthesis of Silver Nanoparticles Using *Aspergillus terreus*. *Int. J. Mol. Sci.* **2012,** *13,* 466–476.

Li, Z. Z.; Chen, J. F.; Liu, F.; Liu, A. Q.; Wang, Q.; Sun, H. Y.; Wen, L. X. Study of UV-Shielding Properties of Novel Porous Hollow Silica Nanoparticle Carriers for Avermectin. *Pest Manag. Sci.* **2007,** *63,* 241–246.

Mandal, D.; Bolander, M. E.; Mukhopadhyay, D.; Sarkar, G.; Mukherjee, P. The Use of Microorganisms for the Formation of Metal Nanoparticles and Their Application. *Appl. Microbiol. Biotechnol.* **2006,** *69,* 485–492.

McKnight, T. E.; Melechko, A. V.; Griffin, G. D.; Guillorn, M. A.; Merkulov, V. I.; Serna, F.; Hensley, D. K.; Doktycz, M. J.; Lowndes, D. H.; Simpson, M. L. Intracellular Integration of Synthetic Nanostructures With Viable Cells for Controlled Biochemical Manipulation. *Nanotechnology* **2003,** *14,* 551–556.

Min, J. S.; Kim, K. S.; Kim, S. W.; Jung, J. H.; Lamsal, K.; Kim, S. B.; Jung, M.; Lee, Y. S. Effects of Colloidal Silver Nanoparticles on Sclerotium-Forming Phytopathogenic Fungi. *Plant Pathol. J.* **2009,** *25,* 376–380.

Mitter, N.; Worrall, E. A.; Robinson, K. E.; Li, P.; Jain, R. G.; Taochy, C.; Fletcher, S. J.; Carroll, B. J.; Lu, G.; Xu, Z. P. Clay Nanosheets for Topical Delivery of RNAi for Sustained Protection Against Plant Viruses. *Nat. Plants* **2017,** *3,* 16207.

Mukherjee, P.; Ahmad, A.; Mandal, D.; Senapati, S.; Sainkar, S. R.; Khan, M. I. R.; Parishcha, P.; Ajaykumar, V.; Alam, M.; Kumar, R.; Sastry, M. Fungus-Mediated Synthesis of Silver Nanoparticles and Their Immobilization in the Mycelia Matrix. A Novel Biological Approach to Nanoparticle Synthesis. *Nano Lett.* **2001,** *1(*10), 515–519. DOI: 10.1021/nl0155274.

https.//www.scirp.org/(S(i43dyn45teexjx455qlt3d2q))/reference/ReferencesPapers.aspx?ReferenceID=810601

Musarrat, J.; Dwivedi, S.; Singh, B. R.; Al-Khedhairy, A. A.; Azam, N. A. A Production of Antimicrobial Silver Nano-Particles in Water Extracts of the Fungus *Amylomyces rouxii* Strain KSU-09. *Biores. Technol.* **2010,** *101,* 8772–8776.

Navarro, E.; Baun, A.; Behra, R.; Hartmann, N. B.; Filser, J.; Miao, A. J.; Quigg, A.; Santschi, P. H.; Sigg, L. Environmental Behavior and Ecotoxicity of Engineered Nanoparticles to Algae, Plants, and Fungi. *Ecotoxicology* **2008,** *17,* 372–386.

Oh, S. D.; Lee, S.; Choi, S. H.; Lee, I. S.; Lee, Y. M.; Chun, J. H.; Park, H. J. Synthesis of Ag and $AgSiO_2$ Nanoparticles by y-irradiation and Their Antibacterial and Antifungal Efficiency Against Salmonella Enteric Serovar *Typhimurium* and *Botrytis cinerea. Coll. Surf. A* **2006,** *275,* 228233.

Owolade, O. F.; Ogunleti, D. O.; Adenekan, M. O. Titanium Dioxide Affects Disease Development and Yield of Edible Cowpea. *Electron. J. Environ. Agric. Food Chem.* **2008,** *7*(50), 2942–2947.

Panacek, A.; Kolar, M.; Vecerova, R.; Prucek, R.; Soukupova, J.; Krystof, V.; Park, H. J.; Kim, S. H.; Kim, H. J.; Choi, S. H. A New Composition of Nanosized Silica-Silver for Control of Various Plant Diseases. *Plant Pathol. J.* **2006,** *22,* 295302.

Prabhu, S.; Poulose, E. K. Silver Nanoparticles: Mechanism of Atimicrobial Action, Synthesis, Medical Applications, and Toxicity Effects. *Int. Nano Lett.* **2012,** *2,* 32. DOI: 10.1186/2228-5326-2-32. https.//link.springer.com/article/10.1186/2228-5326-2-32#citeas

Rai, M.; Deshmukh, S.; Gade, A.; Elsalam, K. A. Strategic Nanoparticles-Mediated Gene Transfer in Plants and Animals A Novel Approach. *Curr. Nano* **2012,** *8,* 170–179.

Rajput, V. D.; Tatiana, M.; Svetlana, S.; Tsitsuashvili, V.; Mandzhieva, S.; Gorovtsov, A.; Nevidomskyaya, D.; Gromakova, N. Effect of Nanoparticles on Crops and Soil Microbial Communities. *J. Soils Sediments* **2017,** *18,* 179–187.

Rajput, V. D.; Minkina, T.; Kumari, A.; Harish, Singh.; Verma, V. K.; Verma, K. K.; Mandzhieva, S.; Sushkova, S.; Srivastava, S.; Keswani, C. Coping With the Challenges of Abiotic Stress in Plants: New Dimensions in the Field Application of Nanoparticles. Plants (Basel) **2021,** *10*(6), 1221.

Rai, M.; Ingle, A. Role of Nanotechnology in Agriculture With Special Reference to Management of Insect Pests. *Appl. Microbiol. Biotechnol.* **2012,** *94,* 287–293.

Retchkiman-Schabes, P. S.; Canizal, G.; Becerra-Herrera, R.; Zorrilla, C.; Liu, H. B.; Ascencio, J. A. Biosynthesis and Characterization of Ti/Ni Bimetallic Nanoparticles. *Opt. Mater.* **2006,** *29,* 95–99.

Rickman, D.; Luvall, J. C.; Shaw. J.; Mask, P.; Kissel, D.; Sullivan, D. Precision Agriculture, Changing the Face of Farming. [Online], 1999. wwwghccmsfcnasagove/ precisionag/

Roe, D.; Karandikar, B.; Bonn-Savage, N.; Gibbins, B.; Roullet, J. B. Antimicrobial Surface Functionalization of Plastic Catheters by Silver Nanoparticles. *J. Antimicrob. Chemother.* **2008,** *61,* 869876.

Ruffolo, S. A.; La, Russa.; M. F.; Malagodi, M.; Oliviero, R. C.; Palermo, A. M.; Crisci, G. M. ZnO and $ZnTiO_3$ Nanopowders for Antimicrobial Stone Coating. *Appl. Phys. A* **2010,** *100,* 829834.

Saharan, V.; Mehrotra, A.; Khatik, R.; Rawal, P.; Sharma, S. S.; Pal, A. Synthesis of Chitosan Based Nanoparticles and Their In Vitro Evaluation Against Phytopathogenic Fungi. *Int. J. Biol. Macromol.* **2013,** *62,* 677–683.

Shankar, S. S.; Ahmad, A.; Sastry, M. Geranium Leaf Assisted Biosynthesis of Silver Nanoparticles. *Biotechnol. Prog.* **2003,** *19*(6), 1627–1631. DOI: 10.1021/bp034070w. PMID. 14656132. https.//pubmed.ncbi.nlm.nih.gov/14656132/

Shankar, S. S.; Rai, A.; Ahmad, A.; Sastry, M. Rapid synthesis of Au, Ag, and Bimetallic Au core-Ag Shell Nanoparticles Using Neem (*Azadirachta indica*) Leaf Broth. *J. Coll. Interf. Sci.* **2004,** *275*(2), 496–502. DOI: 10.1016/j.jcis.2004.03.003. PMID. 15178278. https.//pubmed.ncbi.nlm.nih.gov/15178278/

Sinha, K.; Ghosh, J.; Sil, P. C. New Pesticides: A Cutting-Edge View of Contributions From Nanotechnology for the Development of Sustainable Agricultural Pest Control A2—Grumezescu, Alexandru Mihai. In *New Pesticides and Soil Sensors*; Academic Press: Cambridge, MA, 2017.

Singh, A.; Rajput, V.; Mehrotra, R.; Pal, N.; Singh, V.; Chokheli, V. A.; Singh, R. K. Modern Sustainable Techniques for Enhancing Crop Production. In *Sustainable Soil Fertility Management*, 2021b; pp. 73–100. ISBN: 978-1-53619-055-7.

Singh, A., Rajput, V.; Singh, A. K.; Sengar, R. S.; Singh, R. K.; Minkina, T. Transformation Techniques and Their Role in Crop Improvements: A Global Scenario of GM Crops. In *Policy Issues in Genetically Modified Crops*, 2021a; pp 515–542.

Sondi, I.; Salopek-Sondi, B. Silver Nanoparticles as Antimicrobial Agent: A Case Study on *E. coli* as a Model for Gram-negative Bacteria. *J. Coll. Interf. Sci.* **2004,** *275,* 177–182.

Song, M. R.; Cui, S. M.; Gao, F.; Liu, Y. R.; Fan, C. L.; Lei, T. Q.; Liu, D. C. Dispersible Silica Nanoparticles as Carrier for Enhanced Bioactivity of Chlorfenapyr. *J. Pestic. Sci.* **2012,** *37,* 258–260.

Stephenson, G. R. Pesticide Use and World Food Production: Risks and Benefits. ACS Publications: Washington, DC, 2003.

Taniguchi, N. On the Basic Concept of 'Nano-technology'. In *Proceedings of the International Conference on Production Engineering Tokyo*; ; Japan Soc Precision Engineering: Tokyo, 1974; pp 18–23.

Taylor, T. M.; Davidson, P. M.; Bruce, B.D.; Weiss, J. Liposomal Nanocapsules in Food Science and Agriculture. *Crit. Rev. Food Sci. Nutr*. **2005,** *45,* 587–605.

Torney, F.; Trewyn, B. G.; Lin, S. Y.; Wang, K. Mesoporous Silica Nanoparticles Deliver DNA and Chemicals Into Plants. *Nat. Nanotechnol*. **2007,** *2,* 295–300.

Woo, K. S.; Kim, K. S.; Lamsal, K.; Kim, Y. J.; Kim, S. B.; Jung, M.; Sim, S. J.; Kim, H. S.; Chang, S. J.; Kim, J. K.; Lee, Y. S. An In Vitro Study of the Antifungal Effect of Silver Nanoparticles on Oak Wilt Pathogen *Raffaelea* sp. *J. Microbiol. Biotechnol.* **2009,** *19,* 760–764.

Yeo, S. Y.; Lee, H. J.; Jeong, S. H. Preparation of Nanocomposite Fibers for Permanent Antibacterial Effect. *J. Mater. Sci*. **2003,** *38,* 2143–2147.

Zeng, F.; Hou, C.; Wu, S. Z.; Liu, X. X.; Tong, Z.; Yu, S. N. Silver Nanoparticles Directly Formed on Natural Macroporous Matrix and Their Anti-Microbial Activities. *Nanotechnology* **2007,** *18,* 1–8.

Zhao, P.; Cao, L.; Ma, D.; Zhou, Z.; Huang, Q.; Pan, C. Synthesis of Pyrimethanil-Loaded Mesoporous Silica Nanoparticles and Its Distribution and Dissipation in Cucumber Plants. *Molecules* 2017, *22,* 8.

PART III

Improving Soil Fertility and Crop Protection

CHAPTER 4

Possible Prospects of Nanotechnology in Sustainable Agriculture and Their Response on Soil Health and Plant Growth

PRAMOD U. INGLE[1],VAISHNAVI S. PARMA[2], and ANIKET K. GADE[1]

[1]*Department of Biotechnology, Sant Gadge Baba Amravati University, Amravati, Maharashtra, India*

[2]*Department of Agriculture and Food Engineering, Indian Institute of Technology, Kharagpur, West Bengal, India*

ABSTRACT

Agriculture has always been a global concern whether its productivity or fertility of soil. The productivity mostly relies on the quality, health, and soil fertility. Soil fertility also depends upon macro and micronutrient constituent as well as the population of soil microbiota. However, for better crop production and protection, sustainable agriculture practices are carried out with the aid of advancing technology. Nanotechnology is gaining much importance in agriculture in recent time for application of nanoparticles (NPs), nonmaterials, and nanoformulations, which has benefited agriculture by promoting soil organic matter, enhancing soil fertility, and microbial activity. Conversely, it also perturbs the microbial community when this applicant accumulates into the soil beyond certain limits. This chapter will provide account on mode of entrance of NPs into the soil, its positive and negative effect on the soil and its microbiome, interaction of soil microbial community.

Nanotechnology for Sustainable Agriculture: An Innovative and Eco-Friendly Approach. Vishnu D. Rajput, Abhishek Singh, Tatiana M. Minkina, Krishan K. Verma, and Awani Kumar Singh (Eds.)

4.1 INTRODUCTION

Agriculture acts as an interface between humans and the environment and is responsible for the modification of soil and the ecosystem. Soil quality and ecosystem have greatly impacted because of an unhealthy way of agriculture practices, which leads to vulnerability of agriculture and natural resources. Therefore, there is a requisite need for agricultural development and this can be ensured by adapting a new format of safe agricultural produce, which also helps to bring down environmental pollution (Singh et al., 2021). Sustainable agriculture is the key solution for the holistic future of mankind's survival. These are deliberately practiced to overcome global population demand for food and are mainly adept to protect the environment by preventing its natural resources and also to retain or improve its soil fertility (Rajput et al., 2021). Moreover, the feasible enhancement of soil fertility, agricultural growth, and productivity can be strengthened by the implication of pioneering technologies. Nanotechnology is often documented as a promising next-generation technology. Nanotechnology is one such field whose principal techniques are based on the nanoscale dimension of about 1–100 nm (Klaine et al., 2008). This nanosize range molecule bears specific, unique chemical, and optical properties. Along with those nanoparticles (NPs) also possesses specific surface plasma resonanceand high surface-to-volume ratios that add to its potential for various applications (Pramanik et al., 2020). Whereas the counterpart bulk materials have a dimension greater than 100 nm and possess lower reactivity as they do not have a high surface-to-volume ratio (Hermes Pérez-Hernández et al., 2019). Nanotechnology has traded its progress across the broad spectrum including biological and nonbiological areas and this is mainly due to their size-structure-dependent physicochemical properties. This advanced technology catalyzes enhanced agricultural productivity and growth rate. Undoubtedly, nanotechnology has a remarkable role in the second green revolution. Since it contributes to the sustainability of agriculture as a provider of nanosensors, nanopesticides (NPs), and nanofertilizers (NFs), nanocarriers, nanochips, also nanopackaging or nanocoating that increases shelf life and durability of the food products. While using these nanotechniques a better level of understanding about its potential toxic impacts on soil and other living systems are usually required.

Various forms of newly designed NFs are made commercially available very recently. These NFs are composed of various formulations developed through thorough experimentation and optimization. The main components of these NFs are metal, metal oxides, and quantum dots. Metal NPs includes

In, Mo, Ni, Ti, W, Zn, Al, Bi, Si, Ag, Sn, Co, Cu, Au, and Fe. The core of metal oxide NPs is made of NiO, MgO, SiO_2, SnO_2, ZnO, TiO_2, ZrO_2, Al_2O_3, CeO_2, CuO, Cu_2O, In_2O_3, and La_2O_3 (Ebbs et al., 2016; Olkhovych et al., 2016; Ebrahimi et al., 2016; Van et al., 2016; Singh and Kumar, 2016; Cvjetko et al., 2017; Priesteret et al., 2017; McGee et al., 2017; You et al., 2017). NPs have proven their widespread applications in various fields including agriculture due to their multifunctional ability (Yadav et al., 2014; Keller and Lazareva, 2014; Servin et al., 2017; Rajput et al., 2018). An online database (www.nanodb.dk) has enlisted thousands of nano-based products and huge quantities of these NPs and their ever-increasing production due to continuous consumption by various industries (Yadav et al., 2014; Rajput et al., 2020a).

This unceasing production, use, and continuous disposal have led to the release of tons of NPs into the soil, water, and air. Major amount is accumulated in soil through drainage, landfills, sludge, and waste. At the same time the amount released into water and air cannot be neglected (Keller et al., 2013). This ever-increasing accumulation of NPs into the environment may lead to their constant exposure with the various life forms present at the site of release and disposal, causing harmful effects on the biotic components of the environment (Yadav et al., 2014).

This chapter recapitulates the various modes of NPs entry into the soil, its positive and negative effect onto the soil, fertility, and its microbiome, interaction of soil microbial community with NPs.

4.2 TYPES AND ROLE OF SOIL MICROBIAL COMMUNITY

Soil microbial populations have a significant role in preserving the ecosystem, soil health, and biological production. Soil productivity and soil fertility are influenced by the presence of soil microbial communities. Soil is a habitat of a wide range of distinct microbial populations also refers as soil microbiome, and this diversity is usually because soil environmental conditions are heterogeneous (Bollen, 1956). The diverse microbial population depicts complexity and variability within the community, and this is due to genetic variation, functional groups difference, and relative abundance within taxons of the community. This distinct soil microbiome attributes a healthy soil–plant relationship, while those with low microbial diversity are often regarded as unhealthy soil. Slight alteration or any effect in microbial population significantly results in environmental change. Thus, the active microbial population is inextricably associated with soil health. Aforesaid, the soil is swarmed with a zillion of a living organism comprising of micro and macroflora and fauna. This diverse entity includes every other type of bacteria, actinomycetes, fungi, algae, and fauna like protozoa, nematodes, mites, collembola, earthworm, etc. It also provides shelters to many invertebrates such as worms, insects, and mammals like rabbits, rodents, and badgers. However, among these microbial communities, rhizospheric soil bacteria like N_2-fixing rhizobacteria or plant growth-promoting rhizobacteria, and other symbiotic rhizospheric microorganisms like cyanobacteria, which inhabits the rhizosphere of all plants are a more prominent biological tool in the sustainability of agriculture. Since these bacteria are capable to stimulate plant growth and other biocontrol activities (Meena et al., 2016). Soil microorganisms contribute to soil fertility for sustainable agriculture by influencing efficiency of nutrient cycling, soil carbon sequestration, modifying soil physical structure, greenhouse gas emission, and water regimes, enhancing the effectiveness of nutrient acquisition by the flora and improving plant health, and regulating the soil organic matter dynamics by mediating cation-exchange capacity, the soil S-N-P reserves, toxicity, and soil acidity (Singh et al., 2011). Even though, soil enzymes covalently contribute along with the soil microorganism in metabolic processes and biochemical processes of soil that involve formation and decomposition of organic compounds and humus; various oxidation-reduction reactions, hydrolysis, and transformation of plants, animals, and microorganism residues (You et al., 2017). To a certain extent, these enzymes provide aid to soil microbial processes. This soil biota remains viable as long as there is a carbon source for energy. Moreover, the activity of microbial biomass and their size is mainly governed by the entry of substrate into the soil (Rajput et al., 2020a).

4.3 IMPORTANCE OF SOIL QUALITY IN AGRICULTURE

Soil forms the basis of an ecosystem where it plays a vital function for living system that indorses environmental quality, sustains biological productivity, and maintain plant and animal health. Sustainable developments of agriculture productivity are subject to the concept of soil health. Soil is an intricate arrangement of organic and inorganic matter, air, soil microbial biomass, water, and other biotic and abiotic components. In general, "soil fertility" is the capability of the soil to provide water in adequate amounts and essential plant nutrients to nurture the plant for its growth and development. It is well documented that healthy soil has an abundance of a nutrient element like C, N, O, S, P, Ba, Mg, Mn, Ca, Fe, Zn, Mo, Si, Al, K, Na, Sr, Rb, and nanosized minerals along with this it is also a hub of carbonates, silicates, sulfates, hydroxides, oxides, and phosphates, which somehow play a crucial role in maintaining soil fertility. Any organic matter present in soil is degraded by the microorganisms present and making it available for the plants and other life forms associated. Thus, ultimately increasing the fertility of soil (Bollen, 1956).

Moreover, a fertile soil not only provides vital nutrients for the growth of crop plant but, also supports a miscellaneous and active biotic components, a typical soil structure maintenance, and allows for an uninterrupted decomposition (Bünemann et al., 2018). Soil quality is important both for agro-ecosystems and natural ecosystems as it helps in maintaining environmental quality and also promote biodiversity conservation. A soil status is typically defined by the presence of a microbiological and biochemical ailment in the soil, which is directly related to its natural soil fertility. Although, soil productivity depends upon several factors but microbes have a dominant role. Healthy soil is an indicator of active soil microbial population, which bears the potential to enhance agricultural production (Bharti et al., 2017). Soil quality is also assessed by the presence of biological indicators such as the microbial biomass, soil organic matter, soil respiration, ATP content, soil fatty acid profiles, potential nitrogen (N) mineralization capacity, soil enzyme activities, and DNA characterization (Igalavithana et al., 2017). But the constantly accumulating chemical through various modes may hamper the different components of soil. They may have a deliberate effect on soil microbes leading to the reduction in microbial biomass and ultimately reducing the soil fertility. In contrast to the chemical fertilizers and soil applicants, NPs can be a reliable alternative for improving soil fertility and aid the sustainable agricultural practices. These NPs act as NFs and are emerging as an alternative to chemical fertilizers. NFs can be designed to encapsulate nutrients and growth promoters and their controlled and targeted release as desired. Through controlled release of nutrients and increase in the nutrient utilization efficiency NFs may exert positive effect both in soil and present microflora (Manjunatha et al., 2016; Zulfiqar et al., 2019). For this purpose, the biologically synthesized nanomaterials are applied and the NPs are fabricated so as to cause minimal losses to the soil microbiota and the environment. Still the market availability of the NFs is lacking due to some limitations.

4.4 VARIOUS ELEMENTS OF AGROECOSYSTEM AFFECTED BY NPs

Besides all the reported benefits of NPs, exaggerated use causes diverse effects on the various components of environment. The NPs applied to crops and fields ultimately run-off to the soil, water, and atmosphere (Ali et al., 2020). Figure 4.1 shows the various elements of agroecosystem, which are impacted after exposure with the NPs.

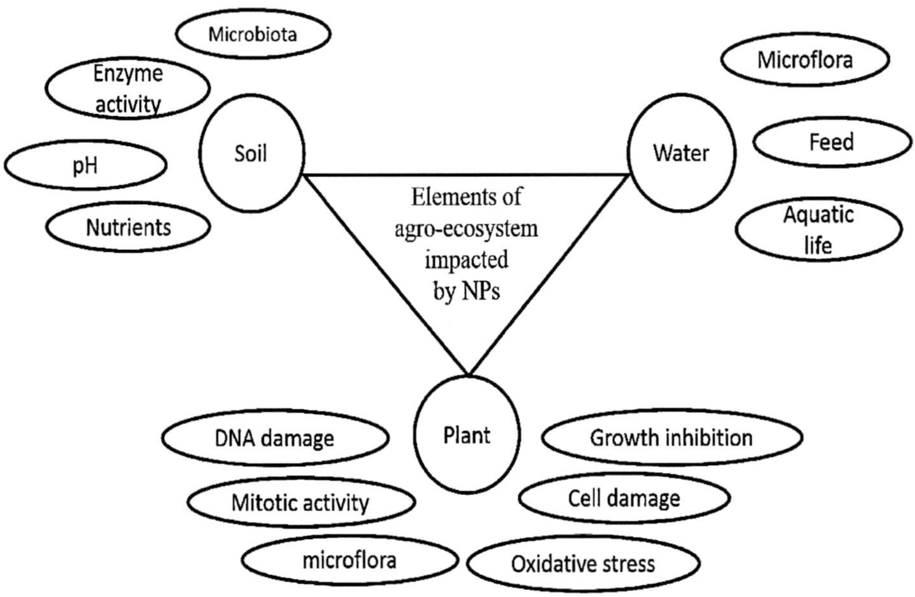

FIGURE 4.1 Different elements of agroecosystem affected by nanoparticles.

4.5 MODES OF NPs ENTRY INTO ENVIRONMENT

Post application fate of NPs is precisely unpredictable. The NPs may be either get absorbed by microbial flora of soil, uptaken by plant roots and leaves, and sometimes carried away with the air currents as an aerosol. As far as the release of NPs in soil is taken into account, there are several ways including leaching, surface run-off, etc. (Gladkova and Terekhova, 2013; Ali et al., 2020; Rajput et al., 2020b). Figure 4.2 represents the various routs of NPs entrance and impacting the environment.

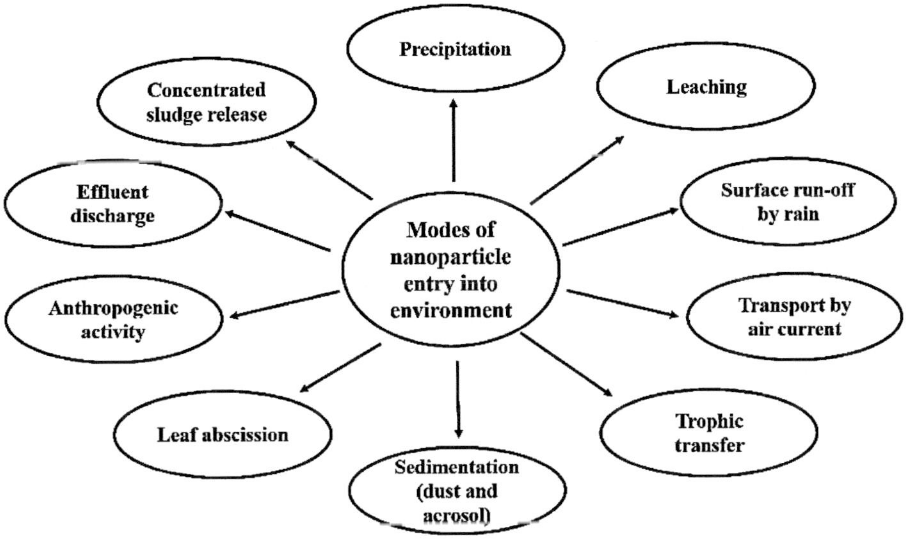

FIGURE 4.2 Various modes of nanoparticle entry into environment.

NFs are responsible for increased biosource of nutrients and are ecofriendly. They can improve organic carbon uptake and soil aggregation (Manjunatha et al., 2016). A good quantity of literature is available with respect to the impact of NPs on plants. But the reports indicating effect of NPs on soil microbes and their application as a soil applicant is scanty. Still, it is important to study the soil microbial communities for determining hazardous effects of NFs of NPs (Holden et al., 2014; Rajput et al., 2020a).

4.6 APPLICATION OF NPs IN ENHANCING SOIL MICROBIAL AND ENZYMATIC ACTIVITY

NPs are found to increase the microbial activity and also supports soil organic carbon cycle. This will help in addressing the infertile lands, wastelands, and support the vegetation and farming on barren lands. NPs are responsible for crop growth and increasing fertility and ultimately enabling farmers for coping up with the increasing food demands of population (Javed et al., 2019). For the sustainable use of soil, it is required to protect microbial biomass and its diversity (Torsvik and Øvreås, 2002). Impact of NPs is a function of its type, concentration, and enzymatic activity of soil. Ultimately, the larger surface availability of NPs for different enzyme activities including photosynthesis gives more yield and protects plants from biotic and abiotic stress thereby increasing the nutrient uptake efficiency (Preetha and Balakrishnan, 2017; Duhan et al., 2017).

But, the higher concentrations of NPs are reported to have negative effects on the soil enzyme activity. A report by Jośko et al. (2014) indicated reduction in dehydrogenase enzyme activity of soil by high concentration of NPs. Ability of soil for self-cleaning, balancing nutrient, and mediating the processes of plant nutrition, fertility, and microbial diversity are hampered due to higher NPs concentration (Janvier et al., 2007; Suresh et al., 2013; Bondarenko et al., 2013). Thus it is needed to study the toxicity effect of NPs in various types of soils for better understanding of their impact on soil pollutants and their mobility (Calvarro et al., 2014). In excess quantities, the NPs show significant negative effects in soil (Grün et al., 2018; Grün and Emmerling, 2018; Rajput et al., 2020a).

4.7 FATE OF NPs AND CONTRIBUTION TO FERTILITY, YIELD, AND REDUCING NPs TOXICITY

4.7.1 APPLICATION OF NPs IN MAINTAINING SOIL FERTILITY

The template of environmental soil is naturally concentrated with NPs, such as primary particles, aggregates, and agglomerates. The presence of nanosize particles and microparticle along with humic substance provide soil a high porosity and specific surface area. The natural compositions of soil are because of continual chemical and physical weathering and re-arrangement of its geogenic constituents associated with a higher bioactivity that transform both minerals and dead organic matter. However, environmental constraint and human action together are responsible for change in biophysical condition and structure of soil. This change results in depletion, erosion modification, or disturbance of soil that generally considered as harmful or undesirable. This has led to the phenomenon of soil degradation that results in the disappearance of nutrients and mineral needed for plant growth. Thus, the aim of nanotechnology is to make soils more fertile and reproducible by improving its nutrient efficiency, which can be used for better environmental security and greater productivity. NPs facilitate nutrient management by ensuring the availability of nutrients to plant as per their requirement. It also makes sure that ions present in soil system must be available according to plant needs, as exchange of ions (e.g., NH_4^+, HPO_4^{2-}, $H_2PO_4^-$, PO_4^{3-}, Zn^{2+}) is the base of nutrient transport in soil-plant systems, adsorption, and desorption (e.g., phosphorus nutrients and solubility and precipitation of iron) reactions

(Mukhopadhyay, 2014). Moreover, the nutrients in soil can be buildup with application of NFs. NFs are an emerging alternative to conventional fertilizer, which aims at supplying one or more micro and macronutrients to the plant–soil system and enhances the growth and productivity. These nutrient-embedded NPs or NFs can be used in aqueous suspension. These are dissolved in desired solvent and, release the nutrient/s as soluble ions. These soluble nutrient ions are comprehensively absorbed by the plants. The application of NFs not only eliminates the contamination of drinking water but also has a potential of low eutrophication and also addresses soil fixation issues (Liu et al., 2015). Soil degradation is one such severe environmental issue, which has impacted soil productivity, and fertility and also a major cause of soil pollution. However, this issue can be ameliorated by the application nanozeolites. Soil acidity could possibly neutralize with the use of nanozeolite. Similarly, iron NPs are used for the remediation of dirtied soil (Baragano et al., 2020). More likely, the contamination is due to heavy application of pesticides, heavy metal, and radionucucliodes. These zerovalent iron NPs have strong affinity against heavy metals and organic compounds (Galdames et al., 2020). Like iron NPs, calcium carbonate NPs bear an excellent soil binding properties that help in formation of soil micro aggregates and macro aggregate. All engineered NPs have myriad application in restoring soil fertility. Nonetheless, it is highly needed to understand the behavior of NPs as this provide apt about toxicity in soil ecosystem.

4.7.2 CONTROLLED RELEASE OF NUTRIENT TO SOIL

The zest of applying nanotechnology in the field of agriculture resulted in the upliftment of the agricultural production. This technology engineered the material that adds value to agriculture NPs and nanomaterials in the form of NFs has been an alternative approach to provide mineral nutrients to the plant instead of traditional fertilizers. These nanostructured micronutrient fertilizers contribute to plant nutrition through either incorporating nanoelements by adsorption or absorption or by encapsulation on NPs. The application of these NFs to soil is mainly as a substrate or by foliar spray. The hybrid NPs with metal ions encapsulated in their core can be used for the meticulous release of micro and macronutrients (Tarafder et al., 2020). However, these NFs must have greater control over the speed and release time of the nutrient element.

4.7.3 TRANSLOCATION OF NPs INTO PLANTS

Aforesaid, NPs are applied as NFs to soil nutrient-deficient plant, exogenously either through irrigation or by foliar spray. In either of the form, the accumulated nanonutrients are released from NFs and translocated into plant through roots to leaves or vice-versa via plant vascular system. NPs applied to soil are firstly adsorbed on the root surface followed by translocation via symplastic and apoplastic pathways into the xylem, passing the endodermis, and then locating to the whole plant through the vascular bundles. While foliar-applied NPs up taken and translocated from leaves to roots via stomatal pathway, where NPs travel distantly through the vascular system of plants when entered through leaf apoplast. Endocytosis or pores or channels can mediate the entry of various NPs and are transported through the cell. However, multiple factors such as the size of NPs, composition of NFs, exposure conditions, morphology and physiology of plants, and the pore-diameter of the plant cell wall affects the uptake and translocation of NPs (Seleiman et al., 2021).

4.7.4 EFFECT OF NPs ON PLANT GROWTH

Nanotechnology accelerated almost every sector of agriculture science, but the applications of nanostructured micronutrient fertilizers, have great effect on the development and plant growth. However, the effect of NPs either positive or negative on plant mainly relies upon the concentration, structure, physical, and biochemical properties of NPs and also depends on the plant species. NP positively brings out various physiological changes in plant, which indirectly or directly influenced the growth and development. Applications of sulfur and zinc oxide metal NPs with appropriate concentration have significantly enhanced the total lipids, amino acids, proteins, thiol, and chlorophyll contents. Even, it has been reported that the enzymatic activity including catalase, peroxidase, superoxide dismutase, ferric reductase activity, malondialdehyde content, root-apoplastic iron, and the root activity has also been increased. Moreover, growth and development of plant species show major response when the photosynthetic activity get altered as electron transport rates is improved and chlorophyll content is increased this generally occurs when plant species are treated with metal NPs (Javed et al., 2019).

4.7.5 DECREASE IN TOXICITY LEVEL IN SOIL

Soil proclaims to be the largest inheritor of both natural and artificial NPs. Wide applications such as NFs, nanopesticides, seed treatment, and applicants, are imparting toxic effect on soil and agro-environment if susceptible to large agglomeration and exposure time. These NPs and/or NFs applied to the plant–soil system, it undergoes certain transformation that leads to accumulation into the soil, and when interacted with plants, these NFs are translocated to other plant part where it starts accumulating in cellular or subcellular organelle. These correspond to change in the physiological process like photosynthesis and transpiration rate, growth, and also the cellular and subcellular integrity of organelles of plant (Morales-Díazet al., 2017; Rajput et al., 2020a, 2020b). As these NPs are getting translocated their concentration at the sink, that is, soil is reducing. Thus, it can be claimed that their potential toxicity is also decreasing with the continuous translocation

4.8 CONCLUSIONS AND FUTURE PROSPECTIVE

The NFs are the formulation designed to supplement the soil or foliar application in agricultural crops. They have the micronutrients encapsulated in their core. These NP formulations are applied for soil applicants, foliar sprays, seed treatment, etc. The unused portion of these ultimately goes to the soil as a residue. This residual content starts accumulating into the soil. The continuous use and ever-increasing amount in the soil may lead to the toxic effect on the elements of environment and more pronounceable on agro-ecosystem. This may hamper the macro and microflora of the soil without discriminating the useful and harmful ones. Toxicity impact on the useful microorganisms that help in micronutrient availability for plants through biodegradation of organic and inorganic contents of soil, will reduce their load from rhizosphere of plants. Resulting in the scarcity of nutrients for the plant roots.

The NPs enter environment through various modes as discussed above. The NPs applied to the crops and soil will stay unavailable for microbes as well as for the plants. Thus, contributing to the losing soil vigor and higher toxicity content of soil. The increase in the threshold concentration of NPs in soil will also decrease the microbial and soil enzymatic activity. To counteract these problems the nanoformulations and NFs applied to the agricultural field are needed to be designed in such a way that they are readily available to the plants, easily enacted by the beneficial microbial counterparts without hampering their number and activity. At the same time, it should be kept into consideration that there is a control over the retrieval of nutrients from nanoformulations applied and their accessibility and translocation into plants is monitored properly. The precise guidelines and policies are needed to be designed for the minimal loss and higher availability and efficient utilization of nutrients from NFs to the desired purpose.

KEYWORDS

- **microbial community**
- **microbiome**
- **soil fertility**
- **soil microbiota**
- **sustainable agriculture**

REFERENCES

Ali, M. A.; Ahmed, T.; Wu, W.; Hossain, A.; Hafeez, R.; Islam Masum, M. M.; Li, B. Advancements in Plant and Microbe-Based Synthesis of Metallic Nanoparticles and Their Antimicrobial Activity Against Plant Pathogens. *Nanomaterials* **2020,** *10*(6), 1146. DOI: 10.3390/nano10061146.

Bharti, V. S.; Dotaniya, M. L.; Shukla, S, P.; Yadav V. K. Managing Soil Fertility Through Microbes. Prospects, Challenges and Future Strategies. In *Agro-Environmental Sustainability*; 2017, chapter 5. DOI: 10.1007/978-3-319-49724-2_5.

Bollen, W. B. *Microorganisms and Soil Fertility*; Oregon State Monograph. Studies in Bacteriology: Oregon, 1956, vol 1.

Bondarenko, O.; Juganson, K.; Ivask, A.; Kasemets, K.; Mortimer, M.; et al. Toxicity of Ag, CuO and ZnO Nanoparticles to Selected Environmentally Relevant Test Organisms and Mammalian Cells *In Vitro*: A Critical Review. *Arch. Toxicol.* **2013,** *87,* 1181–1200.

Bünemann, E. K.; Bongiorno, G.; Bai, Z.; Creamer, R. E.; De Deyn, G.; Goede, R.; Fleskens, L.; Geissen,V.; Kuyper, T.; Mäder, P.; Pulleman, M.; Sukkel, W.; Groenigen, J. W. V.; Brussaard, L. Soil Quality – A Critical Review. *Soil Biol. Biochem.* **2018,** *120,* 105–125. https.//doi.org/10.1016/j.soilbio.2018.01.030.

Calvarro, L. M.; de Santiago-Martín, A.; Gómez, J. Q.; González-Huecas, C.; Quintana, J. R.; et al. Biological Activity in Metal-Contaminated Calcareous Agricultural Soils. The Role of the Organic Matter Composition and the Particle Size Distribution. *Environ. Sci. Pollut. Res.* **2014,** *21*, 6176–6187.

Cvjetko, P.; Milošić, A.; Domijan, A. M.; VinkovićVrček, I.; Tolić, S.; Peharec Štefanić, P.; Letofsky-Papst, I.; Tkalec, M.; Balen, B. Toxicity of Silver Ions and Differently Coated Silver Nanoparticles in *Allium cepa* Roots. *Ecotoxicol. Environ. Saf.* **2017,** *137,* 18–28.

Duhan J. S.; Kumara, R.; Kumara, N.; Kaura, P.; Nehrab, K.; Duhan, S. Nanotechnology: The New Perspective in Precision Agriculture. *Bio-tech. Rep.* **2017,** *15*, 11–23.

Ebbs, S. D.; Bradfield, S. J.; Kumar, P.; White, J. C.; Musante, C.; Ma, X. Accumulation of Zinc, Copper, or Cerium in Carrot (*Daucuscarota*) Exposed to Metal Oxide Nanoparticles and Metal Ions. *Environ. Sci. Nano.* **2016,** *3,* 114–126.

Ebrahimi, A.; Galavi, M.; Ramroudi, M.; Moaveni, P. Effect of TiO2 Nanoparticles on Antioxidant Enzymes Activity and Biochemical Biomarkers in Pinto Bean (*Phaseolus vulgaris* L.). *J. Mol. Biol. Mol.* **2016,** *6*(1), 58–66.

Gladkova, M. M.; Terekhova, V. A. Engineered Nanomaterials in Soil. Sources of Entry and Migration Pathways. *Moscow Univ. Soil Sci. Bull.* **2013,** *68*, 129–134.

Grün, A. L.; Emmerling, C. Long-Term Effects of Environmentally Relevant Concentrations of Silver Nanoparticles on Major Soil Bacterial Phyla of a Loamy Soil. *Environ. Sci. Eur.* **2018,** *30,* 1–13.

Grün, A. L.; Straskraba, S.; Schulz, S.; Schloter, M.; Emmerling, C. Long-Term Effects of Environmentally Relevant Concentrations of Silver Nano-Particles on Microbial Biomass, Enzyme Activity, and Functional Genes Involved in the Nitrogen Cycle of Loamy Soil. *J. Environ. Sci.* **2018,** *69,* 12–22.

Holden, P. A.; Schimel, J. P.; Godwin, H. A. Five Reasons to Use Bacteria When Assessing Manufactured Nanomaterial Environmental Hazards and Fates. *Curr. Opin. Biotechnol.* **2014,** *27,* 73–78.

Igalavithana, A. D.; Lee, S. S.; Niazi, N. K.; Lee, Y. H.; Kim, K.; Park, J.; Moon, D. K.; Ok, Y. S. Assessment of Soil Health in Urban Agriculture: Soil Enzymes and Microbial Properties. *Sustainability* **2017**, *9*(2), 310. DOI: 10.3390/su9020310.

Janvier, C.; Villeneuve, F.; Alabouvette, C.; Edel-Hermann, V.; Mateille, T.; et al. Soil Health Through Soil Disease Suppression: Strategy From Descriptors to Indicators. *Soil Biol. Biochem.* **2007,** *39,* 1–23.

Javed, Z.; Dashora, K.; Mishra, M.; Fasake, V. D.; Srivastva, A. Effect of Accumulation of Nanoparticles in Soil Health- A Concern on Future. *Front Nanosci. Nanotech.* **2019,** *5*, 1–9. DOI: 10.15761/FNN.1000181.

Jośko, I.; Oleszczuk, P.; Futa, B. The Effect of Inorganic Nanoparticles (ZnO, Cr2O3, CuO and Ni) and Their Bulk Counterparts on Enzyme Activities in Different Soils. *Geoderma* **2014,** *232*, 528–537.

Keller, A. A.; Lazareva, A. Predicted Releases of Engineered Nanomaterials: From Global to Regional to Local. *Environ. Sci. Technol. Lett.* **2014,** *1*, 65–70.

Keller, A. A.; McFerran, S.; Lazareva, A.; Suh, S. Global Life Cycle Releases of Engineered Nanomaterials. *J. Nanopart. Res.* **2013,** *15*, 1692.

Manjunatha, S. B.; Biradar, D. P.; Alada-katti, Y. R. Nanotechnology and its Applications in Agriculture: A review. *J. Farm. Sci.* **2016,** *29*(1), 1–13.

McGee, C. F.; Storey, S.; Clipson, N.; Doyle, E. Soil Microbial Community Responses to Contamination With Silver, Aluminium Oxide and Silicon Dioxide Nanoparticles. *Ecotoxicology* **2017,***26*(3), 449–458.

Meena, V. D.; Dotaniya, M. L.; Rajendiran, S.; Coumar, M, V.; Kundu, S.; Rao, A. S. A Case for Silicon Fertilization to Improve Crop Yields in Tropical Soils. *Proc. Natl. Acad. Sci. India Sect. B Biol. Sci.* **2013,** *84*(3), 505–518.

Morales-Díaz, A. B.; Ortega-Ortíz, H.; Juárez-Maldonado, A.; Cadenas-Pliego, G.; González-Morales, S.; Benavides-Mendoza, A. Application of Nanoelements in Plant Nutrition and its Impact in Ecosystems. *Adv. Nat. Sci. Nanosci. Nanotechnol.* **2017,** *8*(1), 013001. DOI: 10.1088/2043-6254/8/1/013001.

Mukhopadhyay, S. Nanotechnology in Agriculture: Prospects and Constraints. *Nanotechnol. Sci. Appl.* **2014,** *7,* 63–71. http.//dx.doi.org/10.2147/NSA.S39409.

Olkhovych, O.; Volkogon, M.; Taran, N.; Batsmanova, L.; Kravchenko, I. The Effect of Copper and Zinc Nanoparticles on the Growth Parameters, Contents of Ascorbic Acid, and Qualitative Composition of Amino Acids and Acylcarnitines in *Pistiastratiotes* L. (Araceae). *Nanoscale Res. Lett.* **2016,** *11*, 218.

Pramanik, P.; Krishnan, P.; Maity, A.; Mridha, N.; Mukherjee, A.; Rai, V. Applications of Nanotechnology in Agriculture. In *Environmental Nanotechnology*; Dasgupta, N., et al., Eds.; *Chem. Sustain. World* **2020,** *4*, 32. https.//doi.org/10.1007/978-3-030-26668-4_9

Priester, J. H.; Moritz, S. C.; Espinosa, K.; Ge, Y.; Wang, Y.; Nisbet, R. M.; Schimel, J. P.; Goggi , S. A.; Gardea-Torresdey, J. L.; Holden, P. A. Damage Assessment for Soybean Cultivated in Soil With Either CeO2 or ZnO Manufactured Nanomaterials. *Sci. Total Environ.* **2017,** *579*, 1756–1768.

Rajput, V.; Minkina, T.; Mazarji, M.; Shende, S.; Sushkova, S.; Mandzhieva, S.; Jatav, H. Accumulation of Nanoparticles in the Soil-Plant Systems and Their Effects on Human Health. *Ann. Agric. Sci.* **2020,** *65*(2), 137–143. DOI: 10.1016/j.aoas.2020.08.001.

Rajput, V. D.; Singh, A.; Singh, V. K.; Minkina, T. M.; Sushkova, S. Impact of Nanoparticles on Soil Resource. *Nanomat. Soil Remediat.* **2020a,** 65–85. DOI. https.//doi.org/10.1016/B978-0-12-822891-3.00004-9.

Rajput, V., Minkina, T., Feizi, M., Kumari, A., Khan, M., Mandzhieva, S., Sushkova, S., El-Ramady, H., Verma, K., Singh, A., Hullebusch, E., Singh, R., Jatav, H.; Choudhary, R. Effects of Silicon and Silicon-Based Nanoparticles on Rhizosphere Microbiome, Plant Stress and Growth. *Biology* **2021,** *10*(8), 7–9.

Seleiman, M. F.; Almutairi, K. F.; Alotaibi, M.; Shami, A.; Alhammad, B. A.; Battaglia, M. L. Nano- Fertilization as an Emerging Fertilization Technique. Why Can Modern Agriculture Benefit From Its Use? *Plants* **2021,** *10*, 2. https.//dx.doi.org/10.3390/plants10010002.

SelvaPreetha, P.; Balakrishnan, N. A Review of Nano Fertilizers and Their Use and Functions in Soil. *Int. J. Curr. Microbiol. App. Sci.* **2017,** *6*(12), 3117–3133.

Servin, A. D.; De la Torre-Roche, R.; Castillo-Michel, H.; Pagano, L.; Hawthorne, J.;Musante, C.; Pignatello, J.; Uchimiya, M.; White, J. C. Exposure of Agricultural Crops to Nanoparticle CeO2 in Biochar-Amended Soil. *Plant Physiol. Biochem.* **2017,** *110,* 147–157.

Singh, D.; Kumar, A. Impact of Irrigation Using Water Containing CuO and ZnO Nanoparticles on *Spinach oleracea* Grown in Soil Media. *Bull. Environ. Contam. Toxicol.* **2016,** *97,* 548–553.

Singh, J, S.; Pandey, V. C.; Singh D. P. Efficient Soil Microorganisms. A New Dimension for Sustainable Agriculture and Environmental Development. *Agric. Ecosyst. Environ.* **2011,** *140*, 339–353. DOI: 10.1016/j.agee.2011.01.017.

Singh, A.; Rajput, V. D., Mehrotra, R.; Pal, N.; Singh, V. K.; Minkina, T.; Chokheli, V. A.; Singh, R. K. In *Sustainable Soil Fertility Management*; Nova Science Publishers Inc., 2020; vol 1, pp 73–100.

Singh, A.; Rajput, V. D., Rawat, S.; Sharma, R.; Singh, A. K.; Singh, A. K.; Tomar, R. S. In *Emerging Tools for SustainableAgriculture and Food Security*; Rajput, Deepika Book Agency: New Delhi, Delhi, 2021; vol 1, pp 1–15.

Suresh, Y.; Annapurna, S.; Bhikshamaiah, G.; Singh, A. K. Characterization of Green Synthesized Copper Nanoparticles: A novel approach. In *International Conference on Advanced Nanomaterials Emerging Engineering Technologies*, 2013; pp 63–67.

Tarafder, C.; Daizy, M.; Alam, M. M.; Ali, M. R.; Islam, M. J.; Islam, R.; Khan, M. Z. H. Formulation of a Hybrid Nanofertilizer for Slow and Sustainable Release of Micronutrients. *ACS Omega* **2020**, *5*(37), 23960–23966. DOI: 10.1021/acsomega.0c03233

Torsvik, V.; Øvreås, L. Microbial Diversity and Function in Soil: From Genes to Ecosystems. *Curr. Opin. Microbiol.* **2002,** *5*(3), 240–245.

Van, N. L.; Ma, C.; Shang, J.; Rui, Y.; Liu, S.; Xing, B. Effects of CuO Nanoparticles on Insecticidal Activity and Phytotoxicity in Conventional and Transgenic cotton. *Chemosphere* **2016a**, *144*, 661–670.

Yadav, T.; Mungray, A. A.; Mungray, A. K. Fabricated Nanoparticles: Current Status and Potential Phytotoxic Threats. *Rev. Environ. Contam. Toxicol.* **2014,** *230*, 83–110.

You, T.; Liu, D.; Chen, J.; Yang, Z.; Dou, R.; Gao, X.; Wang, L. Effects of Metal Oxide Nanoparticles on Soil Enzyme Activities and Bacterial Communities in Two Different Soil Types. *J. Soils Sediments* **2017,** *18,* 211–221. DOI: 10.1007/s11368-017-1716-2.

Zulfiqar, F.; Navarro, M.; Ashraf, M.; Akram, N. A.; Munné-Bosch, S. Nanofertilizer Use for Sustainable Agriculture: Advantages and Limitations. *Plant Sci.* **2019,** *289,* 110270. DOI: 10.1016/j.plantsci.2019.

CHAPTER 5

Impact of Nanoparticles on Soil Microbes for Enhancing Soil Fertility and Productivity

NAVODITA MAURICE

Prophyl Ltd., Dózsa György út 18, Mohács, Hungary

ABSTRACT

Nanotechnology has revolutionized agricultural practices by providing vast applications for instance, metal, and non-metal-based nanoparticles (NPs) (carbon nanotubes, TiO_2, Ag, ZnO, CuO NPs, etc.), nanopesticides, and nanofertilizers for targeted release. NPs offer vast applications as they have variable shapes, sizes, are highly stable, have strong reactivity with larger surface areas for interacting with plants and microbes. NPs gain access into the soil by nanoagrochemicals, animal faeces, atmospheric deposition, plant litter, and industrial wastes. The excessive usage of NP is not considered safe for plants as well as for microbes. Distinct NPs have distinct iveramifications on the plant and microbial relations. Antimicrobial properties of NPs against the microbial populations (bacteria and fungi) have been well studied. NPs can cause cellular deformation in bacteria and alteration of hyphal structure in fungi. The effect of NPs on microbes is dose dependent. NPs, in general, show positive effects on microbes in lower concentrations; however, they inhibit microbial enzyme activity as well as microbial growth at elevated levels. Some bacteria can grow well at elevated concentrations of NPs showing their tolerance toward the toxicity of NPs. The following chapter

Nanotechnology for Sustainable Agriculture: An Innovative and Eco-Friendly Approach. Vishnu D. Rajput, Abhishek Singh, Tatiana M. Minkina, Krishan K. Verma, and Awani Kumar Singh (Eds.)

focuses on how NPs influence the microbes thriving in the soil followed by their effects on their physiological traits.

5.1 INTRODUCTION

Nanotechnology can be exemplified as an inventive, multifaceted scientific way of establishing, shaping, and utilization of substances at the nano-grade (Ali et al., 2019). Nano is the one billionth unit and therefore the substances are evaluated as the billionth of a meter in nanotechnology. Nanometer is commensurate to 10 hydrogen atoms set side by side. Nanotechnology despite being confined to minuscule elements also finds its utilization in multidisciplinary approaches of biology, physics, and chemistry and allied sciences. Nanoparticles (NPs) measuring smaller in size than 100 nm lie in an intermediary state of an atom and its associated elements, with the possibility of changing physicochemical properties (conductivity, sensitivity) (Mishra and Kumar, 2009).

Nanotechnology has found its immense application in the agricultural sector as to feed a population of above seven billion is a real challenge (Singh et al., 2015; Singh et al., 2020). Nanotechnology not only finds its application in the agricultural field, but scientists all over the world are frequently studying and modifying the characteristics of the nanoparticles for the improvement of growth and developmental processes of the plants (Singh et al., 2020). Materials of nano-size can easily take up by bacteria and are capable of penetrating the plant cells but can result in phytotoxic behavior if the dose of these nanomaterials is elevated to higher levels (Liu et al., 2009). Consequences of agrochemicals generated after the utilization of nanotechnological approaches in relation to the developmental processes of agricultural plants are being regularly tested by scientists all over the globe. The major advantages of nanotechnology in the current scenario encompass plant treatment with nanocides, disease prevention, and maintenance of nutrient value with the application of nano-fertilizers.

Distinctive nanomaterials for instance, carbon nanotubes, magnetic particles, polymers, quantum dots, metals, and nonmetals are under investigation all around the world (Rico et al., 2015). The exclusive features of nanoparticles include disease checkup, detection of pathogens, and enhanced nutrient adsorption by plant surface. These remarkable features boost up the future prospects of agricultural and food industries. NPs have the potential to detect the disease-causing agents (Prasad et al., 2014). The surface structure, smaller surface/volume proportion, agglomeration, as well as enhanced reaction rate of NPs have emphasized their use as modern tools in various fields, for example, targeted drug delivery, nano-pharmacology, nano-medicine, and agrochemical processes. Currently, nanomaterials are being tested both in vivo and ex vivo studies (Agrahari and Dubey, 2020) (Fig. 5.1).

5.2 NANOPARTICLES

Nanoparticles are engineered and designed in such a manner that they have peculiar chemical and surface characteristics. A diverse array of NPs has been produced, for example, magnetic, metallic oxide and gold nanoparticles, silica-based NPs, carbon-based NPs (nanotubes, grapheme, and fullerenes), and quantum dots (Wang et al., 2016). Carbon nanotubes enable faster seed germination as they can penetrate seed husks. Usage of optimal concentrations, targeted delivery with low levels of toxicity is possible with nanoparticles for example, silica-based NPs loaded with pesticides protect from photodegradation (Wang et al., 2016). Quantum dots measuring 2–10 nm in diameter are effective semiconductor elements that are the tools of cellular imaging. The exclusive delivery system of nanoparticles has made them suitable to be used in the biotic forms of life ranging from human life to the plant systems. However, the usage of nanoparticles is an effective and fresh techniquc that demands intensive search to achieve the targeted goals. The usage of nanoparticles in the agricultural systems has however increased, but it has also posed some potential serious threats to the environment (Kreyling et al., 2006). Nanomaterials have been used worldwide as they are now easily available and can be easily designed from a vast array of segments (Geisler-Lee et al., 2012). The gathering propensity followed by less solubilizing efficiency (especially water) of the nanoparticles limits their entry in most of the living organisms. Few nanoparticles can gain access into humans and animals through food, skin, water, and air (Jain et al., 2016). They can also circulate in the trophic levels and may concentrate within the top predators resulting in enhanced toxicity among the animals of noticeable levels of food web (Krysanov et al., 2010). Metal oxide nanoparticles, for example, carbon-based fullerenes and titanium oxide nanoparticles result in enhanced toxicity among the microbes. Fullerenes are low mobility in soil and water. The toxicity of nanoparticles is dependent on its physicochemical characteristics and size. It is assumed that the quantities of the nanoparticles that can gain access into the body of organisms are liable to increase in the future, especially among the higher trophic level organisms. The higher accumulation of nanoparticles in the tissues and organs can alter the cellular structure. The usage of conventional methods of pest control by the farmers has already altered the soil's natural flora (Thul and Sarangi, 2015) and releasing these chemicals loaded within the nanoparticles has solved the problem a bit (Gruère et al., 2011). The elevated levels of pernicious and agglomerating nature of these NPs have opened ways for the usage of biodegradable nanoparticles including their impact on organisms (Krysanov et al., 2010). The awareness regarding NP accumulation in surroundings as well as in the organisms is at the preliminary stage at present.

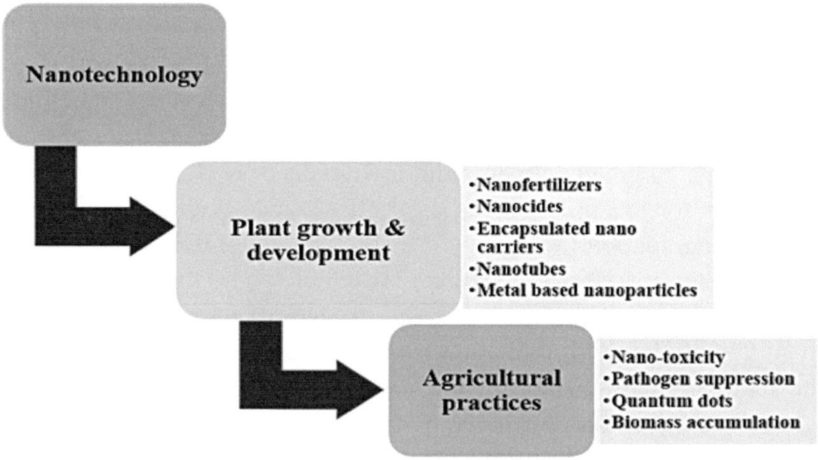

FIGURE 5.1 Application of nanotechnology (modified from Agrahari, S.; Dubey, A. Nanoparticles in plant growth and development. In *Biogenic Nano-Particles and Their Use in Agro-Ecosystems*; Ghorbanpour, M., Bhargava, P., Varma, A., Choudhary, D., Eds.; Springer: Singapore, 2020; pp 9–37).

5.2.1 NATURAL ORIGIN OF NPs

Diverse applications of NPs increase environmental contamination issues. NPs can enter the atmosphere by intentional as well as accidental ways. Releasing these NPs in water, soil, or air is a menace as they are very tiny and can remain in the air or are transported through water. Their accumulation of soil for a prolonged period can alter its physicochemical properties as well as pollute the water present in the soil (Tripathi et al., 2012). Nanoparticle-coated painting materials followed by artificial pigments, however, are more treacherous as they can gain entry into the soil and water. Nanomaterial-coated optical and electronic goods are the major pollutants of landfills. Cosmetics containing NPs are also the major contaminants of soil and water. The processes used in the manufacturing of nanoparticle-coated goods affect the leaching rate. Nanoparticle sources can be point-based NPs and non-point-based NPs. Point agents include research laboratories, production followed by storage sections, as well as wastewater plants (Tiede et al., 2016). These NPs enter the surroundings as sludge and treated water from the treatment plants, as it is almost impossible to separate NPs from the wastewater. Sewage sludge contains silver NPs drained out by the cosmetic industry, drinks, food, as well as textile product washing (McGillicuddy et al., 2017). Titanium oxide (TiO_2) NPs are used in wastewater treatment and water purification reactors and around 99% of the TiO_2 NPs accumulate in the sludge of the wastewater after treatment (Yadav et al., 2014). Treatments plants also have different types of microbes that can also convert metals into NPs for example, elemental selenium NPs are being produced by the reduction of selenate (SeO_4^{2-}) and selenite (SeO_3^{2-}) in the active sludge (Jain et al. 2016). NPs can also react with other constituents of the waste and therefore can result in the formation of new compounds during the treatment process. The non-point sources of NPs include paints, cosmetics, etc. The self-cleaning and antimicrobial feature of photocatalytic TiO_2 NPs makes them suitable for the paint industry and paint coatings can also real the NPs in the surroundings (Shandilya et al., 2015). Zinc oxide (ZnO) as well as TiO_2 NPs find their applications in the cosmetic industry. Sanchez-Quiles and Tovar-Sanchez (2014) found an elevated concentration of TiO_2 NPs due to the prolonged accumulation of sunscreen lotion used by the bathers in the Palmira beach water (Majorca Island, Spain). Apart from its negative side, nanotechnology also finds its potential application in the decrease of the organic pollution of soil as well as environmental remediation (Li et al., 2016). Application of NPs against soil remediation, however, adds to more accumulation of NPs and demands more research. Nanofertilizers are the new tools of nutrient delivery systems utilizing carbon, iron, manganese, and zinc oxide. Nanosensors and smart delivery systems play a role in pest control. Vehicle exhaust release, combustion processes, smoke, and soot are the sources of NPs in the air (Kashyap et al., 2015).

5.3 EFFECT OF NPs

NPs have shown beneficial results on the plant developmental processes as they can be sources of waste-energy conversion, nanofertilizers, nanopesticides for targeted delivery. A diverse variety of NPs show different effects on morphology followed by physiology of plants. Their reactivity size, chemical constitution may be positive as well as negative for the plant growth. The widely used metal NPs are derived from gold, silver, copper, zinc, titanium oxide, and cerium oxide. However, ferric oxide, manganese-cobalt-ferric oxide NPs now these days are also being produced. Both metal and carbon-based NPs can accumulate reactive oxygen species (ROS) leading to plant stress. The attributable creatures of carbon-based NPs are due to their chemical or mechanical or electrical features (Singh et al., 2015). Single-walled carbon nanotubes (SWCNTs) now these days are considered nano transporters as they are involved in the transport of DNA and other materials (dyes) across and also from the outer environment into the cells. However, multi-walled carbon nanotubes (MWCNTs) increase water as well as plant cell nutrient uptake and therefore promote plant propagation. Siddiqui et al. (2015) reported efficacy of carbon nanotubes (CNTs) against ROS aggregation followed by peroxidation of lipids in seedling root tips and cell cultures. NPs perforate the plant cell walls, strengthening water uptake and therefore enhance the seed germination (Agrahari and Dubey, 2020). NPs have found their application in medicine, cosmetics, biosensors, environmental remediation, wastewater treatment, catalysts, etc., which has raised a concern about human and environmental health. NPs can negatively affect some microbes, for instance, bacteria, yeasts, fungi, marine diatoms (Karlsson et al., 2009), and some fishes (zebrafish). The major NPs co-receptors are soil microorganisms and growing plants. Soil microbes are the most accurate indicators of altered soil characteristics (Saison et al., 2010).

5.4 INFLUENCE OF NANOPARTICLES ON THE MICROBIAL COMMUNITIES

Application of NPs in the soil environment is risky as it can influence the essential bacterial and fungal populations. NPs are administered in the soil by several activities for example, as nanopesticides, nanofertilizers for crop improvement, wastewater sludge, and sewage treatment, aerial emulsion (Coll et al., 2016). Soil respiration followed by enzyme activity measurement can give us a clear picture about the levels of NPs accumulated in the soil. TiO_2 and CuO NPs decrease the enzyme activity of soil, especially in the paddy fields (Xu et al., 2015). Enzyme activity especially the phosphatase, urease, invertase, and catalase as well as the number of bacteria also undergo decline in the presence of TiO_2, Fe_3O_4, ZnO, as well as CeO_2 NPs present in the black and saline-alkali soils (You et al., 2017). This indicated that NPs can also affect the process of nitrogen fixation. Fe_3O_4 NPs in higher levels can decrease the number of soil bacteria (Jiling et al., 2016). The populations of P and K-solubilizing bacteria and *Azotobacter* are affected in the presence of Zinc oxide and CeO_2 NPs. TiO_2 NPs pose an adverse effect on the microbial diversity, number, and enzyme activity. Biogenic gold (Au-NPs) does not show any harmful effect to plants and soil microbes (Maliszewska, 2016). The synthesis of ZnO-NPs by chemical and biological methods showed antibacterial response against a number of beneficial soil bacteria, for instance, *Escherichia coli*, *Pseudomonas aeruginosa*, *Bacillus subtilis,* and *Staphylococcus aureus* (Lakshmi et al., 2012). Cu-NPs are a major threat to beneficial as well as pathogenic bacterial species (Lofts et al., 2013). CuO-NPs can cause cellular collapse, membrane degradation, cavities, blebs, holes, and cell lysis of bacteria present in the soil and therefore are very toxic (Concha-Guerrero et al., 2014). The repercussion of CuO-NPs on the soil microbes is less understood at present. Inorganic NPs are more lethal to soil microbes than the organic ones. CuO and Ag-NPs even decrease the rate of leaf decomposition by microbes (Frenk et al., 2013).

5.4.1 CARBON NANOMATERIALS (CNMs)

CNMs have gained widespread usage due to their exclusive performance assets and the commonly applied CNMs comprise carbon nanotubes (CNTs), fullerene, and graphene (Zhang et al., 2013). CNT as well as graphene production has undergone tremendous increase in the last decade (Chen et al., 2018). CNTs and graphene are often used as substitutes for equipment made up of copper and silver due to their good electrical conduction (Politou et al., 2016) which in turn increases the environmental pollution (Aitola et al., 2015). Graphene and fullerenes serve as transporters in the drug delivery and have proven to be suitable against disease prevention (Mehra et al., 2018). CNMs also improve performance of catalysts, concrete (Norhasri et al., 2017). The multiwalled carbon nanotubes (MWCNTs) enhance photochemical quantum production and photosynthesis protein expression in plants, thereby promoting growth and rate of photosynthesis (Fan et al., 2018). MWCNTs have a positive outcome on triazine herbicide adsorption in watery phase (D'Archivio et al., 2018). Fullerene discovery has brought new insights in the field of research; they are composed of carbon atoms and have a cavern spherical or tubular layout (Taylor and Walton, 1993). C60 is the first discovered fullerene molecule (Kroto et al., 1985) with a 32-faced football shape and 0.7-nm diameter (David, 1992). The graphite arc method is widely used for the fullerene production (Gao et al., 2019). C60 has wide applications, for instance, it is used as a lubricant in batteries and superconductors (Guldi and Prato, 2000). CNTs are composed of single or multilayered graphite sheets curled at a certain angle around the center with diameters from picometers to nanometers, smooth hollow nanotubes. They have low density, high strength, strong hydrophobic nature and are major one-dimensional CNMs at the moment. CNTs find application in the lithium-ion batteries and electronic devices (Baughman et al., 2002). Graphene is a well-known carbon allotrope characterized by high mechanical strength, excellent mobility, high thermal conductivity, and excellent light transmission. These excellent features of graphene make it an important part of semiconductors, batteries, etc., and also a tool in biology, chemistry, physics, energy, and environmental science (Zhu et al., 2010). Despite having great economic benefits for the humans and plants CNMs have been found to be a threat for the environment and various organisms. CNMs can result in early embryonic death of animals as well as can inhibit plant and animal development. CNTs from composite materials are agents of pollution of the environment (Laux et al., 2018). It is already established that microorganisms have an important part in ecological balance restoration. The role of microbes in the maintenance of pollution-free environments is remarkable and safe in comparison with chemical followed by physical methods (Chen et al., 2017) (Table 5.1).

TABLE 5.1 Effects of Different CNMs on Microbes. (Chen et al., 2017)

Carbon nanomaterials	Positive effects	Negative effects
SWNTS	Promote growth and metabolic activity of tolerant bacteria **(Lower concentrations)**	Cause cell rupture, suppressed microbial diversity, inhibit microbial enzyme activity **(Higher concentrations)**
MWNTs	Promote growth of tolerant microbes, elevated populations of atrazine degrading microbes **(Lower concentrations)**	Reduce microbial enzyme activity and biomass, microbial cell rupture **(Higher concentrations)**
Graphene	Increases microbial biomass, enhanced microbial enzyme activity **(Lower concentrations)**	Reduces number of nitrogen fixing bacteria, iron-reducing bacteria, and destruction of cellular structure **(Higher concentrations)**
Fullerenes	Increase biocarbon content **(Higher concentrations)**	Reduce microbial number, cell viability, membrane integrity **(Higher concentrations)**

5.4.1.1 EFFECT OF FULLERENE ON THE MICROBIAL POPULATIONS

Several studies have indicated that fullerene can alter the microbial structure and soil biomass. The aftereffect of toxicity of fullerenes on microbes present in a fertile potted soil was studied by Hao et al. (2018). The composition of the bacterial community was tested 30 days after the treatment and it turned out that the amount of *Planctomyces*, *Nitrospira,* and *Lysobacter* declined considerably. Higher concentrations of fullerene also increase the bio-carbon content of the soil that can be utilized as an equipment for the detection of balance of the organic matter of the soil as bio-carbon determines soil balance (Thakur et al., 2015). Tong et al. (2007) applied independently suspensions of fullerenes and granular fullerenes to the soil of the corn field and tested the structural and functional changes in the microbial communities and found very minor changes in the two soil samples. Different doses of fullerenes were added to the clay loam by Johansen et al. (2008) and their results suggested that small doses have a very little effect on the microbial respiration, while increasing the dose can alter populations of bacteria with a fast-growing rate. Fullerenes assimilated in the organic matter of the soil can be absorbed by plants that do not decompose easily and therefore can be accumulated in the plants which in turn can alter the microbial populations of the soil (Avanasi et al., 2014).

Several reports have indicated that fullerenes can indirectly influence microorganisms of the soil for instance, although they promote crop production by enhancing the nutrient uptake of inorganic elements followed by organic content by crops, they restrain soil microbe endurance (Monica and Cremonini, 2014). The colonial composition of bacteria and eukaryotes grown in the anaerobic digesters with and without fullerenes showed no toxic effect of fullerenes on the microbes (Nyberg et al., 2008). Soils rich in organic matter and clay content are slightly impacted by fullerenes and therefore the natural colonization of microbes can be maintained in such soils (Berry et al., 2017). Fullerene has the potential to modify composition of the aquatic microbe aggregation. C60 and SWCNT toxicity on the microflora (aquatic) was examined by Blaise et al. (2008). Their studies indicated that C60 casted very little effect on the microbes in comparison with other eight nanomaterials. Fullerenes casted little effect on the aquatic microbial populations of the river water in comparison with the CNTs (Lawrence et al., 2016). Direct and long-term treatment of bacteria with fullerenes can cause cellular deformation as well as reduced cell viability followed by membrane rectitude (Kang et al., 2011). Fullerenes have shown a strong negative effect on *Escherichia coli* when compared with *Bacillus subtilis* in the aquatic media that can be due to their anti-peptidoglycan surface layers (Lyon et al., 2007). Addition of *E. coli* to a solution of fullerenes along with humic acid and fulvic acid (natural organic matter) in different concentrations resulted in slower growth and lower carbon dioxide production by the bacteria, but increasing the organic matter content resulted in elevated growth and carbon dioxide production (Li et al., 2008). In general, fullerenes have a little effect on microbes in comparison with other nanomaterials as organic content have the potential of mitigating fullerene effect on the microbial populations (Lawrence et al., 2016). Johansen et al. (2008) tested the pristine C60 fullerene ecotoxicological effect on growth activity followed by genetic diversity of soil microbes, especially bacteria and protozoans for 2 weeks. They found that the bacterial colony-forming units (CFUs) dropped instantly after incorporation of fullerenes on the day of onset of the experimental trial, especially for the fast-growing bacteria where the declination was 3–4-fold higher. This finding indicates that C60 fullerene treatment affects the bacteria with a fast-growing rate as C60 can directly interact with the cells resulting in the formation of ROS that deters DNA and membrane lipids (Isakovic et al., 2006). C60 fullerene serves as an antioxidant in vivo, but this behavior is strongly dependent upon its functionalization (Gharbi et al., 2005). Fullerenes can indirectly inhibit growth of bacteria by adsorbing crucial growth elements like vitamins or minerals. The indirect and direct effects of C60 fullerenes are strictly confined to fast-growing bacteria. Microbes are rather flexible for the changes in the soil environment and bacterial species that can tolerate toxicity will grow after a period of time (Nannipieri et al., 2003). It can be predicted from the studies of Johansen et al. (2008) bacterial cultures obtained from the C60 treated soils when incubated on the DNP medium showed three- to four-fold increase in the CFUs at the end of 2 weeks indicating a larger proportion of slow growing bacteria that have lower nutrient demands. Fullerenes do not affect the microbial respiration, suggesting that soil organic matter (SOM) mineralization rate remains unaltered (Ekelund et al., 2003). The microbial biomass, however, remains unaffected by fullerenes, but Tong et al. (2007) found that increasing the concentration of fullerenes lowers the amount of microbial biomass. Higher levels of fullerenes in the soil can chemically change the soil matrix (Thostenson et al., 2001). C60 fullerenes did not show any effect on the population of the protozoans in general, but higher fullerene concentration resulted in a slight decrease (Johansen et al., 2008). Fullerene exposure for a short time does not alter the bacteria as well as the protozoan community of the soil. Protozoans ingest soil and other small latex particles (Matz et al., 2002). Protozoans can ingest C60 particles suspended with the clay particles (Boenigk et al., 2005). The average diameter of C60 fragments lies below 200 nm and therefore it is hard to estimate whether these particles caused any effect on the cellular organelles of the protozoans after ingestion although they have food preferences (Rønn et al., 2001). $C_{63}(COOH)_6$ (carboxylic acid-fullerene derivative) affects the gram-positive bacteria in a denial manner, but no effect has been found on gram-negative bacteria (Tsao et al., 2002).

5.4.1.2 EFFECT OF MWCNTS ON THE MICROBIAL POPULATIONS

Several studies have emphasized the efficacy of MWCNTs on soil microorganisms. Ge et al. (2016) added MWCNTs to the dry soil and found a drop in the microbial biomass. PCR sequencing results predicted that MWCNTs immensely altered the bacterial flora of the soil; however, the fungal community remained unharmed. A short-term exposure of MWCNTs can reduce the microbial enzyme activity as well as microbial biomass (Chung et al., 2011). Addition of lower amounts of MWCNTs to the agricultural topsoil can affect the vegetative growth of plants as well as microbial populations as at lower levels bacterial population increases (Ge et al., 2018). The soil enzyme potential, soil respiration, as well as microbe number of the sandy loamy soil remained unaffected after the addition of lower doses of MWCNTs as found by Shrestha et al. (2013). Presence of higher MWCNT content in the soil can cause declination of the population of specific bacteria (*Waddlia*, *Opitutus*), while the number of some bacterial species can increase (*Pseudomonas*, *Nocardia*, *Rhodococcus*). Higher MWCNT content can transform the dominant microbial groups into microbial communities that can tolerate higher MWCNT contents in the soil. Kerfahi et al. (2015) added natural and acid-treated MWCNTs in different concentrations to overgrown soil samples and found no effect on the diversity of the bacteria; however, the acid-treated MWCNTs significantly altered the soil bacteria community. The bacterial species that were more resistant to the acid-treated MWCNTs were found to be increased. MWCNTs can also affect the microbes of the aquatic environment. Wastewater containing higher concentrations of MWCNTs can immensely affect the chemical oxygen demand (COD) and nutrient content resulting in reduction of microbial biomass and bacterial population that can effectively treat the sewage water. MWCNTs in lower concentrations increase glycogen accumulating microbial populations when treated for a long time (Hai et al., 2014). MWCNTs can result in rupture of microbial cells resulting in alteration of the cellular structure and their increased levels can result in cytoplasmic loss (Yadav et al., 2016). MWCNTs have variable structures and a sharp one can kill the microbial cells (Hai et al., 2014). Lower concentrations of MWCNTs promote growth of atrazine-degrading microbes by elevating gene expression, cell division, and growth (Zhang et al., 2015). Studies have indicated that MWCNTs fail to cast an effect on microorganisms that display a great role in the bioremediation of petroleum, but they can boost these microbial populations as well (Abbasian et al., 2016). Organic biomass dissolved in an aquatic medium lowers MWCNT toxicity against microbes as mobile CNTs are lost. The more the organic biomass the lesser is the effect of CNTs (Kang et al., 2011). MWCNTs can inhibit as well as promote growth of microbes and this effect of MWCNTs is related to the microbial species, concentration and structure of MWCNTs as well as the external environment. Toxicity of MWCNTs is governed by their different size and functionalization. MWCNTs have the potential to form hollow structures that allow larger surface areas in the aqueous medium facilitating pollutant adsorption. The sorption capacity of MWCNTs is dependent upon their size; the larger the surface area, the larger is the pollutant adsorption (Fan et al., 2018). The acid-functionalized MWCNTs hold better adsorption potential as compared with the simple MWCNTs. MWCNT's adsorption capacity is directly related to the functionalization time (D'Archivio et al., 2018). The loose three-dimensional structure of MWCNTs not only provides sites for pollutant adsorption but also nabs a wide array of microbes resulting in the reduction of microbial count (Akasaka and Watari, 2009).

5.4.1.3 EFFECT OF SWCNTS ON THE MICROBES

SWCNTs have the potential to amend biomass as well as soil microbial constitution. Jin et al. (2014) added SWCNTs to the top soil samples and found that with increasing the concentration of SWCNTs the number of major microbes declined. Gram-positive bacteria can modify the composition of their intracellular lipids in response to the SWCNTs. *E. coli* biomass undergoes rapid declination in the presence of SWCNTs (Rodrigues and Elimelech, 2010). The extent of impact of SWCNTs on microbial population in the paddy soil is proportional to the amount of SWCNTs (Zhou et al., 2013). Microbial biomass and enzyme activity undergo rapid declination with increasing the amount of SWCNTs in the soil; however, a very minimal effect appears if the content of SWCNTs is low. SWCNTs can be categorized as natural SWCNTs, m-polyaminobenzene sulfonic acid functionalized ones, and polyethylene glycol functionalized ones at different concentrations when added to soil samples with more and lesser organic content caused reduced respiration rate (soil) and microbial biomass whereas functionalized SWCNTs had more drastic effect than the natural SWCNTs. Soil samples with lower organic matter showed minute change in the microbial biomass (Tong et al., 2012). The feedback of different microbes to carboxyl functionalized SWCNTs is different for example, major populations switch back to their original state after interacting with carboxyl-functionalized SWCNTs within a short period of time; however, the fungal populations do not respond the same (Rodrigues et al., 2013). Carbon and phosphorus cycle of soil also determines the effect of SWCNTs on microbes. The special conformational changes can reduce the degradation of SWCNTs. SWCNTs can adapt between the lignin peroxidase (LiP) and manganese peroxidase (MnP) residues and therefore restrain its degradation (Chen et al., 2016) as it can either suppress or enhance the stability of the hydrogen bonding and hydrophobic nature of the ligand as well as biological enzyme (Chen et al., 2016). SWCNTs during catalytic oxidation may disrupt water molecules as well as catalytic enzymes, thereby hindering the normal pollutant and lignin degradation. Natural and functionalized SWCNTs can modify some microbial enzymes in the soil and they are toxic to aquatic microbes (Chen et al., 2017). The toxicity of CNMs against the microbes can be correlated to their diameters (Liu et al., 2009). SWCNTs can inhibit microbial respiration in wastewater treatment plants within a short period of time (Luongo and Zhang, 2010). SWCNTs in lower doses in the activated sludge systems usually support bacterial growth promoting pollutant degradation (Goyal et al., 2010). Qu et al. (2016) observed that SWCNTs shield phenol degrading microorganisms in phenol-containing wastewater treatment plants as they are capable of increasing the phenol removal rate in the wastewater. Parise and Zhang (2014) examined the influence of short and long, functionalized short, and long SWCNTs on microbial populations of sewage plants and concluded that long SWCNTs are more toxic than the short SWCNTs, while functionalized SWCNTs casted more pronounced effects on microbial populations than simple ones as long and functionalized SWCNTs greatly affect the microbial respiration in comparison to the natural and short SWCNTs. Lower concentrations of SWCNTs in the soil promote microbial growth and degradation of organic matter thereby increasing soil fertility (Qian et al., 2018).

5.4.1.4 EFFECT OF GRAPHENE ON MICROBES

The smaller diameter, larger surface area along with a strong surface antibacterial ability of graphene in comparison to fullerenes, MWCNTs and SWCNTs makes it a friendly nanomaterial for the microbes (Qu et al., 2016). Ren et al. (2015) studied the influence of graphene on microbes thriving in the soil with the help of quantitative real-time polymerase chain reaction (qPCR) as well as pyrosequencing technology. They reported that graphene can significantly affect the number and microbial community structure depending upon the exposure time and concentration. Graphene can positively benefit the enzyme activity of soil microbes and the number of bacteria degrading pollutants in the soil if its concentrations are lower in the soil as after a period of time microbes regain their original levels. Higher doses of graphene reduce the number of soils ammoxidating microbes for example, digestive spirochete and black pine fungus resulting in alterations in the soil nitrogen cycle. Lower doses of graphene oxide promote cell growth as it increases cellular proliferation and attachment (Ruiz et al., 2011). However, graphene oxide has the potential to change soil bacterial composition and structure by resulting in an increased population of iron-reducing and nitrogen-fixing bacterial populations (Du et al., 2016). Hao et al. (2018) observed that by increasing graphene oxide concentration of paddy fields soil microbial counted also reduced as it disrupts oxidative balance, growth ratio, as well as rate of plant photosynthesis (Du et al., 2016). Migration of graphene into the soil particles decreases with increased ion intensity (Lanphere et al., 2013). Addition of inorganic salts and organic matter to the soil also affects graphene migration due to graphene encapsulation by inorganic and organic biomass (Dinesh et al., 2012). Graphene influence on aquatic microbes is analogous to the populations of the soil. Graphene is undoubtedly more toxic to microbes lacking membranes than microbes with membranes (Akhavan and Ghaderi, 2010). The polymer–graphene oxide composites in aquatic environments are more toxic than the natural ones as it can alter the structure of intracellular materials by destroying the cellular structure of microbes resulting in death (Mejias Carpio et al., 2012). This nanomaterial is also not easily degradable due to its highly stable structure (Kurapati et al., 2015). Chen et al. (2017) studied graphene and graphene composite degradation by microbial with enzymatic degradation methods. They reported that microbes could degrade graphene and graphene composite much better in comparison with the enzymes. Graphene has low solubility in water that is why their effects are more pronounced in the soil as compared to that in water. Graphene oxide is more toxic than graphene and is very difficult to degrade (Chen et al., 2019).

5.4.1.5 COMPARISON BETWEEN DIFFERENT CNMS AND THEIR EFFECT ON MICROBES

It has been proven that different CNMs cast different effects on microbial populations, for example, fullerenes have a less impact (Hao et al., 2018) while graphenes have significant influence on soil microbes (Qu et al., 2016). SWCNTs cast higher effect than MWCNTs on soil microbes due to the differences in their diameters, surface area (Qu et al., 2016). However, both are toxic as they cause cell membrane damage, proliferate ROS formation, cellular structure, and composition alterations (Magrez et al., 2006). Damage to the microbial cell membrane poses the cell for further negative effects. Organic matter mollifies the influence of CNMs on microbes, thereby protecting the microbial cells. MWCNTs decrease the counts of beneficial bacterial species, for example, *E. coli* as well as *P. aeruginosa* in wastewater treatment plants (Srivastava et al., 2004). They also mitigate counts of bacterial and viral strains that override the populations of microbes harmful to humans in water (Brady-Estevez et al., 2008); however, they also reduce efficiency of sewage treatment systems (Luongo and Zhang, 2010). MWCNTs and SWCNTs both enhance the growth of microbes participating in the crude oil and phenol decomposition in lower concentrations (Abbasian et al., 2016). Global researches have concluded that composite nanomaterials are more superior than the ordinary ones and can be applied for bioremediation purposes (Mauter and Elimelech, 2008). Fullerenes incorporated in the soil decrease the counts of gram-negative bacterial strains (Tong et al., 2007). Graphenes enhance microbial enzyme activity in soil resulting in material transfer and removal of pollutants; however, they can alter microbial aggregations (Ren et al., 2015). Although fullerenes have little toxic effect on microbes in comparison with different types of CNMs, their higher accumulation in the soil can drop the pollutant degradation rate (Tong et al., 2007). Impact of CNMs on microbial populations is strongly dependent upon the type of microbial strains. CNM toxicity can harm the cell membrane or electrostatic equilibrium of the cells, causing oxidative stress resulting in altered lipid and protein structures, abnormal cell metabolism, and later death.

5.4.2 TITANIUM OXIDE (TIO_2) NPs

Augmentation of NPs to soil changes the prokaryotic community structure, but the fungal communities remain unchanged indicating susceptible nature of prokaryotes to TiO_2 as well as $ZnSO_4$ NPs than fungal species (Moll et al., 2017). NPs mesh directly with the surfaces of bacteria (Neal, 2008). The sensitivity of bacteria and fungi toward the NPs is due to their different surface interactions as observed with TiO_2 NPs (Shah et al., 2014). A larger part of the research has targeted the TiO_2 NPs interaction with bacterial communities rather than fungal species (Burke et al., 2015). It is a well-established fact that different NPs cast affect the bacterial communities of the soil differently, but exact mechanism of interaction between plant exudates, bacteria, and NPs is not known yet (Ge et al., 2014). TiO_2 NPs negatively affect the abundance of some prokaryotic operational taxonomic units (OTUs) while positively increasing the ampleness of other prokaryotes and fungi belonging to Ascomycota. Longer exposure of NPs shows a pronounced effect on the soil bacterial (Ge et al., 2012). TiO_2 NPs have shown to strongly affect the population of 25 microbial taxa that serve as "bio-indicators" (van der Heijden and Hartmann, 2016). The explicit nature of these microbial species is unknown and further technological advancements are required (Schlaeppi et al., 2016). The dominion of NPs on micobes is determined by their crystal structure and size of the particle (Moll et al., 2017). The crystal structures of anatase and rutile TiO_2 NPs affect the structure of the bacterial community. Anatase particles (5–10 nm) have few effects than rutile (55 nm) (Shah et al., 2014). Influence of NPs on microorganisms of the soil can also be studied by physical interactions for instance, hetero-aggregation with soil (Shah et al. (2014). Microbial populations are also affected by soil texture that determines the effect of TiO_2 NPs (Simonin et al., 2015). Bacterial populations are lesser in loamy soils in comparison with silty clay and sandy loam soils. Farming practices like different levels of fertilization also shape the microbial abundance and their reaction with NPs (Leff et al., 2015). Arbuscular mycorrhizal fungi (AMF) comprise a crucial group of fungi which is not affected by TiO_2 NPs as the root colonization remains untouched (Burke et al., 2015). Du et al. (2011) found that wheat shoot biomass decreases with long time exposure of TiO_2 NPs as during the long exposure there are changes in the soil texture and composition. Treating the soil with zinc sulfate ($ZnSO_4$) increases the wheat biomass (Lindsay, 1972) and has a positive effect on soil bacterial abundance (Gogos et al., 2016). Research conducted by several researchers have reported that TiO_2 NPs have distinct aftereffects on distinctive plant types (Giorgetti et al., 2019).

The significance of NPs on the microbial populations related to crop productivity has gained much interest in the last decade as NPs as micronutrients are used as fertilizers (Wang et al., 2019). The crystalline phases of TiO_2 namely anatase and rutile significantly affect the plant parameters, especially biochemical, physiological, and genotoxic (Tan et al., 2018). Anatase is significantly more toxic to plants than rutile. A mixture of NPs is less toxic than anatase as it affects germination of seeds as well as permeability of membranes (Silva et al., 2016). Anatase alone or mixed with anatase + rutile induces elevated levels of oxidative stress along with damage in the ultrastructure of roots (Giorgetti et al., 2019). Anatase showed preferential translocation in rice plants when a mixture of anatase and rutile was given to rice plants (Cai et al., 2017). Bellani et al. (2020) investigated the anatase and rutile TiO_2 NP effect singly and mixed on soil bacteria and mineral nutrition of *Pisum sativum* L. Both anatase and rutile TiO_2 NPs under the transmission electron microscope (TEM) look highly compact with a prism-like shape. Rutile is rod-like with a minor axis of 20–25 nm and a major axis of 30–100 nm while anatase measures 20–80 nm in size. Both of them together appear spherical in shape. Addition of TiO_2 NPs significantly influences the availability of iron (Fe) and phosphorus (P) levels of the soil. About 18–30% declination has been observed in the manganese and 15% iron contents of soil on treatment with TiO_2 NPs. Phosphorus availability has also been found to decline with TiO_2 NPs. However, little effect has been observed for the availability of potassium (K), calcium (Ca), and zinc (Zn) contents of soil. The dose, treatment followed by association of TiO_2 NPs with soil has shown a positive enhancement in the availability of potassium, calcium, magnesium, and zinc. TiO_2 NPs in low doses cause bacterial community biostimulation (Siracusa, 2018). TiO_2 NPs form aggregates in the soil with the soil particles (Conway and Keller, 2016). Carboxylic acids, phenols, amines, and alcoholic functional groups cover TiO_2 NPs (hydrophilic coating) and therefore affect their interaction with particulates of soil (Rowley et al., 2018). Manganese oxy or hydroxides occupy a larger surface area and so are chemically effective as they partake either in oxidation-reduction reactions or in cation-exchange with TiO_2 NPs or other active substances (Gasparatos, 2013). TiO_2 NPs can reduce the availability of Mn in the soils resulting in changes of the Mn oxidation state altering plant growth (Bellani et al., 2020). The lowering of rhizosphere pH and exudate excretion by the roots can combat with the TiO_2 NPs to make Mn available to the plants. TiO_2 NPs can also reduce the available fraction of Fe in the soil by the mechanism of adsorption or complexation. TiO_2 NPs adsorb phosphate ions and therefore form adsorption surface complexes, thereby limiting their availability to the plants (Kang et al., 2011). Chen et al. (2016) concluded that anatase has higher adsorption potential in the aqueous phosphate solution in comparison with rutile. Inorganic ions and organic matter of the soil can recast the P-TiO_2 interaction. Lower doses of TiO_2 cause biostimulation and stress response in the bacterial communities. Higher doses of TiO2 NPs, however, reduced the microbial abundance without affecting any definitive bacterial community. Plants and microbial populations usually compete for the available nutrients in the soil and therefore, amplify the negative effects of NPs present in high amounts (Ge et al., 2013). The reduction in soil microbial diversity can be assigned to the fact that NPs induce the formation of ROS by alleviating oxidative stress; they are adsorbed by the cell membrane and also cause osmotic accentuation (Sohm et al., 2015). Reduction in root length and the increased shoot length has been reported when higher doses of TiO_2 NPs were given to pea plants by Lyu et al. (2017). TiO_2 NPs are toxic even if added in smaller doses as they form big aggregates with the soil particles and later penetrate the cell membrane of the roots causing cellular deformation (Clément et al., 2013). Mn, Zn, and P are highly influenced nutrients when soils are treated with TiO_2 NPs for example, in the nonavailability of Mn in the soils treated with TiO_2 NPs there was no Mn in the plant roots followed by plant shoots. Mn is essential for plant growth; it is a catalyst in the photosystem II (PS II) and also participates in several physiological processes of the plants. There are no visible symptoms that show its deficiency in plants. Application of TiO_2 NPs in the soil, however, shows increased nitrogen (N) content in the plant shoots indicating their translocation and assimilation, but the performance of N-fixing symbiotic bacteria in the treated soil needs further attention. Roots of the plants growing in the soils treated with TiO_2 NPs show declined growth and highly depleted mineral nutrients, especially P, K, Mn, and Zn in comparison with their aerial parts. Changes in the ultrastructure of roots have been noticed, for example, changes in the integrity of the membrane, alterations in the mineral nutrition, and transport efficiency of roots (Giorgetti et al., 2019). Micronutrients mollify toxic-free radicals and initiate production of hormones and enzymes in plants affecting the physiological functions of plants concerned with biotic/abiotic stress as well as uptake of nutrients. The amino acid composition of wheat and peanut crops was found to be different when soils were treated with TiO_2 NPs. Among wheat grains a uniform expansion in the overall amino acid content without any marked effect on the plant growth was observed while the grain weight and total amino acid content of peanuts was remarkably reduced indicating an adverse effect of TiO_2 NPs (Wang et al., 2019). Contemporary factors can affect the uptake of minerals in plants, especially when concerned with TiO_2 NPs as allied microbes and root membranes can undergo modification (Bellani et al., 2020). A delicate but convincing effect of NPs has been observed in the AMF community (Burke et al., 2015). TiO_2 NPs altered the composition of AMF in the rhizosphere (Feng et al. 2013). TiO_2 NP affects the AMF community more than the bacterial ones as AMF colonizes within the roots system and therefore is at higher risks of TiO_2 (Seeger et al., 2009). AMF are essential plant mutualists as they enhance the nutrient uptake and growth of plants (Smith and Read, 2008) and changes produced by TiO_2 NPs can also affect these essential parameters (Avio et al., 2006).

5.4.3 COPPER OXIDE NANOPARTICLES

Soil serves as a large reservoir that can accumulate a huge quantity of NPs gathering from different sources for example from the nano agrochemicals and wastewater treatment systems (Zhao et al., 2018). Influence of NPs on terrestrial ecosystem has been thoroughly examined (Gao et al., 2019). Minor differences have been reported between the natural and ionic NPs, for example, transfiguration of soluble Cu (copper acetate or copper citrate) and copper oxide nanoparticles (CuO-NPs) in the paddy soils shows very minor differences (Gao et al., 2019). Copper (II) hydroxide ($Cu(OH)_2$) nanopesticide is regarded as an alternative to copper sulfate ($CuSO_4$) used in the organic farming practice. The usage of $Cu(OH)_2$ nanopesticide is increasing every year and therefore there is a greater demand of understanding the side effects of this nanopesticide on the terrestrial ecosystems than the natural $CuSO_4$. Momentous changes in the metabolic profiling of the spinach plants have been observed when Zhao et al. (2018) tested the biochemical behavior of spinach plants with Cu $(OH)_2$ nanopesticides where nitrogen metabolism shows a pronounced effect. Soil bacterial communities are sensitive to the pollutants alternating the soil environment (Wu et al., 2019). The bacterial populations and their enzyme activities have an important role in the soil as they participate in the nutrient cycling, degradation of pollutants, and crop production (Gámiz et al., 2019). The indicators of soil health are many enzymes for instance, catalase, urease, acid phosphatase, and invertase (Zhang et al., 2020). The reduction in the amenable soil carbon and clay particles has been observed if the urea hydrolysis undergoes declination. Metal-associated NPs namely, CuO-NPs, Ag-NPs, and ZnO-NPs release metal ions in the soil in the presence of oxygen and can pose an effect on the bacterial communities, thereby modifying the enzyme activity of acid phosphatase which relies on organic matter composition and content of the soil (Gallo et al., 2018). Slight changes have been observed in the urea hydrolysis of organic carbon-rich pasture soils when incorporated with Cu as Cu bioavailability was increased (Adhikari et al., 2018). Impact of different Cu nanofertilizers (commercial nano-$Cu(OH)_2$, $Cu(OH)_2$ nanopesticide formulation (NPF) as well as commercial unformulated $Cu(OH)_2$ nanorods (NR), AI of NPF (AI-NPF), and NPF dispersing agent) soil enzyme potency namely, invertase, acid phosphatase followed by catalase were tested by Zhang et al. (2020). Since enzyme activity of soil directly influences bacteria (Zeng et al., 2019) the consequences of nano-$Cu(OH)_2$ as well as $Cu(OH)_2$ nanopesticides resembling $CuSO_4$ on soil bacteria were tested by Zhang et al. (2020). They observed that AI-NPF resulted in a significant boost of the bacterial communities as compared to $CuSO_4$ suggesting that this NPF can enhance the number of bacteria in the soil during a shorter exposure. Both nano-$Cu(OH)_2$ and $Cu(OH)_2$ nanopesticide caused declination in the bacterial counts. CuSO4 and dispersing agents showed no effect on the soil bacteria. Differences in the bacterial OTUs have been observed when the soils were treated with $CuSO_4$ and NPF respectively indicating that $CuSO_4$ application can temporarily increase the bacterial populations that later decline; however, very minute changes have been noticed if ionic copper was augmented in the soil (Bernard et al., 2009). NPF can alter the bacterial composition during long exposures. Dominance of Proteobacteria has been reported with the soils brushed with $CuSO_4$, NPF, and NR, whereas nano-$Cu(OH)_2$ and $Cu(OH)_2$ nanopesticides reduced the populations of these bacteria. Similarly, a declination in the populations of Actinobacteria has been observed in the soils treated with $Cu(OH)_2$ nanopesticides, suggesting that they can alter the composition of microbial communities. Replacement of traditional copper materials by $Cu(OH)_2$ nanopesticides to prevent microbial plant diseases seems to be a promising tool for reducing copper load in the soil. AI-NPF and $CuSO_4$ show reduction in the urease activity but elevated catalase activity. They not only reduce the dominance of Actinobacteria but also affect its community composition. This indicates that $Cu(OH)_2$ nanopesticides can alter the bacterial population as well as structure of the community.

5.4.4 SILVER NPs

Silver nanoparticles (AgNPs) have magnificent antimicrobial characteristics and therefore they are suitable for several purposes (Kim et al., 2017). AgNPs have a diameter between 1 and 50 nm and they have a larger surface area along with higher reactive potentials in comparison with other NPs (Wijnhoven et al., 2009). Due to their excessive usage the soil is now highly loaded with them (Sun et al., 2014) and therefore they are risky to the soil environment as well as to the microbial populations (Kumar et al., 2015). Many researchers have supported that AgNPs hinder plant growth (Pokhrel and Dubey, 2013), obstruct microbial diversity and their activity (He et al., 2016), and therefore change the composition of microbial populations (Kumar et al., 2015). AgNPs can accumulate in plants and therefore can gain access at the trophic levels (Nair et al., 2010). They can also result in induced oxidative stress (Dimkpa et al., 2013). Soil microbes maintain the ecological balance by enhancing the fertility of soil and degrading the waste matter (Camenzind et al., 2018). Enzymes produced by microbes are essential components of mineral cycles (Bastida et al., 2008). If microbes in the soil are under stress the biogeochemical cycles are also affected (McGee et al., 2017). Scant knowledge is accessible regarding the interaction of AgNPs with soil microbes (Grün et al., 2019). Currently, it is known that AgNPs cause toxicity on microbes as they can concentrate in the soil (Samarajeewa et al., 2017) and their toxicity is dependent upon their amount in soil (Peyrot et al., 2014). Majority of pronounced effects of AgNPs in relation to soil microbes have been observed at higher concentrations (Grün et al., 2019). Montes de Oca-Vásquez et al. (2020) tested AgNPs at the real levels on the diversity, biomass, and microbial activity in a *Coffea arabica* cultivated soil and found no pronounced effect of AgNPs on the microbial populations. Their results were similar to other researchers who also did not find any significant change in the soil microbial populations (Asadishad et al., 2018). However, AgNPs affect the abundance of some fungal species, namely, *Mortierella* (opportunistic decomposer), phylum Mortierellomycota showing increase in their populations (Montes de Oca-Vásquez et al., 2020). Natural dominance of *Mortierella* in soil is relatively significant as it acts as a decomposer of organic content (Brabcová et al., 2018). *Mortierella* uptakes organic carbon from the soil essential for its growth (López-Mondéjar et al., 2018). These silver NPs can alter bacterial and fungal colonies resulting in the relatively higher populations of *Morteriella*. The relative abundance of *Fusarium* and *Fusicolla* usually decreases if the content of *Mortierella* increases in soil after the application of AgNPs. *Fusarium* spp. are responsible for causing disease in the crop plants (barley, wheat, triticale, oats, and rye) all over the world and it is now well known that NPs have a toxic nature (Tarazona et al., 2019). The arbuscular mycorrhizal fungi (AMF) correlating about 80% of the plants are beneficial in increasing the phosphorus (P) uptake in plants and also serve as a barrier against metal stress (Cornejo et al., 2013). AMF are also involved in the metal transport by activating the associated genes (Chen et al., 2019). AMF not only decreases the cadmium transporter gene expression resulting in lower cadmium concentration in rice but also affects the gene expression of P and arsenite transporters (Glassop et al., 2005). AMF mitigate AgNP and zinc oxide nanoparticles (ZnONPs) phytotoxicity by decreasing metal accumulation in plants via reducing metal bioavailability or increasing glomain secretion (Siani et al., 2017). AMF also interacts with soil microbes altering the microbial abundance in the mycorrhizosphere or rhizosphere (Rodriguez-Caballero et al., 2017). They also interact with phosphate-solubilizing as well as nitrogen-fixing bacteria to elevate uptake of nutrients as they are deficient in secretion of enzymes associated with organic matter degradation (Zhang et al., 2018). AMF can also collaborate with bacterial populations and therefore enhance decomposition of organic matter and obtain nutrients for their growth (Xu et al., 2018).

Cao et al. (2020) tested the effect of AgNPs in different concentrations on AMF affecting maize-microbe response and found that the positive effects of AMF are not plant limited (Diagne et al., 2018). Soils treated with AgNPs have a reduced P concentration. Under the influence of AMF usually an adverse effect of AgNPs on metabolic reactions of the soil microbes followed by P availability is mollified. AMF also mitigates the adverse effect of cadmium on soil microbes (Aghababaei et al., 2014). However, under the influence of AgNPs AMF alters the bacterial populations of the soil resulting in the abundance of those bacteria that can help the plant to resist the stress generated by AgNPs (Chen et al., 2019). AMF releases carbon-rich substances that have a positive effect on soil bacterial growth (Zhang et al., 2018). AMF in general do not show a positive effect on all soil microbes, but they have a specific preference for some as they lower the counts of few soil microbes (Marschner and Timonen, 2006). *Bacilli* bacteria have important roles in the decomposition of organic P to make it available to plants (Artursson et al., 2006) and they usually associate with AMF serving as "mycorrhizal helper bacteria" (Bonfante and Anca 2009). In the soils treated with AgNPs the populations of *Bacilli* were found to be higher in the presence of AMF and AMF positively influenced *Anaerolineae*populations

(Cao et al., 2020). *Anaerolineae* belong to the phylum Chloroflexi and are anaerobic bacteria that participate in the decomposition of amino acids and carbon (Yamada and Sekiguchi, 2009). The relatively high population of *Anaerolineae* and *Bacilli* under the influence of AMF, however, decreased the relative abundance of *Nitrospira* (Rodriguez-Caballero et al., 2017). *Nitrospira* are nitrite-oxidizing bacteria having a role nitrogen cycle (Daims et al., 2001) and AMF also partakes in nitrogen transport in the host plants resulting in a competition for nitrogen between AMF and bacteria (Nuccio et al., 2013). Xu et al. (2018) found that AMF causes increase in the number of *Nitrospira* colonies as well as rate of organic matter decomposition. AMF can affect the bacterial population in soil by affecting plant growth (Philippot et al., 2013). The physiological properties of the plants can also shape the bacterial populations in association with AMF (Rodriguez-Caballero et al., 2017). Global research has indicated that AgNPs lose their antibacterial properties in the solution of concentrated electrolytes. AgNPs inhibit aerobic respiration in *Paracoccus denitrificans* in comparison with *E. coli* and *B. subtilis* (Lok et al., 2007). The AgNP effect on the microbial and enzyme activity along with soil microbial biomass for four months was examined by Hänsch and Emmerling (2010). They reported that AgNPs reduced the microbial biomass and they had a dose-dependent effect; however, the levels of nitrogen were not affected at all. AgNPs affect the different taxonomic groups of soils (bacteria, fungi, and grasser protozoans).

5.4.5 ZINC OXIDE NPs (ZNO-NPs)

The ZnO-NPs rank third in the utilization scale of the metal-based NPs in the world (Peng et al., 2017). ZnO has a photocatalytic and photo-oxidizing effect on biological species and is biologically safe (Sirelkhatim et al., 2015). Earlier several studies reported the decisive influences of ZnO-NPs, but their continuous usage has increased their toxicity in the environment (Rajput et al., 2017). However, the toxic nature of these NPs is size and shape dependent (Khare et al., 2015). Hanna et al. (2013) investigated toxic behaviors of highly used NPs, namely, NiO, ZnO, CuO on *Leptocheirus plumulosus*, an estuarine amphipod, and reported that higher levels of Zn dissolved in the sediment in comparison with other NPs indicating higher dissolution rate of ZnO-NPs. It is now known about the toxicity of ZnO-NPs in relation to the aquatic organisms (Jośko et al., 2016). Researchers have concluded that NPs alone or in combination with metals can alter the microbial community and structure (Xu et al., 2015). Several researchers have examined the toxicity of NPs (ZnO-NPs and CuO-NPs) and have found that they have antimicrobial characteristics in the in vitro experiments (Dinesh et al., 2012). Ecotoxic nature of ZnO-NPs on soil microbes in response to ammonification, fluorescent diacetate hydrolase (FDAH) activity, microbial respiration, and dehydrogenase (DH) activity caused decreased number of phosphorus and potassium-solubilizing bacteria and *Azotobacter*, reduced levels of fluorescein diacetatehydrolysis, catalase, and urease activities, as well as reduced thermogenic metabolism. Ammonification and respiration processes also showed a drop (Shen et al., 2015). ZnO-NPs have been found to be toxic against gram-positive as well as gram-negative bacterial species as they completely inhibit their growth by altering their biochemical processes (Reddy et al., 2007). ZnO-NPs are gifted with antimicrobial properties on *E. coli*, *S. aureus*, as well as *B. subtilis* bacterial species along with some fungal species also, for example, *Aspergillus fumigatus* and *A. flavus* (Navale et al., 2015). ZnO-NPs' antibacterial property on bacterial pathogens (food as well as water-borne) (*C. jejuni*and *V. cholerae*) shows alterations in their cellular structure resulting in death of these species (Manzoor et al., 2016). ZnO-NPs also show antimicrobial activity with *P. putida* (an essential soil bacterium) (Zeng et al., 2016). ZnO-NPs show a bacteriostatic effect in comparison with the CuO and AgO-NPs that have bactericidal activity (Gajjar et al., 2009). ZnO-NPs suppress indole acetic acid (IAA) production in *P. chlororaphis*, plant growth promoting rhizobacteria (PGPR) while CuO-NPs increase phytohormone production (Dimkpa et al., 2015). ZnO-NPs display bactericidal effects on *Sinorhizobium meliloti, a* symbiotic diazotrophic soil bacterium (Bandyopadhyay et al., 2012). However, ZnO-NPs are extensively used in bioremediation due to their antibacterial activity (Ahmed et al., 2017). ZnO-NPs have been found to be toxic when tested against the marine algae *Chlorella vulgaris* as the cell viability of the algae shows correlation with the concentration and exposure time of ZnO-NPs (Suman et al., 2015). Antibacterial activity of ZnO-NPs toward *E. coli*, *B. subtilis*, and *K. pneumoniae* indicates that it is superior to TiO_2 NPs (Ge et al., 2011). Shi et al. (2020) examined the influence of ZnO-NPs on soil microbes. They reported the abundance of 11 phyla, namely, Verrucomicrobia, Proteobacteria, Planctomycetes, Acidobacteria, Patescibacteria, Actinobacteria, Gemmatimonadetes, Bacteroidetes, Firmicutes, Chloroflexi, and Cyanobacteria. Proteobacteria were the champions in the relative abundance from the classes, viz., Deltaproteobacteria, Alphaproteobacteria, and Gammaproteobacteria with lower fractions of Actinobacteria. However, the dominance of the phyla was variable with times of exposure and dosage concentrations. The abundance of cyanobacteria, however, did not show any significant change. It can be assumed that ZnO-NPs are not toxic to Cyanobacteria, Patescibacteria, and Verrucomicrobia but are toxic to Bacteroidetes and Gemmatimonadetes. ZnO-NPs inhibit polycyclic aromatic hydrocarbons (PAH) as well as organic matter degradation and therefore affect the relative abundance of Bacteroidetes which have an essential role in their degradation (Li et al., 2014). Dominance of bacterial communities at the genus level in response to the ZnO-NPs was investigated by Shi et al. (2020). They found higher populations of *Acinetobacter*, *Azospirillum,* and *Ohtaekwangia* while a declination of *Piscinibacter* and *Sphingomonas* on the day 0 of ZnO-NPs application. However, after 30 days the populations of *Sphingomonas*, *Lysobacter*, *Flavisolibacter,* and *Gemmatimonas* showed declined numbers, while the numbers of *Altererythrobacter*, *Ramlibacter*, *Lacunisphaera*, *Ohtaekwangia,* and *Opitutus* showed increase. *Altererythrobacter*, *Ohtaekwangia,* and *Massilia* have important roles in the carbon cycle and their populations usually increase with increasing the ZnO-NPs levels (Li et al., 2014). *Terrimonas* partakes in breaking of methyl tert-butyl ether (MTBE) as well as phenanthrene and is also sensitive to the elevated levels of ZnO-NPs (Wu and Xu, 2016). However, populations of *Piscinibacter* (MTBE-degrading bacteria) showed significant increase with the increased concentrations of ZnO-NPs, indicating that hydrocarbon-degrading bacteria can better tolerate the elevated levels of heavy metals as well as antibiotic concentrations (multidrug resistant) (Hamidat et al., 2016). This can be stated that ZnO-NPs cast a pressure on the essential soil microbes to develop resistance against the heavy metals as well as other contaminants (Hemala et al., 2014). The microbial populations of Burkholderiales, Gammaproteobacteria, and Actinobacteria increase with increasing doses of ZnO-NPs (Li et al., 2015). ZnO-NPs may reduce the risk of soil plant diseases and therefore can reduce pesticide usage. The populations of *Pseudomonas* (a phosphate-solubilizing Gammaproteobacteria) show declination within 30 days of exposure time due to the low availability of free phosphorus. ZnO-NPs in general can suppress the populations of bacteria participating in phosphorus cycle, soil metabolic activity, and organic pollutant degradation but can enhance the communities of bacteria suppressing the growth of plant pathogens. Low levels of ZnO-NPs have an adverse influence on the polyphenol oxidase (PPO) activity (Altantuya, 2015).

5.4.6 CERIUM OXIDE (CEO_2) AND CHROMIUM TRIOXIDE (CR_2O_3) NPs

Scant amount of researches has been executed till now to determine the influence of cerium and chromium trioxide nanoparticles in relation with the soil microbial populations. In the last decade the usage of these two NPs has undergone significant increase posing a threat for the environment and organisms (Hu et al., 2015). Luo et al. (2020) tested cerium oxide ($nCeO_2$) and chromium trioxide (nCr_2O_3) NPs on bacterial strains at increased as well as ambient atmospheric CO_2. Toxicity of Cr_2O_3 NPs on mammalian (Alarifi et al., 2016) followed by plant cells (Kumar et al., 2015) along with the aquatic environment is already known, but their effect on the long-term elevated CO_2 (eCO_2) and soil microbes is not known yet. Increase in the levels of $nCeO_2$ and nCr_2O_3 showed decreased enzyme activity of the soil microbes (Luo et al., 2020). Toxicity of $nCeO_2$ and nCr_2O_3 is dependent upon the free metal particles and they can damage the membrane of the cell as well as the DNA (Jośko et al., 2016). Elevated levels of $nCeO_2$, nCr_2O_3, and eCO_2 decrease the soil pH, increase accumulation of ROS causing DNA damage, enzyme inactivation (Wang et al., 2016b). The $nCeO_2$ and nCr_2O_3 can also enhance ROS generation within the microbes which are toxic to them. nCr_2O_3 is more toxic than $nCeO_2$ (Jośko et al., 2016). eCO_2 also enhances the negative effect of NPs on soil microbes affecting the abundance of alpha and beta communities similar to other NPs (Yadav et al., 2014). The enzymatic potential recovery by microbes under the effect of $nCeO_2$ and nCr_2O and eCO_2 can occur due to the decreased levels of available Ce and Cr or by growth of tolerant bacteria (Blagodatskaya et al., 2010). It is proposed that eCO_2 can change the availability of soil carbon to microbes (Carrillo et al., 2018). Soil microbial biomass (MBC) is an essential source of organic carbon and any change in it can alter microbial enzyme activity (Dai et al., 2004). nCr_2O_3 and $nCeO_2$ also pose an inhibitory effect on MBC formation leading to stress in order to ensure survival (Gupta and Diwan, 2017). nCr_2O_3 and $nCeO_2$ also increase the alpha-bacterial community suggesting the fast adaptation of microbes against elevated levels of NPs as similar results have been obtained with Ag-NPs (Sheng et al., 2018). Research has indicated an explicit correlation among NPs as well as bacteria must be present to notice a trend in the microbial number declination (Gómez-Sagasti et al., 2019). In the presence of eCO_2 situations nCr_2O_3 and $nCeO_2$ can be useful as they reduce the toxicity of NPs in soil (Luo et al., 2020).

5.4.7 GOLD NPs

Changing climate and natural resource depletion has oriented the synthesis of gold nanoparticles or (Au-NPs) Ag-NPs from plant extracts opening the way for agro-industrial laboratories (Awad et al., 2019). Extensive research has been done to test the influence of Au-NPs on elongation of roots, germination of seeds followed by plant response against NPs (Ribeiro et al., 2019). Distinctive types of plant extracts have been tested for the production of Au-NPs and it has been reported that Au-NPs have efficient antimicrobial property against fungal and bacterial species (Ali et al., 2019). The differences between the Au-NPs prepared from different plant extracts show higher levels of sapogenins, flavonoids, carbohydrates, and steroids, therefore ensuring the stability of NPs (Yugay et al., 2020). Au-NPs have a precise size and shape that enables them to interact with the cell wall and cell membrane of microbes by changing the osmoregulation, permeability, respiration, as well as electron transport within the cell causing death (Rashmi et al., 2020). The true phenomenon behind the antimicrobial nature of Au-NPs is still hidden. It is believed that Au-NPs have a firm preference toward phosphorus and sulfhydryl groups and therefore they can seriously harm the cellular ingredients of the bacterial cell. According to one hypothesis Au-NPs can gain access within bacterial cell membrane and therefore can attach with the NADH dehydrogenases and generate ROS leading to respiratory chain interruption. These ROS can also associate with phosphorus or sulfur containing cell constituents, DNA, and proteins (Ahmad et al., 2018). Biocidal feature of Au-NPs is dependent on the dose as at higher levels the NPs interact with the nucleic acids as well as with the cytoplasmic organelles (Chauhan et al., 2013). Au-NPs show higher preference toward gram-negative bacteria rather than gram-positive ones. Gram-positive bacteria have a thick peptidoglycan layer that makes the penetration of NPs inside the bacterial cell difficult, while the layer is thinner in gram-negative bacteria due to which the NPs can gain access within the bacterial cell (Muthuvel et al., 2014). Au-NP antimicrobial assay shows significantly good results for *Proteus vulgaris*, *S. aureus*, *K. pneumoniae*, and *E. coli* (Annamalai et al., 2013). The antimicrobial activity of Au-NPs synthesized from *Annona muricata* extract is higher against *Clostridium sporogenes* (Folorunso et al., 2019). Au-NPs produced from *Jasminum auriculatum* leaf extract show antifungal property against *Aspergillus fumigatus* (Balasubramanian et al., 2020).

5.5 INFLUENCE OF NPs ON MICROBIAL GROWTH

5.5.1 PLANT PATHOGENS

NPs exhibit antimicrobial responses in the medical microbiological processes (Seil and Webster, 2012). NPs can influence plant pathogens (bacteria as well as fungi) (Dimpka et al., 2013). Influence of NPs on oomycetes is also under investigation. CuO-NPs can control the growth of *Phytophthora infestans* infecting the tomato plants where necrotic and chlorotic spots were noticed only on the leaves (Giannousi et al., 2013). These signs were not observed on lime plants. The symptoms of powdery mildew (infectious fungal strains are *Golovinomycescichoracearum* as well as *Sphaerotheca fusca* fungi) infecting cucumber and pumpkin plants are minimized by the use of Ag-NPs (Lamsal et al., 2011). However, Ag-NPs affect the conidial germination and mycelial growth in the in vitro tests. A wide array of research has been conducted on the antimicrobial activity of ZnO-NPs. ZnO-NPs also display antimicrobial properties. ZnO-NPs can suppress infectious plant fungal species from growing. ZnO-NPs also suppress the growth of *Botrytis cineria* and *Penicillum expansum* fungi responsible for causing gray mold in fruits (pears, apples) (He et al., 2012). ZnO-NPs (light activated) are more lethal to *B. cinerea* as conidial growth and development is hindered (Kairyte et al., 2013). *Aspergillus flavus* as well as *A. niger* (fruit molds) growth is also suppressed by the use of ZnO-NPs (Jayaseelan et al., 2012). *Fusarium graminearum* growth has been found to be suppressed by ZnO-NPs in vitro studies (Dimpka et al., 2013). Similarly, *F. verticillioides* growth was inhibited by ZnO-NPs as ROS formation was observed while fungal hyphae underwent deformation (Savi et al., 2013). *Fusarium* spp. produce a toxin, fumonisin in cereals. ZnO NPs also inhibit mycelial growth of *Pythium ultimum*.

5.5.2 PROFITABLE SOIL MICROORGANISMS

NPs not only find applications in pest control but they are also used in soil remediation, improvement of agricultural crop production (Dimpka, 2014). Extensive research has already been done so as to understand NP interaction with microbes especially bacteria (Ingle et al., 2014). It has been reported that metal-based NPs namely CuO, ZnO, Fe_3O_4 or SnO_2, CeO_2, Ag NPs alter the soil microbial populations (Gitipour et al., 2013). TiO_2 and ZnO-NPs reduce diversity as well as microbial biomass (Ge et al., 2011). Fe_3O_4-NPs enhance the growth of microbes, modify their structure thereby resulting in increased enzyme activity (acid phosphatase, amylase, urease, and invertase) (Fan et al., 2018). Fe_3O_4 and SnO_2 NPs increase the microbial C/N_2 ratio in the soil and also modified metabolism of microbes in the polluted soils (Antisari et al., 2013). Si, Au, Cu, and Pd NPs failed to influence populations of bacterial species (Shah and Belozerova, 2009). SiO_2 NPs not only augmented protein content and microbial biomass of bacteria in the soil but also boosted maize germination (Karunakaran et al., 2014). CeO_2 and ZnO NPs are lethal to *Sinorhizobium meliloti* as they inhibit cell count and growth. These two NPs alter polysaccharides followed by cell wall's protein configuration of the extracellular polymeric substances (EPS). CeO_2-NPs are bacteriostatic to *S. melba* while ZnO-NPs have bactericidal effect on the same strain (Bandyopadhyay et al., 2012). TiO_2-NPs suppress *Azotobater vinelandii*, *B. megaterium*, *B. brevis*, as well as *Pseudomonas fluorescens* growth, while ZrO_2 (zirconia) NPs show different effect on these strains (Karunakaran et al., 2014). ZnO and Ag-NPs with higher concentrations suppress the growth of *P. aeruginosa*, *B. barbaricus,* and *B. subtilis* (Dhas et al., 2013). *P. fluorescens* and *B. subtilis* are sensitive to the lethal effects of TiO_2, ZnO, SiO_2, and Al_2O_3-NPs. ZnO-NPs cause total mortality in these strains in comparison with other NPs (Jiang et al., 2009). *B. subtilis* growth was inhibited by ZnO, CuO, Sb_2O_3, and NiO_2 NPs (Baek et al., 2011). Growth rate of *P. putida* KT2440 (a biosensor strain) and *P. chlororaphis* O6 (PcO6) declines when Ag as well as CuO-NPs are applied (Gajjar et al., 2009). IAA production decreases by ZnO-NP application while siderophore production increases; however, an opposite effect has been noticed with CuO-NPs (Fang et al., 2013).

5.5.3 NPs AND MICROBES OF AGRICULTURAL IMPORTANCE

Microbes have a very characteristic role to play in the biogeochemical cycles followed by organic matter decomposition despite forming symbiotic collaborations with plants. Some microbes attach with the plant species in a positive manner where they increase the availability of essential nutrients to plants thereby increasing plant growth rate. Fungal strains usually associate with plants and increase their tolerance levels against heat, drought and also increase the protection mechanism against different infectious pathogens as well as insects. However, further research needs to be done (Table 5.2) for utilization of NPs for the betterment of crop production.

5.5.3.1 BIOLOGICAL NITROGEN FIXATION

NP concentration has an essential role to play in maintaining the health of soil microbes. ZnO-NPs can change the morphological structure of *Rhizobium leguminosarum bv. viciae* 3841 community. Such a morphological alteration can influence the nodulation of roots and therefore changes in the nitrogen-fixing cycle (Huang et al., 2014). CeO_2 NPs can fully inhibit the microbial activity of nitrogen-fixing bacteria at 1000 ppm (Priester et al., 2012). The nitrogen-fixing potential of *Rhizobium* is inhibited at higher levels of Ag-NPs (Kumar et al., 2015). However, lower doses of NPs positively affect microbes of the soil, for instance, Ag-NPs in lower concentrations promote root nodulation cowpea (Mehta et al., 2016). The nitrogenase activity of cluster bean, green gram, moth bean, and cowpea shows an increase at lower concentrations of ZnO-NPs; however, decrease is noticed at higher doses (Kumar et al., 2015). Ag-NPs do not cast any negative alteration on gene expression of the denitrifying (*Pseudomonas stutzeri. napB*, *nirH*, *norB*, as well as *narG*), nitrifying (*Nitrosomonas europaea. amoC2*, and *amoA1*), and nitrogen fixing (*Azotobacter vinelandii. nifH*, *nifD*, *anfD*, and *vnfD*) bacteria but inhibit the gene expression at higher concentrations (Yang et al., 2013). Gene expression in *Nitrosococcus oceani* and *Nitrosospira multiformis* is also affected by Ag-NPs when used at higher doses (Beddow et al., 2014). Ag, TiO_2 and ZnO NPs in combination at higher levels suppressed the frequency of nodulation in *S.* and *M. trancatula* (Judy et al., 2016). The potential of ammonia-oxidizing bacteria is reduced at higher doses of Ag and Fe NPs which affects the nitrification process. These NPs repress the growth of bacteria by associating with the bacterial cell surface and therefore hinder wastewater treatment by biological means (Michels et al., 2017). The biological nitrogen-fixing efficiency of *Anabaena variabilis* undergoes declination under the influence of TiO_2-NPs with long exposures (Cherchi and Gu, 2010). Declination in the bacterial growth enhances production of stress indicators, namely, cyanophycin grana proteins (CGPs) against NPs. TiO_2 and ZnO-NPs also decrease the population of nitrogen-fixing bacteria (Ge et al., 2012). However, TiO_2-NPs increased nutrient uptake in clover plants (Moll et al., 2017). The root nodule development and therefore the nitrogen fixation was slowed in peas grown in hydroponics with TiO_2-NPs (Fan et al., 2018). CeO_2 NPs also alter microbial populations at higher doses. ZnO-NPs not only inhibited the growth of *Sphingomonas* and *Rhizobium* but also caused a negative effect on other bacterial species (*Azotobacter, Ensifer*, *Clostridium*, and Rhodospirillaceae) regardless the host plants (Ge et al.,

TABLE 5.2 Influence of NPs on the Beneficial Traits of Agriculturally Important Microbes.

Concentration	Different types of nanoparticles						
	Ag	**ZnO**	**TiO_2**	**Fe**	**CeO_2**	**Mo**	**Cu**
Low	Increases ecological behavior of AMF	Delay NF	Reduces NF	Reduces AMF colonization	Complete NF shutdown	Increases nodulation	–
	Upregulates expression of nitrifying genes	Negatively affects AMF colonization					
High	Inhibits NF	Enhances nutrient uptake with AMF	Delays nodulation in hydrponics	–	–	–	Inhibits siderophore production
	Suppresses gene expression of nitrifiers and dentrifiers	Inhibits siderophore production	Inhibits P-solubilization				
	Enhances nodulation						

2014). A double-fold enhancement has been observed in the root nodule formation of chickpea seeds when molybdenum (Mo-NPs) was administered in the soil (Taran et al., 2014).

5.5.3.2 ARBUSCULAR MYCORRHIZAL FUNGI (AMF)

AMF have a diverse array of symbiotic associations followed by thoroughly researched plant microbe relations. They mitigate the stress levels in plants, therefore enabling the plants to develop tolerance. They help the plants to fight against ROS, metal lethality, and stress factors by forming a covering around the roots. Several researchers have reported that heavy metals, NPs, and other contaminants enter the plant system via roots (Upadhyaya et al., 2010). Researchers have conducted several studies to figure out the effect of AMF and metal NPs. AMF mollifies the toxicity of Fe_3O_4 NPs in the maize fields on the bacterial populations (Cao et al., 2020). *Gemmatimonadetes*, *Chloroflexi*, and *Nitrospira* have been found in higher proportions in soils treated with Fe_3O_4 NPs remaining unchanged in the presence of AMF. However, increasing the concentrations of Fe_3O_4 NPs AMF also had negative effects as glomalin content, mycorrhizal colonization followed by nutrient uptake by the roots underwent declination (Feng et al., 2013). TiO_2 as well as Ag-NPs also repress mycorrhizal colony formation on the roots of plants (Dubchak et al., 2010). Ag-NPs however under the increased levels increased the stress tolerance behavior in the AMF (Feng et al., 2013). The growth and nutrient uptake increased even at a higher concentration of ZnO-NPs in soybean plants in the presence of AMF (Wang et al., 2016a). These facts indicate that AMF inhibits the toxic influence of NPs toward soil and therefore increases plant growth.

5.5.3.3 SECONDARY METABOLITES

The profitable soil microbes produce secondary metabolites and therefore enhance plant growth. Siderophore is utilized by microbes to capture free iron radicals important for their survival. The metal NP and microbe interaction can enhance siderophore production that regulates plant–microbe interactions. Scant information is available regarding the influence of NPs on secondary metabolite production. CuO-NPs downregulate the gene expression of pyoverdine siderophore in *Pseudomonas chlororaphis* O6 resulting in the suppression of siderophore production (Dimkpa et al., 2013). TiO_2-NPs at higher doses repress siderophore production in *P. fluorescens*, *P. aeruginosa*, and *B. amyloliquefaciens*. ZnO-NPs increase the iron-binding capacity at higher concentrations. TiO_2 and ZnO-NPs at higher concentrations inhibit the phosphate-solubilizing efficiency of *P. fluorescens*, *P. aeruginosa*, and *B. amyloliquefaciens* (Haris and Ahmad, 2018).

5.5.3.4 *BIOCONTROL ACTIVITY*

Plant growth promoting rhizobacteria (PGPRs) residing in the rhizosphere of plants enhance plant growth by regulating the populations of phytopathogens via biocontrol mechanisms (Liu et al., 2017). Rangaraj et al. (2014) investigated biocontrol efficiency of the fluorescent pseudomonads by the application of nano-silica in maize plants. They found that higher silica levels resulted in increased phenolic activity, roughness of the leaves but lesser stress levels. Higher phenol levels lead to accumulation of silica in the leaf epidermis and therefore provide a physical barrier with increased resistance against diseases. Ag and ZnO-NPs increase antifungal activity of *P. protegens* in association with *Candida albicans*. Antifungal compound (pyrrolnitrin) production at higher doses of Ag and SiO_2 NPs in *P. protegens* inhibited its growth (Khan et al., 2018). NP accumulation at sublethal doses can enhance the biocontrol efficiency of useful microbes and therefore it can be an option to enhance crop productivity in a safer manner.

5.6 INTERACTIONS BETWEEN MICROORGANISMS AND NPs AT THE CELLULAR LEVEL

The NPs and microbe interactions have been described as toxic as the majority of the NPs possess antimicrobial properties (Raghunath and Perumal, 2017). The metal NP and microbial interactions are useful at agricultural followed by industrial as well as environmental levels (Guilger et al., 2017). Majority of the metals partake in cellular functions and so therefore are essential for the life activities. Microbes uptake these metal elements in different forms and thereby act as source and reservoirs of essential nutrients. The microbe–plant interactions enrich the chemical, physical, and biological characteristics of soil and therefore increase crop production and soil fertility. For sustainable agriculture complete knowledge regarding the associations of NPs with microbes is mandatory. Metals like Zn, Cu, Fe, and Mn are essential for growth and sustenance of microbial communities (Porcheron et al., 2013), but high levels of these metals are toxic as they can damage the cellular structure leading to death of microbial cells (Lima e Silva et al., 2012). Other metals, for example, Pb, Hg, and Al assemble within the microbial cells and cause cellular malfunctions and DNA damage (Oves et al., 2016). To overcome the higher levels of metals, microbes have tolerance mechanisms that serve as extracellular barriers and therefore prevent the entry of excess metal ions. The ATP-binding cassette transporters (ABC) participate in metal ion movement within as well as out of the cellular compartments so that detoxifcation can occur by specific enzymes (Ianeva, 1993). NPs offer a large surface area with varied chemical reactivity as well as quantum effects (Fig. 5.2). The possible mechanisms of NP interaction with microbial cells include membrane potential loss, membrane disruption, oxidation of proteins, ROS formation, and nucleic acid's integrity disruption (Slavin et al., 2017). Cell wall regulates the susceptibility and tolerance of bacterial species against NPs (Hajipour et al., 2012). NP oxides usually attach to the bacterial cell wall by Van der Waals forces (hydrophobic), but they can also attach via electrostatic interactions or by receptor-ligand methods (Parikh and Chorover, 2006). The electrostatic interactions of NPs with bacterial cell walls not only disrupt the structure of bacterial cell membrane but also attack other essential cell wall components, namely, amide, carboxyl, phosphate, carbohydrate moieties, and hydroxyl groups (Leone et al., 2007). Bacterial cell wall's lipopolysaccharide (LPS) layer serves as an initial interaction site for both bacteria and NPs (Guan et al., 2005). S-layer (surface layer, made up of glycoproteins present in monolayer) also interacts with NPs (Prakash et al., 2011). Some NPs gain entry within the cell by endocytosis where they settle within the endolysosomes and therefore cannot reach cytosol (Conner and Schmid, 2003). The prism-shaped, polygonal, or hierarchical Ag-NPs show inhibitory effects on the growth of both gram-negative (*E. coli*) and gram-positive (*S. aureus*) bacterial species in comparison with both spherical and disc-shaped NPs (Nateghi and Hajimirzababa, 2014). Higher antibacterial efficiency has been noticed in the plate-shaped Ag-NPs against *E. coli* in comparison with the cube-shaped but spherical NPs (Kim et al., 2017). Spherical Ag-NPs exhibit strong antibacterial potential toward *P. aeruginosa* (Raza et al., 2016). It can be suggested that the shape of NPs orients their effects on the microbes as different shapes have different surface areas and therefore different interactions with the microbes are generated (Buszewski et al., 2018).

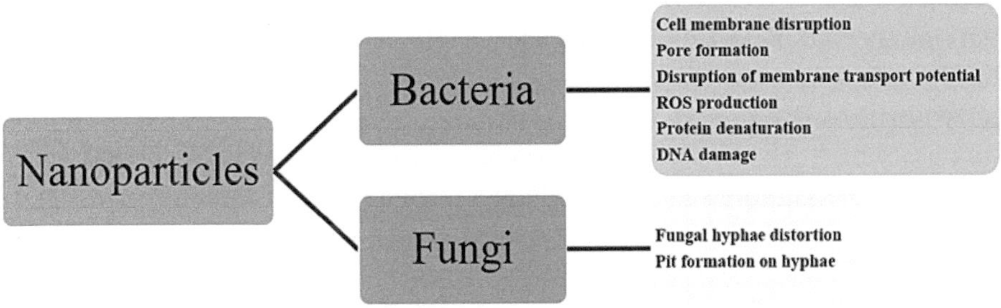

FIGURE 5.2 Effects of nanoparticles on bacterial and fungal cells (modified from Mahawar, H.; Prasanna, R. Prospecting the Interactions of Nanoparticles With Beneficial Microorganisms for Developing Green Technologies for Agriculture. *Environ. Nanotechnol. Monitor. Manag.* 2018, 10, 477–485).

5.7 CONCLUSIONS

NPs due to their special features have attracted the attention of scientists all over the globe as a source for sustainable agriculture leading to crop production. The application of NPs for the improvement of soil activity and crop growth has undergone tremendous increase in the last decade. Excessive increase has increased the risk of environmental pollution and soil contamination. NPs also appeared to pose a threat for the soil microbial communities altering their cellular composition and essential biological processes. Several types of NPs are toxic to microbes if used in higher concentrations; however, at lower levels they show significant effects on microbes as well as on plants. Many NPs have antimicrobial activities that inhibit microbial growth. The deleterious effects of NPs depend upon their size and concentrations applied in the soil. NPs are also toxic to aquatic microbes as well as microbes involved in the sewage and sludge treatment systems. NPs are a boon if used in small amounts and handled with care. They are a threat if used on larger concentrations.

ACKNOWLEDGMENTS

I am grateful to Gábor Draskovits, Laboratory Researcher, Dr. József Marek Animal Health Laboratory, Prophyl Kft., Dózsa György út 18, Mohács-7700, Hungary for his innovative ideas, motivation, and continuous moral support in writing this chapter. I also want to thank Prof. (Dr.) Pramod W. Ramteke (now retired), former Dean PG Studies and Head, Department of Biological

Sciences, Faculty of Science, Sam Higginbottom University of Agriculture, Technology and Sciences, Prayagraj-211007, UP, India. Last, but not the least, wisdom shared by Dr. Pradeep Kumar Shukla, Assistant Professor, Department of Biological Sciences, Faculty of Science, Sam Higginbottom University of Agriculture, Technology and Sciences, Prayagraj-211007, UP, India cannot be ignored as he has always been a source of inspiration to me.

KEYWORDS

- **antimicrobial properties**
- **nanoparticles**
- **nanotechnology**
- **soil microbes**
- **toxicity**

REFERENCES

Abbasian, F.; Lockington, R.; Palanisami, T.; Megharaj, M.; Naidu, R. Multiwall Carbon Nanotubes Increase the Microbial Community in Crude Oil Contaminated Fresh Water Sediments. *Sci. Total Environ.* **2016,** *539,* 370–380.

Adhikari, K. P.; Saggar, S.; Hanly, J. A.; Guinto, D. F.; Taylor, M. D. Why Copper and Zinc are Ineffective in Reducing Soil Urease Activity in New Zealand Dairy-Grazed Pasture Soils. *Soil Res.* **2018,** *56,* 491–502.

Aghababaei, F.; Raiesi, F.; Hosseinpur, A. The Combined Effects of Earthworms and Arbuscular Mycorrhizal Fungi on Microbial Biomass and Enzyme Activities in a Calcareous Soil Spiked With Cadmium. *Appl. Soil Ecol.* **2014**, *75,* 33–42.

Agrahari S.; Dubey A. Nanoparticles in Plant Growth and Development. In *Biogenic Nano-Particles and Their Use in Agro-Ecosystems*; Ghorbanpour, M., Bhargava, P., Varma, A., Choudhary, D., Eds.; Singapore: Springer, 2020**;** pp 9–37.

Ahmad, H. R.; Zia-ur-Rehman, M.; Sohail, M. I.; ul Haq, M. A.; Khalid, H.; Ayub, M. A.; Ishaq, G. Effects of Rare Earth Oxide Nanoparticles on Plants. In *Nanomaterials in plants, algae, and microorganisms*. Academic Press: Pittsburgh, 2018**;** pp 239–275.

Aitola, K.; Zhang, J.; Vlachopoulos, N.; Halme, J.; Kaskela, A.; Nasibulin, A. G.; Kauppinen, E. I.; Boschloo, G.; Hagfeldt, A. Carbon Nanotube Film Replacing Silver in High-Efficiency Solid-State Dye Solar Cells Employing Polymer Hole Conductor. *J. Solid State Electrochem*. **2015,** *19,* 3139–3144.

Akasaka, T.; Watari, F. Capture of Bacteria by Flexible Carbon Nanotubes. *Acta Biomater.* **2009,** *5,* 607–612.

Akhavan, O.; Ghaderi, E. Toxicity of Graphene and Graphene Oxide Nanowalls Against Bacteria. *ACS Nano*. **2010,** *4,* 5731–5736.

Alarifi, S.; Ali, D.; Alkahtani, S. Mechanistic Investigation of Toxicity of Chromium Oxide Nanoparticles in Murine Fibrosarcoma Cells. *Int. J. Nanomed.* **2016,** *11,* 1253–1259.

Ali, J.; Ali, N.; Wang, L.; Waseem, H.; Pan, G. Revisiting the Mechanistic Pathways for Bacterial Mediated Synthesis of Noble Metal Nanoparticles. *J. Microbiol. Meth.* **2019,** *159,* 18–25.

Altantuya, S. Transport and Aggregation Behavior of Zinc Oxide Nanoparticles in the Natural Soil and Water Environment. Doctoral Dissertation, Zhejiang Gongshang University, 2015.

Annamalai, A.; Christina, V. L. P.; Sudha, D.; Kalpana, M.; Lakshmi, P. T. V. Green Synthesis, Characterization and Antimicrobial Activity of Au NPs Using *Euphorbia hirta* L. Leaf Extract. *Coll. Surfaces B Biointer.* **2013,** *108,* 60–65.

Antisari, L. V.; Carbone, S.; Gatti, A.; Vianello, G.; Nannipieri, P. Toxicity of Metal Oxide (CeO_2, Fe_3O_4, SnO_2) Engineered Nanoparticles on Soil Microbial Biomass and Their Distribution in Soil. *Soil Biol. Biochem.* **2013,** *60,* 87–94.

Artursson, V.; Finlay, R. D.; Jansson, J. K. Interactions Between Arbuscular Mycorrhizal Fungi and Bacteria and Their Potential for Stimulating Plant Growth. *Environ. Microbiol.* **2006,** *8,* 1–10.

Asadishad, B.; Chahal, S.; Akbari, A.; Cianciarelli, V.; Azodi, M.; Ghoshal, S.; Tufenkji, N. Amendment of Agricultural Soil With Metal Nanoparticles: Effects on Soil Enzyme Activity and Microbial Community Composition. *Environ. Sci. Technol.* **2018,** *52,* 1908–1918.

Avanasi, R.; Jackson, W. A.; Sherwin, B.; Mudge, J. F.; Anderson, T. A. C60 Fullerene Soil Sorption, Biodegradation, and Plant Uptake. *Environ. Sci. Technol.* **2014,** *48,* 2792–2797.

Avio, L.; Pellegrino, E.; Bonarim, E.; Giovannetti, M. Functional Diversity of Arbuscular Mycorrhizal Fungal Isolates in Relation to Extra-Tadical Mycelial Networks. *New Phytol.* **2006,** *172,* 347–357.

Awad, M. A.; Eisa, N. E.; Virk, P.; Hendi, A. A.; Ortashi, K. M. O. O.; Mahgoub, A. S. A.; Elobeid, M. A.; Eissa, F. Z. Green Synthesis of Gold Nanoparticles: Preparation, Characterization, Cytotoxicity, and Anti-Bacterial Activities. *Mat. Lett.* **2019,** *256,* 126608.

Baek, Y.-W.; An, Y.-J. Microbial Toxicity of Metal Oxide Nanoparticles (CuO, NiO, ZnO, and Sb_2O_3) to *Escherichia coli*, *Bacillus subtilis* and *Streptococcus aureus*. *Sci. Total Environ.* **2011,** *409,* 1603–1608.

Balasubramanian, S.; Kala, S. M. J.; Pushparaj, T. L. Biogenic Synthesis of Gold Nanoparticles Using *Jasminum auriculatum* Leaf Extract and Their Catalytic, Antimicrobial and Anticancer Activities. *J. Drug Deliv. Sci. Technol.* **2020,** *57,* 101620.

Bandyopadhyay, S.; Peralta-Videa, J. R.; Plascencia-Villa, G.; Jose-Yacaman, M.; Gardea-Torresdey, J. L. Comparative Toxicity Assessment of CeO_2 and ZnO Nanoparticles Towards *Sinorhizobium meliloti*, A Symbiotic Alfalfa Associated Bacterium, Use of Advanced Microscopic and Spectroscopic Techniques. *J. Hazard. Mater.* **2012,** *241,* 379–386.

Bastida, F.; Kandeler, E.; Moreno, J. L.; Ros, M.; García, C.; Hernández, T. Application of Fresh and Composted Organic Wastes Modifies Structure, Size and Activity of Soil Microbial Community Under Semiarid Climate. *Appl. Soil Ecol.* **2008,** *40,* 318–329.

Baughman, R. H.; Zakhidov, A. A.; Heer, W. A. D. Carbon Nanotubes—The Route Toward Applications. *Science* **2002,** *5582,* 787–792.

Beddow, J.; Stolpe, B.; Cole, P.; Lead, J. R.; Sapp, M.; Lyons, B. P., Colbeck, I.; Whitby, C. Effects of Engineered Silver Nanoparticles on the Growth and Activity of Ecologically Important Microbes. *Environ. Microbiol. Rep.* **2014,** *6,* 448–458.

Bellani, L.; Siracusa, G.; Giorgetti, L.; Di Gregorio, S.; Castiglione, M. R.; Spanò, C.; Muccifora, S.; Bottega, S.; Pini, R.; Tassi, E.TiO_2 Nanoparticles in a Biosolid-Amended Soil and Their Implication in Soil Nutrients, Microorganisms and *Pisum sativum* Nutrition. *Ecotoxicol. Environ. Saf.* **2020,** *190,* 110095.

Bernard, L.; Maron, P. A.; Mougel, C.; Nowak, V.; Leveque, J.; Marol, C.; Balesdent, J.; Gibiat, F.; Ranjard, L. Contamination of Soil by Copper Affects the Dynamics, Diversity, and Activity of Soil Bacterial Communities Involved in Wheat Decomposition and Carbon Storage. *Appl. Environ. Microbiol.* **2009,** *75,* 7565–7569.

Berry, T. D.; Filley, T. R.; Clavijo, A. P.; Bischoff Gray, M.; Turco, R. Degradation and Microbial Uptake of C60 Fullerols in Contrasting Agricultural Soils. *Environ. Sci. Technol.* **2017,** *51,* 1387–1394.

Blagodatskaya, E.; Blagodatsky, S.; Dorodnikow, M.; Kuzyakov, Y. Elevated Atmospheric CO_2 Increases Microbial Growth Rates in Soil: Results of Three CO_2 Enrichment Experiments. *Global Change Biol.* **2010,** *16,* 836–848.

Blaise, C.; Gagne, F.; Ferard, J. F.; Eullaffroy, P. Ecotoxicity of Selected Nano-Materials to Aquatic Organisms. *Environ. Toxicol.* **2008,** *23,* 591–598.

Boenigk, J.; Wiedlroither, A.; Pfandl, K. Heavy Metal Toxicity and Bioavailability of Dissolved Nutrients to a Bacterivorous Flagellate are Linked to Suspended Particle Physical Properties. *Aq. Toxicol.* **2005,** *71,* 249–259.

Bonfante, P.; Anca, I. A. Plants, Mycorrhizal Fungi, and Bacteria: A Network of Interactions. *Ann. Rev. Microbiol.* **2009,** *63,* 363–383.

Brabcová, V.; Štursová, M.; Baldrian, P. Nutrient Content Affects the Turnover of Fungal Biomass in Forest Topsoil and the Composition of Associated Microbial Communities. *Soil Biol. Biochem.* **2018,** *118,* 187–198.

Brady-Estevez, A. S.; Kang, S.; Elimelech, M. A Single-Walled-Carbon-Nanotube Filter for Removal of Viral and Bacterial Pathogens. *Small* **2008,** *4,* 481–484.

Burke, D. J.; Pietrasiak, N.; Situ, S. F.; Abenojar, E. C.; Porche, M.; Kraj, P.; Lakliang, Y.; Samia, A. C. S. Iron Oxide and Titanium Dioxide Nanoparticle Effects on Plant Performance and Root Associated Microbes. *Int. J. Mol. Sci.* **2015,** *16,* 23630–23650.

Buszewski, B.; Railean-Plugaru, V.; Pomastowski, P.; Rafińska, K.; Szultka-Mlynska, M.; Golinska, P.; Wypij, M.; Laskowski, D.; Dahm, H. Antimicrobial Activity of Biosilver Nanoparticles Produced by a Novel *Streptacidiphilus durhamensis* strain. *J. Microbiol. Immunol. Infect.* **2018,** *51,* 45–54.

Cai, F.; Wu, X.; Zhang, H.; Shen, X.; Zhang, M.; Chen, W.; Gao, Q.; White, J. C.; Tao, S.; Wang, X. Impact of TiO_2 Nanoparticles on Lead Uptake and Bioaccumulation in Rice (*Oryza sativa* L.). *Nano Impact* **2017,** *5,* 101–108.

Camenzind, T.; Hättenschwiler, S.; Treseder, K. K.; Lehmann, A.; Rillig, M. C. Nutrient Limitation of Soil Microbial Processes in Tropical Forests. *Ecol. Monographs* **2018,** *88,* 4–21.

Cao, J.; Feng, Y.; Lin, X.; Wang, J. A Beneficial Role of Arbuscular Mycorrhizal Fungi in Influencing the Effects of Silver Nanoparticles on Plant-Microbe Systems in a Soil Matrix. *Environ. Sci. Pollut. Res.* **2020,** 27, 11782–11796.

Carrillo, Y.; Dijkstra, F.; LeCain, D.; Blumenthal, D.; Pendall, E. Elevated CO_2 and Warming Cause Interactive Effects on Soil Carbon and Shifts in Carbon Use by Bacteria. *Ecol. Lett.* **2018,** *21,* 1639–1648.

Chauhan, R.; Kumar, A.; Abraham, J. A Biological Aproach to the Synthesis of Silver Nanoparticles With *Streptomyces* sp JAR1 and its Antimicrobial Activity. *Sci. Pharm.* **2013,** *81,* 607–624.

Chen, M.; Qin, X.; Zeng, G. Single-Walled Carbon Nanotube Release Affects the Microbial Enzyme-Catalyzed Oxidation Processes of Organic Pollutants and Lignin Model Compounds in Nature. *Chemosphere* **2016,** *163,* 217–226.

Chen, M.; Qin, X.; Zeng, G. Biodegradation of Carbon Nanotubes, Graphene, and Their Derivatives. *Trends Biotechnol.* **2017,** *35,* 836–846.

Chen, M.; Sun, Y.; Liang, J.; Zeng, G.; Li, Z.; Tang, L.; Zhu, Y.; Jiang, D.; Song, B. Understanding the Influence of Carbon Nanomaterials on Microbial Communities. *Environ. Int.* **2019,** *126,* 690–698.

Chen, M.; Zhou, S.; Zhu, Y.; Sun, Y.; Zeng, G.; Yang, C.; Xu, P.; Yan, M.; Liu, Z.; Zhang, W. Toxicity of Carbon Nanomaterials to Plants, Animals and Microbes, Recent Progress From 2015-Present. *Chemosphere* **2018,** *206,* 255–264.

Cherchi, C.; Gu, A. Z. Impact of Titanium Dioxide Nanomaterials on Nitrogen Fixation Rate and Intracellular Nitrogen Storage in *Anabaena variabilis*. *Environ. Sci. Technol.* **2010,** *44,* 8302–8307.

Chung, H.; Son, Y.; Yoon, T. K.; Kim, S.; Kim, W. The Effect of Multi-Walled Carbon Nanotubes on Soil Microbial Activity. *Ecotoxicol. Environ. Saf.* **2011,** *74,* 569–575.

Clément, L.; Hurel, C.; Marmier, N. Toxicity of TiO_2 Nanoparticles to Cladocerans, Algae, Rotifers and Plants – Effects of Size and Crystalline Structure. *Chemosphere* **2013,** *90,* 1083–1090.

Coll, C.; Notter, D.; Gottschalk, F.; Sun, T.; Som, C.; Nowack, B. Probabilistic Environmental Risk Assessment of Five Nanomaterials (nano-TiO_2, nano-Ag, nano-ZnO, CNT, and fullerenes). *Nanotoxicology* **2016,** *10,* 4.

Concha-Guerrero, S. I.; Brito, E. M. S.; Piñón-Castillo, H. A.; Tarango-Rivero, S. H.; Caretta, C. A.; Luna-Velasco, A.; Duran, R.; Orrantia-Borunda, E. Effect of CuO Nanoparticles Over Isolated Bacterial Strains From Agricultural Soil. *J. Nanomater.* **2014,** *13,* (2014).

Conner, S. D.; Schmid, S. L. Regulated Portals of Entry into the Cell. *Nature* **2003,** *422,* 37–44.

Conway, J. R.; Keller, A. A. Gravity-Driven Transport of Three Engineered Nanoparticles in Unsaturated Soils and Their Effects on Soil pH and Nutrient Release. *Water Resour.* **2016,** *98,* 250–260.

Cornejo, P.; Perez-Tienda, J.; Meier, S.; Valderas, A.; Borie, F.; Azcon-Aguilar, C.; Ferrol, N. Copper Compartmentalization in Spores as a Survival Strategy of Arbuscular Mycorrhizal Fungi in Cu-Polluted Environments. *Soil Biol. Biochem.* **2013,** *57,* 925–928.

Dai, J.; Becquer, T.; Rouiller, J. H.; Reversat, G.; Bernhard-Reversat, F.; Lavelle, P. Influence of Heavy Metals on C and N Mineralisation and Microbial Biomass in Zn-, Pb-, Cu-, and Cd-Contaminated Soils. *Appl. Soil Ecol.* **2004,** *25,* 99–109.

Daims, H.; Nielsen, J. L.; Nielsen, P. H.; Schleifer, K. H.; Wagner, M. *In situ* Characterization of *Nitrospira*-Like Nitrite Oxidizing Bacteria Active in Wastewater Treatment Plants. *Appl. Environ. Microbiology* **2001,** *67,* 5273– 5284.

D'Archivio, A. A.; Maggi, M. A.; Odoardi, A.; Santucci, S.; Passacantando, M. Adsorption of Triazine Herbicides From Aqueous Solution by Functionalized Multiwall Carbon Nanotubes Grown on Silicon Substrate. *Nanotechnology* **2018,** *29,* 065701.

David, W. I. F.; Ibberson, R. M.; Dennis, T. J. S.; Hare, J. P.; Prassides, K. Structural Phase Transitions in the Fullerene C60. *Europhys. Lett.* **1992,** *18,* 219.

Dhas, S. P.; Shiny, P. J.; Khan, S. S.; Mukherjee, A.; Chandrasekaran, N. Toxic Behavior of Silver and Zinc Oxide Nanoparticles on Environmental Microorganisms. *J. Basic Microbiol.* **2013,** *54,* 916–927

Diagne, N.; Baudoin, E.; Svistoonoff, S.; Ouattara, C.; Diouf, D.; Kane, A.; Ndiaye, C.; Noba, K.; Bogusz, D.; Franche, C.; Duponnois, R. Effect of Native and Allochthonous Arbuscular Mycorrhizal Fungi on *Casuarina equisetifolia* Growth and its Root Bacterial Community. *Arid Land Res. Manag.* **2018,** *32,* 212–228.

Dimkpa, C. O.; McLean, J. E.; Britt, D. W.; Anderson, A. J. Antifungal Activity of ZnO Nanoparticles and Their Interactive Effect With a Biocontrol Bacterium on Growth Antagonism of the Plant Pathogen, *Fusarium graminearum. Bio Metals* **2013,** *26,* 913–924.

Dimpka, C. O. Can Nanotechnology Deliver the Promised Benefits Without Negatively Impacting Soil Microbial Life? *J. Basic Microbiol.* **2014,** *54,* 889–904.

Dimpka, C. O.; McLean, J. E.; Britt, D. W.; Anderson, A. J. Nano-CuO and Interaction With Nano-ZnO or Soil Bacterium Provide Evidence for the Interference of Nanoparticles in Metal Nutrition of Plants. *Ecotoxicology* **2015,** *24,* 119–129.

Dinesh, R.; Anandaraj, M.; Srinivasan, V.; Hamza, S. Engineered Nanoparticles in the Soil and Their Potential Implications to Microbial Activity. *Geoderma* **2012,***173–174,* 19–27.

Du, S.; Zhang, P.; Zhang, R.; Lu, Q.; Liu, L.; Bao, X.; Liu, H. Reduced Graphene Oxide Induces Cytotoxicity and Inhibits Photosynthetic Performance of the Green Alga *Scenedesmus obliquus. Chemosphere* **2016,** *164,* 499–507.

Dubchak, S.; Ogar, A.; Mietelski, J. W.; Turnau, K. Influence of Silver and Titanium Nanoparticles on Arbuscular Mycorrhiza Colonization and Accumulation of Radio-Caesium in *Helianthus annuus. Spanish J. Agri. Res.* **2010,** *8,* 103–108.

Ekelund, F.; Olsson, S.; Johansen, A. Changes in the Succession and Diversity of Protozoan and Microbial Populations in Soil Spiked With a Range of Copper Concentrations. *Soil Biol. Biochem.* **2003,** *35,* 1507–1516.

Fan, X.; Xu, J.; Lavoie, M.; Peijnenburg, W.; Zhu, Y.; Lu, T.; Fu, Z.; Zhu, T.; Qian, H. Multiwall Carbon Nanotubes Modulate Paraquat Toxicity in *Arabidopsis thaliana. Environ. Pollut.* **2018,** *233,* 633–641.

Fang, T.; Watson, J. L.; Goodman, J.; Dimkpa, C. O.; Martineau, N.; Das, S.; McLean, J. E.; Britt, D. W.; Anderson, A. J. Does Doping With Aluminum Alter the Effects of ZnO Nanoparticles on the Metabolism of Soil Pseudomonads? *Microbiol. Res.* **2013,** *168,* 91–98.

Feng, Y. Z.; Cui, X. C.; He, S. Y.; Dong, G.; Chen, M.; Wang, J. H.; Lin, X. The Role of Metal Nanoparticles in Influencing Arbuscular Mycorrhizal Fungi Effects on Plant Growth. *Environ. Sci. Technol.* **2013,** *47,* 9496–9504.

Folorunso, A.; Akintelu, S.; Oyebamiji, A. K.; Ajayi, S.; Abiola, B.; Abdusalam, I.; Morakinyo, A. Biosynthesis, Characterization and Antimicrobial Activity of Gold Nanoparticles From Leaf Extracts of *Annona muricata. J. Nanostruct. Chem.* **2019,** *9,* 111–117.

Frenk, S.; Ben-Moshe, T.; Dror, I.; Berkowitz, B.; Minz, D. Effect of Metal Oxide Nanoparticles on Microbial Community Structure and Function in Two Different Soil Types. *PLoS One* **2013,** *8,* e84441.

Gajjar, P.; Pettee, B.; Britt, D. W.; Huang, W. J.; Johnson, W. P.; Anderson, A. J. Antimicrobial Activities of Commercial Nanoparticles Against an Environmental Soil Microbe, *Pseudomonas putida* KT2440. *J. Biol. Eng.* **2009,** *3,* 9.

Gallo, A.; Manfra, L.; Boni, R.; Rotini, A.; Migliore, L.; Tosti, E. Cytotoxicity and Genotoxicity of CuO Nanoparticles in Sea Urchin Spermatozoa Through Oxidative Stress. *Environ. Int.* **2018,** *118,* 325–333.

Gámiz, B.; Velarde, P.; Spokas, K. A.; Cox, L. Dynamic Effect of Fresh and Aged Biochar on the Behavior of the Herbicide Mesotrione in Soils. *J. Agri. Food Chem.* **2019,** *67,* 9450–9459.

Gao, X.; Rodrigues, S. M.; Spielman-Sun, E.; Lopes, S.; Rodrigues, S.; Zhang, Y.; Avellan, A.; Duarte, R. M. B. O.; Duarte, A.; Casman, E. A.; Lowry, G. V. Effect of Soil Organic Matter, Soil pH, and Moisture Content on Solubility and Dissolution Rate of CuO NPs in Soil. *Environ. Sci. Technol.* **2019,** *53,* 4959–4967.

Gasparatos, D. Sequestration of heavy metals from soil with Fe-Mn concretions and nodules. *Environ. Chem. Letters***2013,**11, 1–9.

Ge, Y.; Priester, J.H.; Mortimer, M.; Chang, C.H.; Ji, Z.; Schimel, J.P.; Holden, P.A. Long-term effects of multiwalled carbon nanotubes and graphene on microbial communities in dry soil.*Environ.Sci. Technol.***2016,**50, 3965–3974.

Ge, Y.; Priester, J. H.; van de Werfhorst, L. C.; Walker, S. L.; Nisbet, R. M.; An, Y.-J.; Schimel, J. P.; Gardea-Torresdey, J. L.; Holden, P. A. Soybean Plants Modify Metal Oxide Nanoparticle Effects on Soil Bacterial Communities. *Environ. Sci. Technol.* **2014,** *48,* 13489–13496.

Ge, Y.; Schimel, J. P.; Holden, P. A. Identification of Soil Bacteria Susceptible to TiO_2 and ZnO Nanoparticles. *Appl. Environ. Microbiol.* **2012,** *78,* 6749c6758.

Ge, Y.; Shen, C.; Wang, Y.; Sun, Y. Q.; Schimel, J. P.; Gardea-Torresdey, J. L.; Holden, P. A. Carbonaceous Nanomaterials Have Higher Effects on Soybean Rhizosphere Prokaryotic Communities During the Reproductive Growth Phase Than During Vegetative Growth. *Environ. Sci. Technol.* **2018,** *52,* 6636–6646.

Ge, Y. G.; Schimel, J. P.; Holden, P. A. Evidence for Negative Effects of TiO_2 and ZnO Nanoparticles on Soil Bacterial Communities. *Environ. Sci. Technol.* **2011,** *45,* 1659–1664.

Geisler-Lee, J.; Wang, Q.; Yao, Y.; Zhang, W.; Geisler, M.; Li, K.; Huang, Y.; Chen, Y.; Kolmakov, A.; Ma, X. Phytotoxicity, Accumulation and Transport of Silver Nano-Particles by *Arabidopsis thaliana*. *Nanotoxicology* **2012,** *7,* 323–337.

Gharbi, N.; Pressac, M.; Hadchouel, M.; Szwarc, H.; Wilson, S. R.; Moussa, F. (60) Fullerene is a Powerful Antioxidant In Vivo With no Acute or Subacute Toxicity. *Nano Lett.* **2005,** *5,* 2578–2585.

Giannousi, K.; Avramidis, I.; Dendrinou-Samara, C. Synthesis, Characterization and Evaluation of Copper-Based Nanoparticles as Agrochemicals Against *Phytophthora infestans*. *RSC Adv.* **2013,** *3,* 21743.

Giorgetti, L.; Spanò, C.; Muccifora, S.; Bellani, L.; Tassi, E.; Bottega, S.; Di Gregorio, S.; Siracusa, G.; Sanità di Toppi, L.; Ruffini Castiglione, M. An Integrated Approach to Highlight Biological Responses of *Pisum sativum* Root to Nano-TiO_2 Exposure in a Biosolid-Amended Agricultural Soil. *Sci. Total Environ.* **2019,** *650,* 2705–2716.

Gitipour, A.; El Badawy, A.; Arambewela, M.; Miller, B.; Scheckel, K.; Elk, M.; Ryu, H.; Gomez-Alvarez, V.; Domingo, J. S.; Thiel, S.; Tolaymat, T. The Impact of Silver Nanoparticles on the Composting of Municipal Solid Waste. *Environ. Sci. Technol.* **2013,** *47,* 14385–14393.

Glassop, D.; Smith, S. E.; Smith, F. W. Cereal Phosphate Transporters Associated With the Mycorrhizal Pathway of Phosphate Uptake into Roots. *Planta* **2005,** *222,* 688–698.

Gogos, A.; Moll, J.; Klingenfuss, F.; van der Heijden, M.; Irin, F.; Green, M.; Zenobi, R.; Bucheli, T. Vertical Transport and Plant Uptake of Nanoparticles in a Soil Mesocosm Experiment. *J. Nanobiotechnol.* **2016,***14*.

Gómez-Sagasti, M. T.; Epelde, L.; Anza, M.; Urra, J.; Alkorta, I.; Garbisu, C. The Impact of Nanoscale Zero-Valent Iron Particles on Soil Microbial Communities is Soil Dependent. *J. Hazard. Mater.* **2019,** *364,* 591–599.

Goyal, D.; Zhang, X. J.; Rooney-Varga, J. N. Impacts of Single-Walled Carbon Nanotubes on Microbial Community Structure in Activated Sludge. *Lett. Appl. Microbiol.* **2010,** *51,* 428–435.

Gruère, G.; Narrod, C.; Abbott, L. Agricultural, Food, and Water Nanotechnologies for the Poor. International Food Policy Research Institute, Washington, DC, 2011.

Grün, A. L.; Manz, W.; Kohl, Y. L.; Meier, F.; Straskraba, S.; Jost, C.; Drexel, R.; Emmerling, C. Impact of Silver Nanoparticles (AgNP) on Soil Microbial Community Depending on Functionalization, Concentration, Exposure Time, and Soil Texture. *Environ. Sci. Eur.* **2019,** *31,* 15.

Guan, Z.; Breazeale, S. D.; Raetz, C. R. H. Extraction and Identification by Mass Spectrometry of Undecaprenyl Diphosphate-MurNAc-pentapeptide-GlcNAc From *Escherichia coli. Analyt. Biochem.* **2005,** *345,* 336–339.

Guilger, M.; Pasquoto-Stigliani, T.; Bilesky-Jose, N.; Grillo, R.; Abhilash, P. C.; Fraceto, L. F.; De Lima, R. Biogenic Silver Nanoparticles Based on *Trichoderma harzianum* Synthesis, Characterization, Toxicity Evaluation and Biological Activity. *Sci. Rep.* **2017,** *7,* 44421.

Guldi, D. M.; Prato, M. Excited-State Properties of C60 Fullerene Derivatives. *Acc. Chem. Res.* **2000,** *33,* 695–703.

Gupta, P.; Diwan, B. Bacterial Exopolysaccharide Mediated Heavy Metal Removal: A Review on Biosynthesis, Mechanism and Remediation Strategies. *Biotechnol. Rep.* **2017,** *13,* 58–71.

Hai, R.; Wang, Y.; Wang, X.; Du, Z.; Li, Y. Impacts of Multiwalled Carbon Nanotubes on Nutrient Removal From Wastewater and Bacterial Community Structure in Activated Sludge. *PLoS One* **2014,** *9,* e107345.

Hajipour, M. J.; Fromm, K. M.; Ashkarran, A. A.; de Aberasturi, D. J.; de Larramendi, I. R.; Rojo, T.; Serpooshan, V.; Parak, W. J.; Mahmoudi, M. Antibacterial Properties of Nanoparticles. *Trends Biotechnol.* **2012,** *30,* 499–511.

Hamidat, M.; Barakat, M.; Ortet, P.; Chaneac, C.; Rose, J.; Bottero, J.-Y.; Bottero, J. Y.; Heulin, T.; Achouak, W.; Santaella, C. Design Defines the Effects of Nanoceria at a Low Dose on Soil Microbiota and the Potentiation of Impacts by the Canola Plant. *Environ. Sci. Technol.* **2016,** *50,* 6892–6901.

Hanna, S. K.; Miller, R. J.; Zhou, D.; Keller, A. A.; Lenihan, H. S. Accumulation and Toxicity of Metal Oxide Nanoparticles in a Soft Sediment Estuarine Amphipod. *Aq. Toxicol.* **2013,** *143,* 441–446.

Hänsch, M.; Emmerling, C. Effects of Silver Nanoparticles on the Microbiota and Enzyme Activity in Soil. *J. Plant Nut. Soil Sci.* **2010,** *173,* 554–558.

Hao, Y.; Ma, C.; Zhang, Z.; Song, Y.; Cao, W.; Guo, J.; Zhou, G.; Rui, Y.; Liu, L.; Xing, B. Carbon Nanomaterials Alter Plant Physiology and Soil Bacterial Community Composition in a Rice-Soil-Bacterial Ecosystem. *Environ. Pollut.* **2018,** *232,* 123–136.

Haris, Z.; Ahmad, I. Impact of Metal Oxide Nanoparticles on Beneficial Soil Microorganisms and their Secondary Metabolites. *Int. J. Life Sci. Scien. Res.* **2017,** *3,* 1020–1030.

He, L.; Liu, Y.; Mustapha, Z.; Lin, M. Antifungal Activity of Zinc Oxide Nanoparticles Against *Botrytis cinerea* and *Penicillium expansum*. *Microbiol. Res.* **2012,** *166,* 207–215.

He, S. Y.; Feng, Y. Z.; Ni, J.; Sun, Y. F.; Xue, L. H.; Feng, Y. F.; Yu, Y. L.; Lin, X. G.; Yang, L. Z. Different Responses of Soil Microbial Metabolic Activity to Silver and Iron Oxide Nanoparticles. *Chemosphere* **2016,** *147,* 195–202.

Hemala, L.; Zhang, D.; Margesin, R. Cold-Active Antibacterial and Antifungal Activities and Antibiotic Resistance of Bacteria Isolated From an Alpine Hydrocarbon-Contaminated Industrial Site. *Res. Microbiol.* **2014,** *165,* 447–456.

Hu, Z.; Xu, M.; Shen, Z.; Yu, J. C. A Nanostructured Chromium (III) Oxide/Tungsten(VI) Oxide p-n Junction Photoanode Toward Enhanced Efficiency for Water Oxidation. *J. Mater. Chem. A*. **2015,** *3,* 14046–14053.

Huang, Y. C.; Fan, R.; Grusak, M. A.; Sherrier, J. D.; Huang, C. P. Effects of Nano-ZnO on the Agronomically Relevant *Rhizobium*-Legume Symbiosis. *Sci. Total Environ.* **2014,** *497,* 78–90.

Ianeva, O. D. Mechanisms of Bacteria Resistance to Heavy Metals. *Mikrobiolohichnyĭ zhurnal* **1993,** *71,* 54–65.

Ingle, A. P.; Duran, N.; Rai, M. Bioactivity, Mechanism of Action, and Cytotoxicity of Copper-Based Nanoparticles: A Review. *Appl. Microbiol. Biotechnol.* **2014,** *98,* 1001–1009.

Isakovic, A.; Markovic, Z.; Todorovic-Markovic, B.; Nikolic, N.; Vranjes-Djuric, S.; Mirkovic, M.; Dramicanin, M.; Harhaji, L.; Raicevic, N.; Nikolic, Z.; Trajkovic, V. Distinct Cytotoxic Mechanisms of Pristine Versus Hydroxylated Fullerene. *Toxicol. Sci.* **2006,** *91,* 173–183.

Jain, R.; Matassa, S.; Singh, S.; van Hullebusch, E. D.; Esposito. G.; Lens, P. N. Reduction of Selenite to Elemental Selenium Nanoparticles by Activated Sludge. *Environ. Sci. Pollut. Res. Int.* **2016,** *23,* 1193–1202.

Jayaseelan, C.; Rahuman, A. A.; Kirthi, A. V.; Marimuthu, S.; Santhoshkumar, T.; Bagavan, A.; Gaurav, K.; Karthik, L.; Bhaskara Rao, K. V. Novel Microbial Route to Synthesize ZnO Nanoparticles Using Aeromonas Hydrophila and Their Activity Against Pathogenic Bacteria and Fungi. *Spectrochim. Acta A Mol. Biomol. Spectrosc.* **2012,** *90,* 78–84.

Jiang, W.; Mashayekhi, H.; Xing, B. Bacterial Toxicity Comparison Between Nano and Micro-Scaled Oxide Particles. *Environ. Pollut.* **2009,** *157,* 1619–1625.

Jiling, C.; Youzhi, F.; Xiangui, L.; Junhua, W. Arbuscular Mycorrhizal Fungi Alleviate the Negative Effects of Iron Oxide Nanoparticles on Bacterial Community in Rhizospheric Soils. *Front. Environ. Sci.* **2016,** *4,* 10.

Jin, L.; Son, Y.; DeForest, J. L.; Kang, Y. J.; Kim, W.; Chung, H. Single-Walled Carbon Nanotubes Alter Soil Microbial Community Composition. *Sci. Total Environ.* **2014,** *466–467,* 533–538.

Johansen, A.; Pedersen, A.; Jensen, K. A.; Karlson, U.; Hansen, B. M.; Scott-Fordsmand, J. J.; Winding, A. Effects of C60 Fullerene Nanoparticles on Soil Bacterial and Protozoans. *Environ. Toxicol. Chem.* **2008,** 27, 1895–1903.

Jośko, I.; Oleszczuk, P.; Skwarek, E. The Bioavailability and Toxicity of ZnO and Ni Nanoparticles and Their Bulk Counterparts in Different Sediments. *J. Soils Sed.* **2016,** *16,* 1798–1808.

Judy, J. D.; Kirby, J. K.; McLaughlin, M. J.; Cavagnaro, T.; Bertsch, P. M. Gold Nanomaterial Uptake From Soil is not Increased by Arbuscular Mycorrhizal Colonization of *Solanum lycopersicum* (Tomato). *Nanomaterials* **2016,** *6,* 68.

Kairyte, K.; Kadys, A.; Luksiene, Z. Antibacterial and Antifungal Activity of Photoactivated ZnO Nanoparticles in Suspension. *J. Photochem. Photobiol. B Biol.* **2013,** *128,* 78–84.

Kang, S. A.; Li, W.; Lee, H. E.; Phillips, B. L.; Lee, Y. J. Phosphate Uptake by TiO_2: Batch Studies and NMR Spectroscopic Evidence for Multisite Adsorption. *J. Coll. Int. Sci.* **2011,** *364,* 455–461.

Karlsson, H. L.; Gustafsson, J.; Cronholm, P.; Möller, L. Size Dependent Toxicity of Metal Oxide Particles–a Comparison Between Nano- and Micrometer Size. *Toxicol. Lett.* **2009,** *188,* 112–118.

Karunakaran, G.; Suriyaprabha, R.; Manivasakan, P.; Rajendran, V.; Kannan, N. Influence of Nano and Bulk SiO_2 and Al_2O_3 Particles on PGPR and Soil Nutrient Contents. *Curr. Nanosci.* **2014,** *10,* 604–612.

Kashyap, P. L.; Xiang, X.; Heiden, P. Chitosan Nanoparticle-Based Delivery Systems for Sustainable Agriculture. *Int. J. Biol. Macromol.* **2015,** *77,* 36–51.

Kerfahi, D.; Tripathi, B. M.; Singh, D.; Kim, H.; Lee, S.; Lee, J.; Adams, J. M. Effects of Functionalized and Raw Multi-Walled Carbon Nanotubes on Soil Bacterial Community Composition. *PLoS One* **2015,** *10,* e0123042.

Khan, S. T.; Ahmad, J.; Ahamed, M.; Jousset, A. Sub-lethal Doses of Widespread Nanoparticles Promote Antifungal Activity in *Pseudomonas Protegens* CHA0. *Sci. Total Environ.* **2018,** *627,* 658–662.

Khare, P.; Sonane, M.; Nagar, Y.; Moin, N.; Ali, S.; Gupta, K. C.; Satish, A. Size Dependent Toxicity of Zinc Oxide Nanoparticles in Soil Nematode *Caenorhabditis elegans. Nanotoxicology* **2015,** *9,* 423–432.

Kim, D. H.; Park, J. C.; Jeon, G. E.; Kim, C. S.; Seo, J. H. Effect of the Size and Shape of Silver Nanoparticles on Bacterial Growth and Metabolism by Monitoring Optical Density and Fluorescence Intensity. *Biotechnol. Bioproc. Eng.* **2017,** *22,* 210–217.

Kreyling, W. G.; Semmler-Behnke, M.; Möller, W. Health Implications of Nanoparticles. *J. Nanopart. Res.* **2006,** *8,* 543–562.

Kroto, H. W.; Heath, J. R.; O'Brien, S. C.; Curl, R. F.; Smalley, R. E. C60. Buckminsterfullerene. *Nature* **1985,** *318,* 162–163.

Krysanov, E. Y.; Pavlov, D. S.; Demidova, T. B.; Dgebuadze, Y. Y. Effect of Nano-Particles on Aquatic Organisms. *Biol. Bull.* **2010,** *37,* 406–412.

Kumar, D.; Rajeshwari, A.; Jadon, P. S.; Chaudhuri, G.; Mukherjee, A.; Chandrasekaran, N.; Mukherjee, A. Cytogenetic Studies of Chromium (III) Oxide Nanoparticles on *allium cepa* Root Tip Cells. *J. Environ. Sci.* **2015,** *38,* 150–157.

Kurapati, R.; Russier, J.; Squillaci, M. A.; Treossi, E.; Menard-Moyon, C.; Del Rio-Castillo, A. E.; Vazquez, E.; Samori, P.; Palermo, V.; Bianco, A. Dispersibility-Dependent Biodegradation of Graphene Oxide by Myeloperoxidase. *Small* **2015,** *11,* 3985–3994.

Lakshmi, J. V.; Sharath, R.; Chandraprabha, M. N.; Neelufar, E.; Abhishikta, H.; Malyasree, P. Synthesis, Characterization and Evaluation of Antimicrobial Activity of Zinc Oxide Nanoparticles. *J. Biochem. Technol.* **2012,** *3,* S151–S154.

Lamsal, K.; Kim, S.-W.; Jung, J. H.; Kim, Y. S.; Kim, K. S.; Lee, Y. S. Inhibitions Effects of Silver Nanoparticles Against Powdery Mildews on Cucumber and Pumpkin. *Mycobiology* **2011**, *39,* 26–232.

Lanphere, J. D.; Luth, C. J.; Walker, S. L. Effects of Solution Chemistry on the Transport of Graphene Oxide in Saturated Porous Media. *Environ. Sci.Technol.* **2013,** *47,* 4255–4261.

Laux, P.; Riebeling, C.; Booth, A. M.; Brain, J. D.; Brunner, J.; Cerrillo, C.; Creutzenberg, O.; Estrela-Lopis, I.; Gebel, T.; Johanson, G.; Jungnickel, H.; Kock, H.; Tentschert, J.; Tlili, A.; Schäffer, A.; Sips, A. J. A. M.; Yokel, R. A.; Luch, A. Challenges in Characterizing the Environmental Fate and Effects of Carbon Nanotubes and Inorganic Nanomaterials in Aquatic Systems. *Environ. Sci. Nano*. **2018,** *5,* 48–63.

Hao, Y.; Ma, C.; Zhang, Z.; Song, Y.; Cao, W.; Guo, J.; Zhou, G.; Rui, Y.; Liu, L.; Xing, B. Carbon Nanomaterials Alter Plant Physiology and Soil Bacterial Community Composition in a Rice-Soil-Bacterial Ecosystem. *Environ. Pollut.* **2018,** *232,* 123–136.

Lawrence, J. R.; Swerhone, G. D. W.; Dynes, J. J.; Hitchcock, A. P.; Korber, D. R. Complex Organic Corona Formation on Carbon Nanotubes Reduces Microbial Toxicity by Suppressing Reactive Oxygen Species Production. *Environ. Sci. Nano* **2016,** *3,* 181–189.

Leff, J. W.; Jones, S. E.; Prober, S. M.; Barberan, A.; Borer, E. T.; Firn, J. L.; Harpole, W. S.; Hobbie, S. E.; Hofmockel, K. S.; Knops, J. M. Consistent Responses of Soil Microbial Communities to Elevated Nutrient Inputs in Grasslands Across the Globe. *Proc. Nat. Acad. Sci.* **2015,** *112,* 10967–10972.

Leone, L.; Ferri, D.; Manfredi, C.; Persson, P.; Shchukarev, A.; Sjoberg, S.; Loring, J. Modeling the Acid–Base Properties of Bacterial Surfaces: A Combined Spectroscopic and Potentiometric Study of the Gram-Positive Bacterium *Bacillus Subtilis*. *Environ. Sci. Technol*. **2007,** *41,* 6465–6471.

Li, D.; Lyon, D. Y.; Li, Q.; Alvarez, P. J. J. Effect of Soil Sorption and Aquatic Natural Organic Matter on the Antibacterial Activity of a Fullerene Water Suspension. *Environ. Toxicol. Chem.* **2008,** *27,* 1888–1894.

Li, Q.; Chen, X.; Zhuang, J.; Chen, X. Decontaminating Soil Organic Pollutants With Manufactured Nanoparticles. *Environ. Sci. Pollut. Res.* **2016,** *23,* 11533–11548.

Li, X.; Rui, J.; Mao, Y.; Yannarell, A.; Mackie, R. Dynamics of the Bacterial Community Structure in the Rhizosphere of a Maize Cultivar. *Soil Biol. Biochem.* **2014,** *68,* 392–401.

Li, X.; Zhang, Y.; Ding, C.; Jia, Z.; He, Z.; Zhang, T.; Wang, X. Declined Soil Suppressiveness to *Fusarium Oxysporum* by Rhizosphere Microflora of Cotton in Soil Sickness. *Biol. Fert. Soils* **2015,** *51,* 935–946.

Lima e Silva, A. A. D.; Carvalho, M. A.; de Souza, S. A.; Dias, P. M. T.; Silva Filho, R. G. D.; Saramago, C. S.; Bento, C. A.; Hofer, E. Heavy Metal Tolerance (Cr, Ag, and Hg) in Bacteria Isolated From Sewage. *Braz. J. Microbiol.* **2012,** *43,* 1620–1631.

Lindsay, W. L. Zinc in Soils and Plant Nutrition. In *Advances in Agronomy*; Brady, N. C., Ed.; Academic Press, 1972; pp 147–186.

Liu, K.; Newman, M.; McInroy, J. A.; Hu, C. H.; Kloepper, J. W. Selection and Assessment of Plant Growth-Promoting Rhizobacteria (PGPR) for Biological Control of Multiple Plant Diseases. *Phytopathology* **2017,** *107,* 928–936.

Liu, Q.; Chen, B.; Wang, Q.; Shi, X.; Xiao, Z.; Lin, J.; Fang, X. Carbon Nanotubes as Molecular Transporters for Walled Plant Cells. *Nano Lett.* **2009,** *9,* 1007–1010.

Liu, S.; Wei, L.; Hao, L.; Fang, N.; Chang, M. W.; Xu, R.; Yang, Y.; Chen, Y. Sharper and Faster "Nano Darts" Kill More Bacteria: A Study of Antibacterial Activity of Individually Dispersed Pristine Single-Walled Carbon Nanotube. *ACS Nano* **2009,** *3,* 3891–3902.

Lofts, S.; Criel, P.; Janssen, C. R.; Lock, K.; McGrath, S. P.; Oorts, K.; Rooney, C. P.; Smolders, E.; Spurgeon, D. J.; Svendsen, C.; Eeckhout, H. V.; Zhao, F. Z. Modelling the Effects of Copper on Soil Organisms and Processes Using the Free Ion Approach Towards a Multi-Species Toxicity Model. *Environ. Pollut.* **2013,** *178,* 244–253.

Lok, C. N.; Ho, C. M.; Chen, R.; He, Q. Y.; Yu, W. Y.; Sun, H.; Tam, P. K. H.; Chiu, J. F.; Che, C. M. Silver Nanoparticles: Partial Oxidation and Antibacterial Activities. *J. Biol. Inorg. Chem.* **2007,** *12,* 527–534.

López-Mondéjar, R.; Brabcová, V.; Štursová, M.; Davidová, A.; Jansa, J.; Cajthaml, T.; Baldrian, P. Decomposer Food Web in a Deciduous Forest Shows High Share of Generalist Microorganisms and Importance of Microbial Biomass Recycling. *ISME J.* **2018,** *12,* 1768–1778.

Luo, J.; Song, Y.; Liang, J.; Li, J.; Islam, E.; Li, T. Elevated CO_2 Mitigates the Negative Effect of CeO_2 and Cr_2O_3 Nanoparticles on Soil Bacterial Communities by Alteration of Microbial Carbon Use. *Environ. Pollut.* **2020,** *263,* 114456.

Luongo, L. A.; Zhang, X. J. Toxicity of Carbon Nanotubes to the Activated Sludge Process. *J. Hazard. Mater.* **2010,** *178,* 356–362.

Lyon, D. Y.; Brunet, L.; Hinkal, G. W.; Wiesner, M. R.; Alvarez, P. J. J. Antibacterial Activity of Fullerene Water Suspensions (nC60) is not due to ROS-Mediated Damage. *Nano Lett.* **2007,** *8,* 1539–1543.

Lyu, S.; Wei, X.; Chen, J.; Wang, C.; Wang, X.; Pan, D. Titanium as a Beneficial Element for Crop Production. *Front. Plant Sci.* **2017,** *8,* 597.

Magrez, A. E. A.; Kasas, S.; Salicio, V.; Pasquier, N.; Seo, J. W.; Celio, M.; Catsicas, S.; Schwaller, B.; Forró, L. Cellular Toxicity of Carbon-Based Nanomaterials. *Nano Lett.* **2006,** *6,* 1121–1125.

Mahawar, H.; Prasanna, R. Prospecting the Interactions of Nanoparticles With Beneficial Microorganisms for Developing Green Technologies for Agriculture. *Environ. Nanotechnol. Monitor. Manag.* **2018,** *10,* 477–485.

Maliszewska, I. Effects of the Biogenic Gold Nanoparticles on Microbial Community Structure and Activities. *Ann. Microbiol.* **2016,** *66,* 785–794.

Manzoor, U.; Siddique, S.; Ahmed, R.; Noreen, Z.; Bokhari, H.; Ahmad, I. Antibacterial, Structural and Optical Characterization of Mechano-Chemically Prepared ZnO Nanoparticles. *PLoS One* **2016,** *11,* e0154704.

Marschner, P.; Timonen S. Bacterial Community Composition and Activity in Rhizosphere of Roots Colonized by Arbuscular Mycorrhizal Fungi. In *Microbial Activity in the Rhizoshere*; Mukerji, K. G., Manoharachary, C., Singh, J., Eds.; *Soil Biol.* Springer: Berlin, Heidelberg 2006; vol 7.

Matz, C.; Boenigk, J.; Arndt, H.; Jurgens, K. Role of Bacterial Phenotypic Traits in Selective Feeding of the Heterotrophic Nanoflagellate *Spumella* sp. *Aq. Microbial Ecol.* **2002,** *27,* 137–148.

Mauter, M. S.; Elimelech, M. Environmental Applications of Carbon-Based Nanomaterials. *Environ. Sci. Technol.* **2008,** *42,* 5843–5859.

McGee, C. F.; Storey, S.; Clipson, N.; Doyle, E. Soil Microbial Community Responses to Contamination With Silver, Aluminium Oxide and Silicon Dioxide Nanoparticles. *Ecotoxicology* **2017,** *26,* 449–458.

McGillicuddy, E.; Murray, I.; Kavanagh, S.; Morrison, L.; Fogarty, A.; Cormican, M.; Dockery, P.; Prendergast, M.; Rowan, N.; Morris, D. Silver Nanoparticles in the Environment Sources, Detection and Ecotoxicology. *Sci. Total Environ.* **2017,** *575,* 231–246.

Mehra, N. K.; Jain, A. K.; Nahar, M. Carbon Nanomaterials in Oncology: An Expanding Horizon. *Drug Discov. Today* **2018,** *23,* 1016–1025.

Mehta, C. M.; Srivastava, R.; Arora, S.; Sharma, A. K. Impact Assessment of Silver Nanoparticles on Plant Growth and Soil Bacterial Diversity. *3 Biotech* **2016,** *6,* 254–263.

Mejias Carpio, I. E.; Santos, C. M.; Wei, X.; Rodrigues, D. F. Toxicity of a Polymergraphene Oxide Composite Against Bacterial Planktonic Cells, Biofilms, and Mammalian Cells. *Nanoscale* **2012,** *4,* 4746–4756.

Michels, C.; Perazzoli, S.; Soares, H. M. Inhibition of an Enriched Culture of Ammonia Oxidizing Bacteria by Two Different Nanoparticles: Silver and Magnetite. *Sci. Total Environ.* **2017,** *586,* 995–1002.

Mishra, V. K.; Kumar, A. Impact of Metal Nano-Particles on the Plant Growth Promoting Rhizobacteria. *Digest J. Nanomater. Biostruct.* **2009,** *4,* 587–592.

Moll, J.; Klingenfuss, F.; Widmer, F.; Gogos, A.; Bucheli, T. D.; Hartmann, M.; van der Heijden, M. G. A. Effects of Titanium Dioxide Nanoparticles on Soil Microbial Communities and Wheat Biomass. *Soil Biol. Biochem.* **2017,** *111,* 85–93.

Monica, R. C.; Cremonini, R. Nanoparticles and Higher Plants. *Caryologia* **2014,** *62,* 161–165.

Montes de Oca-Vásquez, G.; Solano-Camposc, F.; Vega-Baudrit, J. R.; López-Mondéjar, R.; Odriozola, I.; Vera, A.; Moreno, J. L.; Bastida, F. Environmentally Relevant Concentrations of Silver Nanoparticles Diminish Soil Microbial Biomass but do not Alter Enzyme Activities or Microbial Diversity. *J. Hazard. Mater.* **2020,** *391,* 122224.

Muthuvel, A.; Adavallan, K.; Balamurugan, K.; Krishnakumar, N. Biosynthesis of Gold Nanoparticles Using *Solanum nigrum* Leaf Extract and Screening Their Free Radical Scavenging and Antibacterial Properties. *Biomed. Prevent. Nutr.* **2014,** *4,* 325–332.

Nair, R.; Varghese, S. H.; Nair, B. G.; Maekawa, T.; Yoshida, Y.; Kumar, D. S. Nanoparticulate Material Delivery to Plants. *Plant Sci.* **2010,** *179,* 154–163.

Nannipieri, P.; Ascher, J.; Ceccherini, M. T.; Landi, L.; Pietramellara, G.; Renella, G. Microbial Diversity and Soil Functions. *Eur. J. Soil Sci.* **2003,** *54,* 655–670.

Nateghi, M. R.; Hajimirzababa, H. Effect of Silver Nanoparticles Morphologies on Antimicrobial Properties of Cotton Fabrics. *J. Text. Inst.* **2014,** *105,* 806–813.

Navale, G. R.; Thripuranthaka, M.; Late, D. J.; Shinde, S. S. Antimicrobial Activity of ZnO Nanoparticles Against Pathogenic Bacteria and Fungi. *JSM Nanotechnol. Nanomed.* **2015,** *3,* 1033.

Neal, A. L. What Can be Inferred From Bacterium Nanoparticle Interactions About the Potential Consequences of Environmental Exposure to Nanoparticles? *Ecotoxicology* **2008,** *17,* 362–371.

Norhasri, M. S. M.; Hamidah, M. S.; Fadzil, A. M. Applications of Using Nano Material in Concrete: A Review. *Const. Build. Mater.* **2017,** *133,* 91–97.

Nuccio, E. E.; Hodge, A.; Pett-Ridge, J.; Herman, D. J.; Weber, P. K.; Firestone, M. K. An Arbuscular Mycorrhizal Fungus Significantly Modifies the Soil Bacterial Community and Nitrogen Cycling During Litter Decomposition. *Environ. Microbiol.* **2013,** *15,* 1870–1881.

Nyberg, L.; Turco, R. F.; Nies, L. Assessing the Impact of Carbon Materials on Anaerobic Microbial Communities. *Environ. Sci. Technol.* **2008,** *42,* 1938–1943.

Oves, M.; Khan, S.; Qari, H.; Felemban, N.; Almeelbi, T. Heavy Metals: Biological Importance and Detoxification Strategies. *J. Bioremed. Biodegrad.* **2016,** *7,* 334.

Parikh, S. J.; Chorover, J. ATR-FTIR Spectroscopy Reveals Bond Formation During Bacterial Adhesion to Iron Oxide. *Langmuir* **2006,** *22,* 8492–8500.

Parise, A.; Zhang, J. Activity Inhibition on Municipal Activated Sludge by Single Walled Carbon Nanotubes. *J. Nanopart. Res.* **2014,** *16.*

Peng, C.; Xu, C.; Liu, Q.; Sun, L.; Luo, Y.; Shi, J. Fate and Transformation of CuO Nanoparticles in the Soil-Rice System During the Life Cycle of Rice Plants. *Environ. Sci. Technol.* **2017,** *51,* 4907–4917.

Peyrot, C.; Wilkinson, K. J.; Desrosiers, M.; Sauvé, S. Effects of Silver Nanoparticles on Soil Enzyme Activities With and Without Added Organic Matter. *Environ. Toxicol. Chem.* **2014,** *33,* 115–125.

Philippot, L.; Raaijmakers, J. M.; Lemanceau, P.; van der Putten, W. H. Going Back to the Roots: The Microbial Ecology of the Rhizosphere. *Nat. Rev. Microbiol.* **2013,** *11,* 789–799.

Pokhrel, L. R.; Dubey, B. Evaluation of Developmental Responses of Two Crop Plants Exposed to Silver and Zinc Oxide Nanoparticles. *Sci. Total Environ.* **2013,** *452–453,* 321–332.

Politou, M.; Asselberghs, I.; Soree, B.; Lee, C. S.; Sayan, S.; Lin, D.; Pashaei, P.; Huyghebaert, C.; Raghavan, P.; Radu, I.; Tokei, Z.; De Gendt, S.; Heyns, M. Single- and Multilayer Graphene Wires as Alternative Interconnects. *Microelect. Eng.* **2016,** *156,* 131–135.

Porcheron, G,; Garénaux, A.; Proulx, J.; Sabri, M.; Dozois, C. M. Iron, Copper, Zinc, and Manganese Transport and Regulation in Pathogenic Enterobacteria: Correlations Between Strains, Site of Infection and the Relative Importance of the Different Metal Transport Systems for Virulence. *Front. Cell. Infect. Microbiol.* **2013,** *3,* 90.

Prakash, A.; Sharma, S.; Ahmad, N.; Ghosh, A.; Sinha, P. Synthesis of Ag Nps by *Bacillus Cereus* Bacteria and Their Antimicrobial Potential. *J. Biomater. Nanobiotechnol.* **2011**, *2,* 156–162.

Prasad, R.; Kumar, V.; Prasad, K. S. Nanotechnology in Sustainable Agriculture: Present Concerns and Future Aspects. *Afr. J. Biotechnol.* **2014,** *13,* 705–713.

Qian, H.; Ke, M.; Qu, Q.; Li, X.; Du, B.; Lu, T.; Sun, L.; Pan, X. Ecological Effects of Single-Walled Carbon Nanotubes on Soil Microbial Communities and Soil Fertility. *Bull. Environ. Contam. Toxicol.* **2018,** *101,* 536–542.

Qu, Y.; Zhang, X.; Shen, W.; Ma, Q.; You, S.; Pei, X.; Li, S.; Ma, F.; Zhou, J. Illumina MiSeq Sequencing Reveals Long-Term Impacts of Single-Walled Carbon Nanotubes on Microbial Communities of Wastewater Treatment Systems. *Bioresour. Technol.* **2016,** *211,* 209–215.

Raghunath, A.; Perumal, E. Metal Oxide Nanoparticles as Antimicrobial Agents: A Promise for the Future. *Int. J. Antimicrob. Agents*. **2017,** *49,* 137–152.

Rajput, V. D.; Minkina, T.; Sushkova, S.; Tsitsuashvili, V.; Mandzhieva, S.; Gorovtsov, A.; Nevidomskaya, D.; Gromoakova, N. Effect of Nanoparticles on Crops and Soil Microbial Communities. *J. Soils Sed.* **2017,** 1–9.

Rangaraj, S.; Gopalu, K.; Muthusamy, P.; Rathinam, Y.; Venkatachalam, R.; Narayanasamy, K. Augmented Biocontrol Action of Silica Nanoparticles and *Pseudomonas Fluorescens* Bioformulant in Maize (*Zea mays* L.). *RSC Adv*. **2014,** *4,* 8461–8465.

Rashmi, B. N.; Harlapur, S. F.; Avinash, B.; Ravikumar, C. R.; Nagaswarupa, H. P.; Anil Kumar, M. R.; Gurushantha, K.; Santosh, M. S. Facile Green Synthesis of Silver Oxide Nanoparticles and Their Electrochemical, Photocatalytic and Biological Studies. *Inorg. Chem. Comm.* **2020,** *111,* 107580.

Raza, M. A.; Kanwal, Z.; Rauf, A.; Sabri, A. N.; Riaz, S.; Naseem, S. Size-and Shape Dependent Antibacterial Studies of Silver Nanoparticles Synthesized by Wet Chemical Routes. *J. Nanomater*. **2016,** *6,* 74.

Reddy, K. M.; Feris, K.; Bell, J.; Wingett, D. G.; Hanley, C.; Punnoose, A. Selective Toxicity of Zinc Oxide Nanoparticles to Prokaryotic and Eukaryotic Systems. *Appl. Phys. Lett.* **2007,** *90,* 213902.

Ren, W.; Ren, G.; Teng, Y.; Li, Z.; Li, L. Time-Dependent Effect of Graphene on the Structure, Abundance, and Function of the Soil Bacterial Community. *J. Hazard. Mater.* **2015,** *297,* 286–294.

Ribeiro, C. A. S.; Albuquerque, L. J. C.; de Castro, C. E.; Batista, B. L.; de Souza, A. L. M.; Albuquerque, B. L.; Zilse, M. S.; Bellettini, I. C.; Giacomelli, F. C. One-Pot Synthesis of Sugar-Decorated Gold Nanoparticles With Reduced Cytotoxicity and Enhanced Cellular Uptake. *Colloids Surf. A Physicochem. Eng. Asp.* **2019,** *580,* 123690.

Rico, C. M.; Peralta-Videa, J. R.; Gardea-Torresdey, J. L. Chemistry, Biochemistry of Nano-Particles, and Their Role in Antioxidant Defense System in Plants. In *Nanotechnology and Plant Sciences*; Springer: Cham, 2015, pp 1–17.

Rodrigues, D. F.; Elimelech, M. Toxic Effects of Single-Walled Carbon Nanotubes in the Development of *E. coli* Biofilm. *Environ. Sci. Technol.* **2010,** *44,* 4583–4589.

Rodrigues, D. F.; Jaisi, D. P.; Elimelech, M. Toxicity of Functionalized Single-Walled Carbon Nanotubes on Soil Microbial Communities: Implications for Nutrient Cycling in Soil. *Environ. Sci. Technol.* **2013,** *47,* 625–633.

Rodriguez-Caballero, G.; Caravaca, F.; Fernandez-Gonzalez, A. J.; Alguacil, M. M.; Fernandez-Lopez, M.; Roldan, A. Arbuscular Mycorrhizal Fungi Inoculation Mediated Changes in Rhizosphere Bacterial Community Structure While Promoting Revegetation in a Semiarid Ecosystem. *Sci. Total Environ.* **2017,** *584,* 838–848.

Rønn, R.; Ekelund, F.; Grunert, J. Protozoan Predation on the Bacteria *Mycobacterium Chlorophenolicus* and *Pseudomonas Chlororaphis* in Soil Microcosms. *Biol. Fert. Soils.* **2001,** *33,* 126–131.

Rowley, M. C.; Grand, S.; Verrecchia, É. P. Calcium-Mediated Stabilization of Soil Organic Carbon. *Biogeochemistry* **2018,** *137,* 27–49.

Ruiz, O. N.; Fernando, K. A.; Wang, B.; Brown, N. A.; Luo, P. G.; Mcnamara, N. D.; Vangsness, M.; Sun, Y. P.; Bunker, C. E. Graphene Oxide: A Nonspecific Enhancer of Cellular Growth. *ACS Nano*. **2011,** *5,* 8100–8107.

Saison, C.; Perreault, F.; Daigle, J. C.; Fortin, C.; Claverie, J.; Morin, M.; Popovic, R. Effect of Core Shell Copper Oxide Nanoparticles on Cell Culture Morphology and Photosynthesis (Photosystem II Energy Distribution) in the Green Alga, *Chlamydomonas reinhardtii*. *Aq. Toxicol.* **2010,** *96,* 109–114.

Samarajeewa, A. D.; Velicogna, J. R.; Princz, J. I.; Subasinghe, R. M.; Scroggins, R. P.; Beaudette, L. A. Effect of Silver Nano-Particles on Soil Microbial Growth, Activity and Community Diversity in a Sandy Loam Soil. *Environ. Pollut.* **2017,** *220,* 504–513.

Sanchez-Quiles, D.; Tovar-Sanchez, A. Sunscreens as a Source of Hydrogen Peroxide Production in Coastal Waters. *Environ. Sci.Technol.* **2014,** *48,* 9037–9042.

Savi, G. D.; Vitorino, V.; Bortoluzzi, A. J.; Scussel, V. M. Effect of Zinc Compounds on *Fusarium Verticillioides* Growth, Hyphae Alterations, Conidia, and Fumonisin Production. *J. Sci. Food Agri.* **2013,** *93,* 3395–3402.

Schlaeppi, K.; Bender, S.; Mascher, F.; Russo, G.; Patrignani, A.; Camenzind, T.; Hempel, S.; Rillig, M.; van der Heijden, M. G. A. High-Resolution Community Profiling of Arbuscular Mycorrhizal Fungi. *New Phytologist*. **2016,** *212,* 780–791.

Seeger, E.; Baun, A.; Kastner, M.; Trapp, S. Insignificant Acute Toxicity of TiO_2 Nanoparticles to Willow Trees. *J. Soils Sed.* **2009,** *9,* 46–53.

Seil, J.; Webster, T. J. Antimicrobial Applications of Nanotechnology: Methods and Literature. *Int. J. Nanomed.* **2012,** *7,* 2767–2781.

Shah, V.; Belozerova, I. Influence of Metal Nanoparticles on the Soil Microbial Community and Germination of Lettuce Seeds. *Water Air Soil Pollut.* **2009,** *197,* 143–148.

Shah, V.; Jones, J.; Dickman, J.; Greenman, S. Response of Soil Bacterial Community to Metal Nanoparticles in Biosolids. *J. Hazard. Mater.* **2014,** *274,* 399–403.

Shandilya, N.; Le, B. O.; Bressot, C.; Morgeneyer, M. Emission of Titanium Dioxide Nanoparticles From Building Materials to the Environment by Wear and Weather. *Environ. Sci. Technol.* **2015,** *49,* 2163–2170.

Shen, Z.Y.; Chen, Z.; Hou, Z.; Li, T.T.; Lu, X.X. Ecotoxicological effect of zinc oxide nanoparticles on soil microorganisms.*Front. Environ.Sci. Eng.***2015,** 9, 912–918.

Sheng, Z.; Van Nostrand, J. D.; Zhou, J.; Liu, Y. Contradictory Effects of Silver Nanoparticles on Activated Sludge Wastewater Treatment. *J. Hazard. Mater.* **2018,** *341,* 448–456.

Shi, Y.; Xiao, Y.; Li, Z.; Zhang, X.; Liu, T.; Li, Y.; Pan, Y.; Yan, W. Microorganism Structure Variation in Urban Soil Microenvironment Upon ZnO Nanoparticles Contamination. *Chemosphere* **2021,** *273,* 128565.

Shrestha, B.; Acosta-Martinez, V.; Cox, S. B.; Green, M. J.; Li, S.; Canas-Carrell, J. E. An Evaluation of the Impact of Multiwalled Carbon Nanotubes on Soil Microbial Community Structure and Functioning. *J. Hazard. Mater.* **2013,** *261,* 188–197.

Siani, N. G.; Fallah, S.; Pokhrel, L. R.; Rostamnejadi, A. Natural Amelioration of Zinc Oxide Nanoparticle Toxicity in Fenugreek (*Trigonella foenum-gracum*) by Arbuscular Mycorrhizal (*Glomus intraradices*) Secretion of Glomalin. *Plant Physiol. Biochem.* **2017,** *112,* 227–238.

Siddiqui, M.H.; Al-Whaibi, M.H.; Mohammad, F. Role of nanoparticles in plants. Nanotechnology and Plant Sciences, Springer, Cham, **2015,**19-35.

Silva, S.; Oliveira, H.; Craveiro, S. C.; Calado, A. J.; Santos, C. Pure Anatase and Rutile + Anatase Nanoparticles Differently Affect Wheat Seedlings. *Chemosphere* **2016,** *151,* 68–75.

Simonin, M.; Guyonnet, J. P.; Martins, J. M. F.; Ginot, M.; Richaume, A. Influence of Soil Properties on the Toxicity of TiO_2 Nanoparticles on Carbon Mineralization and Bacterial Abundance. *J. Hazard. Mater.* **2015,** *283,* 529–535.

Singh, A.; Singh, N. B.; Hussain, I.; Singh, H.; Singh, S. C. Plant-Nanoparticle Interaction: An Approach to Improve Agricultural Practices and Plant Productivity. *Int. J. Pharma. Sci. Invent.* **2015,** *4,* 25–40.

Singh, A., Rajput, V., Rawat, S., Kumar Singh, A., Bind, A., Kumar Singh, A., Chernikova, N.; Voloshina, M.; Lobzenko, I. Monitoring Soil Salinity and Recent Advances in Mechanism of Salinity Tolerance in Plants. *Biogeosyst. Tech.* **2021,** *2,* 66–87. https://doi.org/10.13187/bgt.2020.2.66

Singh, A.; Rajput, V.; Mehrotra, R.; Pal, N.; Singh, V.; Chokheli, V. A.; Singh, R. K. Modern Sustainable Techniques for Enhancing Crop Production. In *Sustainable Soil Fertility Management*, 2020**;** pp 73–100

Siracusa, G. Recalcitrant Compounds in Soils and Dredged Sediments. New Biotechnological Approaches for the Recover to the Public Use, Ph.D. Thesis, University of Pisa, Italy 2018.

Sirelkhatim, A.; Shahrom, M.; Azman, S.; Noor, H. M. K.; Chuo, A. L.; Siti, K. M. B.; Habsah, H.; Dasmawati, M. Review on Zinc Oxide Nanoparticles: Antibacterial Activity and Toxicity Mechanism. *Nanomicro. Lett.* **2015,** *7,* 219–242.

Slavin, Y. N.; Asnis, J.; Häfeli, U. O.; Bach, H. Metal Nanoparticles: Understanding the Mechanisms Behind Antibacterial Activity. *J. Nanobiotechnol.* **2017,** *15,* 65.

Smith, S. E.; Read, D. J. Mycorrhizal Symbiosis. Academic Press: New York, 2008.

Sohm, B.; Immel, F.; Bauda, P.; Pagnout, C. Insight into the Primary Mode of Action of TiO2 Nanoparticles on *Escherichia coli* in the Dark. *Proteomics* **2015,** *15,* 98–113.

Srivastava, A.; Srivastava, O. N.; Talapatra, S.; Vajtai, R.; Ajayan, P. M. Carbon Nanotube Filters. *Nat. Mater*. **2004,** *3,* 610–614.

Suman, T. Y.; RadhikaRajasree, S. R.; Kirubagaran, R. Evaluation of Zinc Oxide Nanoparticles Toxicity on Marine Algae *Chlorella vulgaris* Through Flow Cytometric, Cytotoxicity and Oxidative Stress Analysis. *Ecotoxicol. Environ. Saf.* **2015,** *113,* 23–30.

Sun, T.Y.; Gottschalk, F.; Hungerbühler, K.; Nowack, B. Comprehensive probabilistic modelling of environmental emissions of engineered nanomaterials. *Environ. Pollut.* **2014,**185, 69-76.

Tan, W.; Peralta-Videa, J. R.; Gardea-Torresdey, J. L. Interaction of Titanium Dioxide Nanoparticles With Soil Components and Plants: Current Knowledge and Future Research Needs - A Critical Review. *Environ. Sci. Nano*. **2018,** *5,* 257–278.

Taran, N. Y.; Gonchar, O. M.; Lopatko, K. G.; Batsmanova, L. M.; Patyka, M. V.; Volkogon, M. V. The Effect of Colloidal Solution of Molybdenum Nanoparticles on the Microbial Composition in Rhizosphere of *Cicer arietinum* L. *Nanoscale Res. Lett.* **2014,** *9,* 289.

Tarazona, A.; Gómez, J. V.; Mateo, E. M.; Jiménez, M.; Mateo, F. Antifungal Effect of Engineered Silver Nanoparticles on Phytopathogenic and Toxigenic *Fusarium* spp. and Their Impact on Mycotoxin Accumulation. *Int. J. Food Microbiol.* **2019,** *306,* 108259.

Taylor, R.; Walton, D. R. M. The Chemistry of Fullerenes. *Nature* **1993,** *363,* 685–693.

Thakur, M. P.; Milcu, A.; Manning, P.; Niklaus, P. A.; Roscher, C.; Power, S.; Reich, P. B.; Scheu, S.; Tilman, D.; Ai, F.; Guo, H.; Ji, R.; Pierce, S.; Ramirez, N. G.; Richter, A. N.; Steinauer, K.; Strecker, T.; Vogel, A.; Eisenhauer, N. Plant Diversity Drives Soil Microbial Biomass Carbon in Grasslands Irrespective of Global Environmental Change Factors. *Global Change Biol.* **2015,** *21,* 4076–4085.

Thostenson, E. T.; Ren, Z. F.; Chou, T. W. Advances in the Science and Technology of Carbon Nanotubes and Their Composites: A Review. *Comp. Sci. Technol.* **2001,** *61,* 1899–1912.

Thul, S. T.; Sarangi, B. K. Implications of Nanotechnology on Plant Productivity and its Rhizospheric Environment. In: *Nanotechnology and plant sciences*; Springer: Cham, 2015; pp 37–53.

Tiede, K.; Hanssen, S. F.; Westerhoff, P.; Fern, G. J.; Hankin, S. M.; Aitken, R. J.; Chaudhry, Q.; Boxall, A. B. How Important is Drinking Water Exposure for the Risks of Engineered Nanoparticles to Consumers? *Nanotoxicology* **2016,** *10,* 102–110.

Tong, Z.; Bischoff Gray, M.; Nies, L. F.; Applegate, B.; Turco, R. F. Impact of Fullerene (C60) on Soil Microbial Community. *Environ. Sci. Technol.* **2007,** *41,* 2985–2991.

Tong, Z.; Bischoff, M.; Nies, L. F.; Myer, P.; Applegate, B.; Turco, R. F. Response of Soil Microorganisms to as-Produced and Functionalized Single-Wall Carbon Nanotubes (SWNTs). *Environ. Sci. Technol.* **2012,** *46,* 13471–13479.

Tripathi, S.; Champagne, D.; Tufenkji, N. Transport Behavior of Selected Nanoparticles With Different Surface Coatings in Granular Porous Media Coated With *Pseudomonas aeruginosa* Biofilm. *Environ. Sci. Technol.* **2012,** *46,* 6942–6949.

Tsao, N.; Luh, T. Y.; Chou, C. K.; Wu, J. J.; Liu, C. C.; Lei, H. Y. *In Vitro* Action of Carboxyfullerene. *Antimicrob. Agents Chemo.* **2002,** *49,* 641–649.

Upadhyaya, H.; Panda, S. K.; Bhattacharjee, M. K.; Dutta, S. Role of Arbuscular Mycorrhiza in Heavy Metal Tolerance in Plants: Prospects for Phytoremediation. *J. Phytol.* **2010,** *2,* 16–27.

van der Heijden, M.; Hartmann, M. Networking in the Plant Microbiome. *PLoS Biol.* **2016,** *14,* e1002378.

Wang, P.; Lombi, E.; Zhao, F. J.; Kopittke, P. M. Nanotechnology: A New Opportunity in Plant Sciences. *Trends Plant Sci.* **2016,** *21,* 699–712.

Wang, Y.; Jiang, F.; Ma, C.; Rui, Y.; Tsang, D. C. W.; Xing, B. Effect of Metal Oxide Nanoparticles on Amino Acids in Wheat Grains (*Triticum aestivum*) in a Life Cycle Study. *J. Environ. Manag.* **2019,** *241,* 319–327.

Wijnhoven, S. W. P.; Peijnenburg, W. J. G. M.; Herberts, C. A.; Hagens, W. I.; Oomen, A. G.; Heugens, E. H. W.; Roszek, B.; Bisschops, J.; Gosens, I.; Van de Meent, D.; Dekkers, S.; De Jong, W. H.; van Zijverden, M.; Sips, A. J. A. M.; Geertsma, R. E. Nano-silver - A Review of Available Data and Knowledge Gaps in Human and Environmental Risk Assessment. *Nanotoxicology* **2009,** *3,* 109–138.

Wu, F.; You, Y.; Zhang, X.; Zhang, H.; Chen, W.; Yang, Y.; Werner, D.; Tao, S.; Wang, X. Effects of Various Carbon Nanotubes on Soil Bacterial Community Composition and Structure. *Environ. Sci. Technol.* **2019,** *53,* 5707–5716.

Wu, Y. H.; Xu, X. W. Advances in the Taxonomy of Erythrobacteraceae. *Microbiol. China* **2016,** *43,* 1082–1094.

Xu, C.; Peng, C.; Sun, L.; Zhang, S.; Huang, H.; Chen, Y.; Shi, J. Distinctive Effects of TiO_2 and CuO Nanoparticles on Soil Microbes and Their Community Structures in Flooded Paddy Soil. *Soil Biol. Biochem.* **2015,** *86,* 24–33.

Xu, J.; Liu, S. J.; Song, S. R.; Guo, H. L.; Tang, J. J.; Yong, J. W. H.; Ma, Y. D.; Chen, X. Arbuscular Mycorrhizal Fungi Influence Decomposition and the Associated Soil Microbial Community Under Different Soil Phosphorus Availability. *Soil Biol. Biochem.* **2018,** *120,* 181–190.

Yadav, T.; Mungray, A. A.; Mungray, A. K. Fabricated Nanoparticles. *Rev. Environ. Contam. Toxicol.* **2014,** *230,* 83–110.

Yadav, T.; Mungray, A. A.; Mungray, A. K. Effect of Multiwalled Carbon Nanotubes on UASB Microbial Consortium. *Environ. Sci. Pollut. Res. Int.* **2016,** *23,* 4063–4072.

Yamada, T.; Sekiguchi, Y. Cultivation of Uncultured Chloroflexi Subphyla: Significance and Ecophysiology of Formerly Uncultured Chloroflexi 'Subphylum I' With Natural and Biotechnological Relevance. *Microbes Environ.* **2009,** *24,* 205–216.

Yang, Y.; Wang, J.; Xiu, Z. M.; Alvarez, P. J. J. Impacts of Silver Nanoparticles on Cellular and Transcriptional Activity of Nitrogen-Cycling Bacteria. *Environ. Toxicol. Chem.* **2013,** *32,* 1488–1494.

You, T.; Liu, D.; Chen, J.; Yang, Z.; Dou, R.; Gao, X.; Wang, L. Effects of Metal Oxide Nanoparticles on Soil Enzyme Activities and Bacterial Communities in Two Different Soil Types. *J. Soils Sed.* **2017,** *18,* 211–221.

Yugay, Y. A.; Usoltseva, R. V.; Silant'ev, V. E.; Egorova, A. E.; Karabtsov, A. A.; Kumeiko, V. V.; Ermakova, S. P.; Bulgakov, V. P.; Shkryl, Y. N. Synthesis of Bioactive Silver Nanoparticles Using Alginate, Fucoidan and Laminaran From Brown Algae as a Reducing and Stabilizing Agent. *Carbohydr. Polym.* **2020,** *245,* 116547.

Zeng, P.; Guo, Z.; Xiao, X.; Peng, C. Effects of Tree-Herb Co-planting on the Bacterial Community Composition and the Relationship Between Specific Microorganisms and Enzymatic Activities in Metal-(loid)-Contaminated Soil. *Chemosphere* **2019,** *220,* 237–248.

Zeng, X.; Zhang, F.; He, N.; Zhang, B.; Liu, X.; Li, X. ZnO Nanoparticles of Different Shapes and Their Antimycotic Property Against Penicillium and Mucor. *Nanosci. Nanotechnol. Lett.* **2016,** *8,* 688–694.

Zhang, C.; Li, M.; Xu, X.; Liu, N. Effects of Carbon Nanotubes on Atrazine Biodegradation by *Arthrobacter* sp. *J. Hazard. Mater.* **2015,** *287,* 1–6.

Zhang, L.; Shi, N.; Fan, J. Q.; Wang, F.; George, T. S.; Feng, G. Arbuscular Mycorrhizal Fungi Stimulate Organic Phosphate Mobilization Associated With Changing Bacterial Community Structure Under Field Conditions. *Environ. Microbiol.* **2018,** *20,* 2639–2651.

Zhang, M.; Gao, B.; Chen, J.; Li, Y. Effects of Graphene on Seed Germination and Seedling Growth. *J. Nanopart. Res.* **2015,** *17,* 78.

Zhang, Q.; Huang, J. Q.; Qian, W. Z.; Zhang, Y. Y.; Wei, F. The Road for Nanomaterials Industry: A Review of Carbon Nanotube Production, Post-Treatment, and Bulk Applications for Composites and Energy Storage. *Small* **2013,** *9,* 1237–1265.

Zhang, X.; Xu, Z.; Qian, X.; Lin, D.; Zeng, T.; Filser, J.; Li, L.; Kah, M. Assessing the Impacts of Cu(OH)2 Nanopesticide and Ionic Copper on the Soil Enzyme Activity and Bacterial Community. *J. Agri. Food Chem.* **2020,** *68,* 3372–3381.

Zhao, L.; Huang, Y.; Keller, A. A. Comparative Metabolic Response Between Cucumber (*Cucumis sativus*) and Corn (*Zea mays*) to a $Cu(OH)_2$ Nanopesticide. *J. Agri. Food Chem.* **2018,** *66,* 6628– 6636.

Zhou, W.; Shan, J.; Jiang, B.; Wang, L.; Feng, J.; Guo, H.; Ji, R. Inhibitory Effects of Carbon Nanotubes on the Degradation of 14C-2,4-Dichlorophenol in Soil. *Chemosphere* **2013,** *90,* 527–534.

Zhu, Y.; Murali, S.; Cai, W.; Li, X.; Suk, J. W.; Potts, J. R.; Ruoff, R. S. Graphene and Graphene Oxide Synthesis, Properties, and Applications. *Adv. Mater.* **2010,** *22,* 3906–3924.

CHAPTER 6

Nanoparticles: Plant Protective Agents Against Pathogenic Microbes and Pests

IMMANUEL SURESH J. and ISWAREYA LAKSHIMI V.

PG Department of Microbiology, The American College, Madurai, Tamil Nadu, India

ABSTRACT

Population explosion in the world leading to the high requirement for food has obligated the need for optimizing the agriculture methods with minimal loss on fields. Crops are generally lost due to factors like pests, insects, pathogens causing plant diseases, environmental conditions, etc. Crops can be protected from these damages by the application of antimicrobials and pesticides. Due to long-time usage of chemical pesticides and antimicrobials, pests and phytopathogens have developed resistance. So, there is a need to develop innovative, cost-effective agents made possible by nanotechnology. Numerous organic and inorganic nanoparticles have been successfully developed and used against a wide range of agricultural pests and pathogens.

Many nanoparticles like gold, silica, titanium, silver, carbon, titanium dioxide, iron, and zinc oxide are used as pesticides. Metal oxide nanoparticles of CuO and CaO were developed and tested against cotton leafworm (*Spodoptera littoralis*), which showed potential entomo-toxic activity. Many nanoparticles based on the green synthesis method like Iron oxide nanoparticles were developed using *Trigonella foenum-graecum* leaf, against South American leaf miner *Tuta absoluta* larvae. Similarly, many nanoparticles like copper, chitosan-coated fungal metabolites, Copper oxide, sulfur, and nanocomposites showed antibacterial and antifungal activity against

Nanotechnology for Sustainable Agriculture: An Innovative and Eco-Friendly Approach. Vishnu D. Rajput, Abhishek Singh, Tatiana M. Minkina, Krishan K. Verma, and Awani Kumar Singh (Eds.)

notable phytopathogens. One of the important bacterial pathogens of rice, *Xanthomonas oryzae pv. Oryzae,* which is developing resistance to general antimicrobials, had been effectively targeted by Copper nanoparticles of size 28 nm. Silver nanoparticles exhibited significant antifungal activity against phytopathogenic fungi like *Rhizoctonia solani, Sclerotinia sclerotiorum,* and *Sclerotinia minor* and were also confirmed using microscopic observation of hyphae. This review will evidence the applications of nanoparticles as an effective plant protective–nano control agent by protecting the crops from pests and pathogens and preventing heavy agricultural losses.

6.1 INTRODUCTION

Nowadays, population explosion is a major problem globally due to which there is a huge requirement of food (Singh et al., 2021). So agricultural products are in huge demand, and it is necessary to safeguard the plants from various damages so that the unwanted loss of plants can be avoided (Singh et al., 2020a). Plants generally encounter two different stresses, which leads to their loss. They are biotic and abiotic stresses. Abiotic stress includes environmental conditions like drought, salinity, metal toxicity, nutrient depletion, etc. (Singh et al., 2020b). In contrast, biotic stress is caused by living organisms that mainly include pests that damage the plants and pathogens that cause diseases to the plants (Rajput et al., 2021). Many chemical pesticides and antimicrobials have been in use today, but due to inappropriate and excessive usage, pathogens and pests have developed resistance to them, and some of them are toxic to humans and animals. So, there is an immediate need to develop innovative ways to eliminate the pest and pathogen in a specific and eco-friendly mechanism.

Every year 20–40% of crops are damaged by plant pests and pathogens, which cause complete loss of those plants. Chemical antimicrobials are toxic to plants and animals and have other problems like poor solubility, less shelf life, and action against nontarget organisms (Stephenson, 2003; Worrall et al., 2018). Many plant pathogens and pests have developed resistance to conventional chemicals, which must be overcome (Hernández-Díaz et al., 2021). Annual use of chemical pesticides has reached 4.6 million worldwide, but even though after the application of huge quantities of pesticides, about 90% of it runs off into the environment, leaves residue in agricultural products, and interferes with the ecological cycle. Thus, the appropriate use of pesticides and antimicrobials has led to many problems like non-point pollution, eutrophication, resistance development, soil fertility loss, biodiversity loss, and bioaccumulation in the food chain (Malaj et al., 2014; Zhao et al., 2017). These disadvantages are crucial for leading nanopesticides and nanoparticles-based antimicrobials for their effective development and research.

Nanotechnology is developing and manipulating materials with at least one dimension in the nanoscale range of 1–100 nm, which is 10^{-9} m (Ahmed and Lee, 2015; Islam, 2020; Misra et al., 2013). Nanoparticles can be produced by various physical, chemical, and biological methods. Nanoparticles have high surface-to-volume ratio that enhances their reactivity and biochemical activity (Khan et al., 2019). Nanoparticles differ widely in their properties from their bulk material. Copper in its nanoscale is transparent to visible light, whereas its bulk form is entirely impermeable (Gao et al., 2003; Zong et al., 2005). Similarly, gold and silver nanoparticles show inhibitory activity against pathogens, whereas macro form does not. Nanotechnology has applications in various fields like medicine, pharmacology, energy, environment, food, and agriculture (Malaj et al., 2014). It is also reported that nanoparticles synthesized from various sources like marine endophytic fungi exhibit antibacterial, anticancer activity and have various uses in pharmaceutical industries (Lydia et al., 2021).

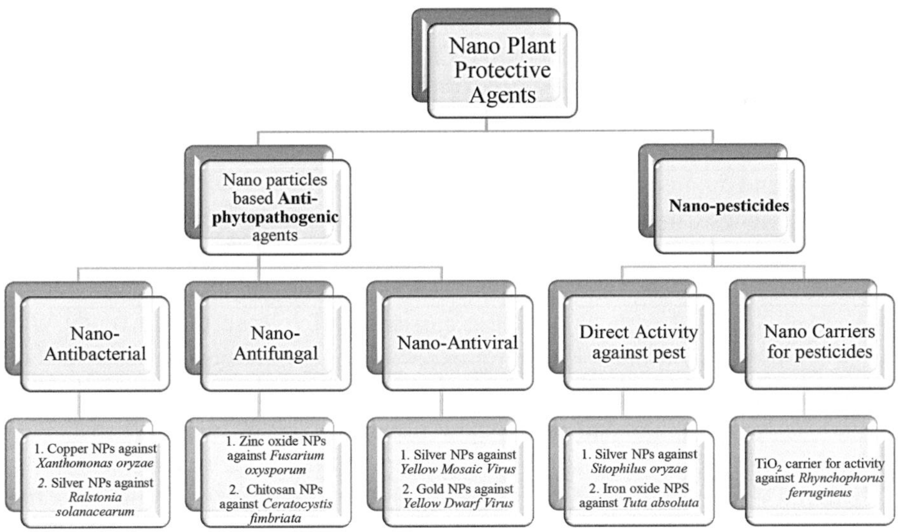

FIGURE 6.1 Overview of plant protective activity of nanoparticles (NPs) against plant pests and pathogens.

There are various methods by which nanoparticles can be synthesized. The two major mechanisms are the top-down and bottom-up approaches. In the top-down approach, nanoparticles are produced from bulk material by physical methods like sonication, radiation, thermal decomposition, and laser ablation. But in the bottom-up approach, nanomaterials are produced by the building up of atom and molecular components (Dantas et al., 2020; Malaj et al., 2014). In the biological synthesis method, nanoparticles are produced by using biological reducing agents like bacterial, fungi, algae, or their metabolites and by using plant extracts. In chemical synthesis, chemical reducing agents like sodium citrate are used (Wuithschick et al., 2015). The biological synthesis method is cost-effective, eco-friendly, safe, and readily available (Huang and Yang, 2004).

In agriculture, nanoparticles are used for various applications like promoting plant growth, yield, and combating biotic and abiotic stresses. Properties of nanoparticles that must be considered for use in pesticides and antimicrobial activity are size, shape, surface-to-volume ratio, and chemical composition, which determines its efficacy, activity, and toxicity (Athanassiou et al., 2018). This review gives a detailed report about various nanomaterials studied for their pesticide activity and activity against plant pathogens. An overview of plant protective effects of nanoparticles against plant pests and pathogens with few examples is given in Figure 6.1.

6.2 PROPERTIES TO BE CONSIDERED FOR DEVELOPING NPs-BASED PESTICIDES AND ANTIPHYTOPATHOGENS

There are various properties to be considered for the nanoparticles to be used as pesticides and against phytopathogens. The chemical structures like geometry, chains, branches, rings, bonded and unbonded atoms, and chemical composition are major factors determining nanoparticles' role. Pesticides should be water-soluble in nature so that they can be less volatile and more reactive. Nano pesticides designed with hydrophilic carriers will release the pesticide with direct contact with water, preventing excessive accumulation of pesticides on unspecific sites. As mentioned above, nanopesticides and anti-phytopathogens should be less volatile with low vapor pressure so that they will evaporate before reaching the site of action. Field temperature, mode of application also affects the rate of the volatility of the nano formulations. Formulations like emulsion, granules, and pellets are better than dust, sprayable liquid, liquid mixtures, and wettable powders as the former can withstand drifts and volatile loss than the latter. So, in brief, nanoparticles used for agricultural applications must have good solubility, less volatility, suitable formulation, and should not be toxic to the plants (Gahukar and Das, 2020).

6.3 NANOPESTICIDES

Pesticides that are developed in the nano-based formulation are nanopesticides. It may be active pesticide ingredients in nano-scale particles or nanoparticles itself exhibiting pesticidal activity. Nanopesticides have various advantages like reducing toxicity, increasing solubility, targeted action, high pesticide activity due to the high surface area-to-volume ratio, and requiring less quantity for application (Nuruzzaman et al., 2016). Many nanomaterials like silica, sulfur, silver, copper, copper oxide, zinc, titanium oxide, graphene, and chitosan-based nanoparticles and nanocomposites have been studied for their pesticidal activity (Pho et al., 2020).

6.3.1 NANOPESTICIDES PRODUCED BY BIOLOGICAL SYNTHESIS METHOD NANOPESTICIDES SYNTHESIZED USING PLANT EXTRACTS

Many nanoparticles that are synthesized based on plant extracts have shown pesticidal activity. *Helicoverpa armigera* is the major pest in Asia, affecting many crops like tomato, pigeon pea, okra, and blackgram. Significant larvicidal activity of silver nanoparticles synthesized from *Aristolochia indica* against third instar larvae of *Helicoverpa armigera* was observed (Siva and Kumar, 2015). The pesticidal activity was exhibited by silver and lead nanoparticles produced using *Avicennia marina* against the grain storage pest *Sitophilus oryzae,* and a significant effect was observed in silver nanoparticles after 4 days (Sankar and Abideen, 2015). Silver nanoparticles synthesized using aqueous leaf extracts of *Euphorbia prostrata* showed pesticidal activity with 100% mortality against *Sitophilus oryzae L.* after 7 days (Zahir et al., 2012). Silver and gold nanoparticles synthesized based on pungamoil compared with commercial neem-based insecticide showed significant pesticide activity against *Pericallia ricini* by reducing the food conversion by pests, thereby affecting its growth (Sahayaraj et al., 2016). Silver nanoparticles generally cause pesticide activity by releasing Ag^{+} ions, which are highly toxic and can disrupt the membrane, kill pests by binding to the cysteine-containing proteins in the plasma membrane (Servin et al., 2015).

Iron oxide nanoparticles produced by photosynthesis using capping and reducing agents from aqueous extract of *Anthocephalous* cadamba exhibited significant pesticide activity against *Sitophilus granaries* compared to chemical nanoparticles. It was also observed that these iron oxide nanoparticles did not show any antimicrobial activity against *E. coli* and *Bacillus* sp. Hence, it can be concluded that iron oxide nanoparticles will not affect plant-promoting microbes (Sivapriya et al., 2018). Iron oxide nanoparticles synthesized from *Trigonella foenum-graecum* were tested for their insecticidal activity against tomato pinworm, *Tuta absoluta,* and 72% mortality was observed after 96 h in 100 μg mL^{-1} concentration (Ramkumar et al., 2020).

The synergistic effect of thiamethoxam nanocomposite and zinc nanoparticles was evaluated against the fourth instar larvae of *Spodoptera litura*. The larvae fed on the composite of zinc nanoparticles with thiamethoxam showed larvae mortality by altering levels of superoxide dismutase, glutathione S transferase, and thiobarbituric acid-reactive substances in the pest (Jameel et al., 2020). Zinc oxide nanoparticles synthesized using *Pongamia pinnata* extract were evaluated against the major pulse beetle *Callosobruchus maculatus*. Pupal, larval, and complete development of *C. maculatus* was delayed, and it caused the pest's mortality exhibiting potential pesticide activity (Malaikozhundan and Vinodhini, 2018). The better pesticide activity was exhibited against pistachio psylla, *Agonoscena pistacie* compared with amitraz (a commercial insecticide) by ZnO nanoparticles (Samih et al., 2011).

Cubic Nickel nanoparticles of size 47 nm synthesized using methanolic extract of *Cocos nucifera* showed pesticidal activity against agricultural pest *C. maculates* with 97.31% mortality (Elango et al., 2016). Nanoenzyme conjugate was prepared by doping Silica nanoparticles with chitinase enzyme, and its synergistic pesticidal activity in the combination of extracts of *Azadirachta indica, Adhatoda vasica, Leucas aspera,* and *Curcuma longa* were studied against *Spodoptera litura*. The enhanced pesticidal activity was observed in the combination of silica nanoparticles, chitinase enzyme conjugate, and *Azadirachta indica* (Narendrakumar and Namasivayam, 2021).

Pimpinella anisum L., Lemongrass, and peppermint-based nanoemulsion showed repellent activity and mortality against *Rhopalosiphum padi L*., an oat aphid at low (0.02 μL/cm^2) and high (0.15 μL/cm^2) concentrations, respectively (Pascual-Villalobos et al., 2017). Similarly, nanoparticles based on essential oil of garlic in a polyethylene glycol carrier were developed with sustained-release mechanism; it showed 80% mortality against *Tribolium castaneum* after 5 months (Yang et al., 2009). Nanoparticles prepared using diatomaceous earth also showed mortality against *Tribolium confusum* and *T. castaneum* than the macro form and caused insecticidal activity by disrupting oviposition, adult infestation, and emergence (Sabbour et al., 2015).

6.3.1.1 NANOPESTICIDES SYNTHESIZED USING MICROBES AND ITS METABOLITES

Nanoconjugate prepared by the combination of the pesticidal fungal metabolite from *Nomuraea rileyi* and silver nanoparticles was studied against the pest *Spodoptera litura,* and significant pesticidal activity was observed against all the larval instars (Namasivayam and Bharani, 2021). Pesticidal fungal metabolite from *Nomuraea rileyi* with chitosan nanoparticles was evaluated for pesticidal activity against Spodoptera litura compared with fungal metabolite and fungal spores. Fungal metabolite coated with chitosan nanoparticles showed better activity than the other two, and pesticidal activity was inversely proportional to the instar stage (Chandra et al., 2013). It is evidenced that nanoparticle conjugation and coating on natural pesticidal metabolites can enhance its activity, and nanoparticles play a major role in improving the pesticidal activity of the fungal metabolite (Ayoub et al., 2018). Titanium oxidenanoparticles synthesized by *Trichoderma viride* showed significant pesticidal activity against the first three instar stages of larvae of *Helicoverpa armigera*. This activity was due to reducing detoxifying enzymes like carboxylesterase and β-glucosidase and increased the level of glutathione S transferase in the pest (Chinnaperumal et al., 2018). Copper nanoparticles treated with cell-free extract of copper-resistant bacteria *Pseudomonas fluorescens MAL2* showed pesticidal activity against *Tribolium castaneum,* grain pest, and significant activity compared with chemically synthesized copper nanoparticles (El-Saadony et al., 2020).

6.3.2 NANOPESTICIDES PRODUCED BY CHEMICAL SYNTHESIS METHOD

Many nanoparticles synthesized by chemical means also have been studied for their pesticidal activity. In Europe, SiO_2 nanoparticles of size greater than 1 μm synthesized by chemical methods are an approved insecticide (Adisa et al., 2019). World Health Organization (WHO) also approved amorphous silica as nano biopesticide since it is safe for humans, and several NPs are considered safe as they do not alter photosynthesis or gene expression in plants and insects, respectively (Athanassiou et al., 2018). *Tinea pellionella*, a cloth moth, causes high weight loss in cotton fiber crops. 20 ppm of colloidal nano silver caused the pest's complete mortality and inhibited weight loss of the cotton fiber (Ki et al., 2007).

Nanoparticles pesticidal activity was also enhanced by coating or tagging it. Au, Ag, and CdS nanoparticles were tagged with pesticidal DNA, and CdS showed maximum activity against *S. litura* larvae. Similarly, nanoparticles coated with Ecdysteroid showed significant pesticidal activity against *Corcyra cephalonica* (Chandrashekharaiah et al., 2015). Nanoparticles of metal oxides like flower structured Copper oxide and hexagonal sheet-like Calcium oxide were analyzed for entomotoxic activity against cotton leafworm *Spodoptera littoralis*. Copper oxide nanoparticles and calcium oxide nanoparticles showed significant pesticidal activity after 3 and 11 days, respectively, by disrupting the mid gut and cuticle layer of the body wall of insects. They also damaged vital enzymes beyond the cell membrane and generated superoxide damaging cell membrane leading to leakage of cellular compounds respectively and caused the death of the pest. Changes in the midgut, cuticle layer, and related biochemical aspects were also observed on the application of pesticidal nanoparticles (Ayoub et al., 2018). Nanoparticles help in overcoming the resistance to commercial insecticides. Chemically synthesized silver nanoparticles showed an effective pesticidal effect in combination with commercial insecticide malathion, whereas most pests like *Tribolium castaneum* are developing resistance to malathion (Alif Alisha and Thangapandiyan, 2019).

Nanomaterials of Chitosan-g-poly (acrylic acid) also reported insecticidal activity against soybean insect pests, namely *Callosobruchus maculatus*, *Aphis gossypii,* and *Callosobruchus maculatus* (Sahab et al., 2015). Similarly, Calcium nanoparticles against *Bactrocera dorsalis* and *Aonidiella aurantia*, silver and silica NPs against *S. litura* and *Achaea Janata*, nanostructured alumina against *Sitophilus oryzae* and *Rhyzopertha dominica*, silica, Al_2O_3, TiO_2 NPs against *S. oryzae* are also studied (Kitherian, 2017).

6.3.3 NANOPARTICLES AS CARRIER FOR PESTICIDES

Besides directly acting as pesticides, nanoparticles help slow and sustain various available pesticides and enhance performance by acting as nanocarriers. This technique is called nanoencapsulation, which mainly helps in the steady release of pesticides, aids in safer handling by less direct exposure, and ensures appropriate concentration to attain bio-efficacy (Athanassiou et al., 2018). Since nanocarriers can penetrate the cells and deliver the biomolecules in various cells and tissues of plants, they form a nanocarrier transport system (Du et al., 2012). Major advantages of nanocarriers include reducing cost and amount of pesticide requirement, enhanced shelf life, delay loss in bioefficacy due to degradation, promoting site-specific action, safety by avoiding direct contact, protection against environmental stress, allow high loading due to large surface area, and capability to load more than one pesticide (Athanassiou et al., 2018; Worrall et al., 2018). Nanocarriers include controlled-release formulations like beads, gels, capsules, and granules (Mishra et al., 2020). They can also be used for directly altering plants to become pest-resistant by delivering genetic materials and other active metabolites (Prasad et al., 2014).

Several nanoparticles have been studied for their nanocarrier property. Silica nanoparticles have the structural flexibility of producing NPs of various sizes and shapes and can also form pores for loading the pesticides (Athanassiou et al., 2018). Chlorfenapyr in silica nanoparticles ensured the slow and sustained release of around 10–20 weeks and showed insecticidal activity twice better than the direct application of chlorfenapyr (Song et al., 2012). Similar results have been observed in various studies in plants like *Cajanus cajan, Vigna mungo, Cicer arietinum* against *Callosobruchus maculatus, Tribolium confusum,* and *R. dominica* (Adisa et al., 2019). 100% mortality was observed against *Rhynchophorus ferrugineus*, the red palm weevil, after 10 days when NPs based on *Moringa olifera* leaf extract were administered with TiO_2 carrier (Hamza, 2015). Similarly, $TiCl_4$-based carrier for neem bar gum extract showed pesticidal activity against *H. armigra* and *Spodoptera litura*. Many activities like reduction of feeding effect, abnormal changes of pest morphology leading to mortality of various instar stages with higher activity against second stage and pupae were also observed (Kamaraj et al., 2018).

The insecticidal protein beauvericin incorporated in chitosan nanoparticles by ionic gelation was tested against groundnut defoliator *Spodoptera litura,* and pesticidal activity against all stages of the pest was observed with 91% release of beauvericin after 24 h (Bharani et al., 2014). Emamectin benzoate, a biopesticide produced by the fermentation of *Streptomyces avermitilis,* is used broadly in controlling pests and is specific in action, but the major drawback is its sensitivity to UV and light. It was overcome by incorporating this biopesticide in cellulose nanocrystals and silicon dioxide nanoparticles, and it was found that it was not affected by UV or light. Emamectin benzoate delivered in nanocarriers showed better pesticidal activity against *Phenacoccus solenopsis* than the biopesticide alone (Elabasy et al., 2019).

Nanocapsules have also shown 68% reduction in root-knot disease, and poly (ε-caprolactone) and poly (lactic-co-glycolic-acid) nanocapsule with ametryn and atrazine also showed better activity due to slow and sustained release (Adisa et al., 2019; Clemente et al., 2014). In the latter nanocapsule, only 15% of atrazine was found to be released even after 72 h of administration (Schnoor et al., 2018).

6.3.4 *ADVANTAGES OF NANOPESTICIDES*

Nanopesticides have been suggested as better than conventional pesticides as they have shown better pesticidal activity by increasing mortality rate and morphological disruptions of the pest. Besides, nanopesticides also have several advantages. They have better stability, specific activity, physical stability (Kah and Hofmann, 2014), increased solubility, improved mobility, larger surface area ensures higher longevity, less toxicity, and sustained delivery (Sasson et al., 2007). Since the loss of pesticides is low, there are less chances of pollution and toxicity to humans and animals (Islam, 2020).

6.4 PHYTOPATHOGENS

Several pathogens like bacteria, fungi, and viruses affect plants causing disease, accounting for a huge loss of agricultural crops. Bacterial pathogens like *Xanthomonas sp.,* fungal pathogens like *Fusarium sp.*, and viral pathogens like *Tomato mosaic virus* have been suppressed by using various nanoparticles in plants. Chitosan nanoparticles derived by ionic gelation of chitosan showed antifungal activity against *Fusarium oxysporum* and antibacterial activity against phytopathogens *Erwinia carotovora subsp. carotovora* and *Xanthomonas campestris pv. Vesicatoria* (Oh et al., 2019). Chitosan guar nanoparticles prepared using sodium tripolyphosphate by the ionic gelation method showed antifungal activity against the rice blast disease pathogen *Pyricularia grisea* and antibacterial rice blight disease pathogen *Xanthomonas oryzae* (Sathiyabama and Muthukumar, 2020). Silver nanoparticles synthesized by algae *Chlorella vulgaris* showed effective activity against phytopathogens like *Erwinia carotovora* and *Alternaria alternata* (El-Moslamy et al., 2016).

6.4.1 *NANOPARTICLES AGAINST BACTERIAL PHYTOPATHOGENS SYNTHESIZED BY BIOLOGICAL AND CHEMICAL METHOD*

Bacterial pathogens are of major concern in plants causing a variety of diseases like bacterial wilt. Many nanoparticles have been studied for their antibacterial activity against bacterial phytopathogens.

6.4.1.1 COPPER NANOPARTICLES

Chemically synthesized copper nanoparticles of various sizes, including 18, 24, 28, and 33 nm, exhibited antibacterial activity against *Xanthomonas oryzae pv. Oryzae* in size and concentration-dependent manner. Antibacterial activity was attributed to the increase in reactive oxygen species, and copper nanoparticles showed better activity than streptomycin sulfate (Majumdar et al., 2019).

6.4.1.2 SLIVER NANOPARTICLES

Antibacterial activity of silver nanoparticles produced by papaya extract against *Ralstonia solanacearum,* which causes bacterial wilt, is reported. Nanoparticles exhibit antibacterial activity by preventing biofilm formation, disturbing ATP production, reducing swarming motility, causing mechanical damage to the cell membrane at the nano level, and downregulating the bacterial genes that play a major role in virulence (Chen et al., 2019). Silver nanoparticles synthesized by the green method using *Mentha longifolia* showed antibacterial activity against plant pathogens like *Pectobacterium carotovorum, Xanthomonas oryzae, Xanthomonas vesicatoria,* and *Ralstonia solanacearum*. It was also biocompatible against RBC and suggested that *Mentha longifolia*-based silver nanoparticles can be used as anti-phyto-pathogenic and consumed by humans due to its biocompatibility (Javed et al., 2021). Silver nanoparticles synthesized using *Taraxacum officinale* as reducing agent showed significant antibacterial activity against plant pathogens *Xanthomonas axonopodis* and *Pseudomonas syringae.* Green synthesized silver nanoparticles showed better activity than chemically synthesized nanoparticles (Saratale et al., 2018). Silver nanoparticles synthesized using the stem extract of *Gossypium hirsutum* showed antibacterial activity against *Xanthomonas axonopodis pv. malvacearum* and *Xanthomonas campestris pv. Campestris* and no phytotoxicity was detected, suggesting that the produced silver nanoparticles are nontoxic to the plants and specific to plant pathogens (Vanti et al., 2019).

Synergistic antibacterial effects of silver nanoparticles with commercial antibiotics are also studied against *Pseudomonas solanocearum, Pseudomonas syringae, Xanthomonas malvacearum,* and *Xanthomonas campestris* (Mala et al., 2012). Similarly, silver nanoparticles on treatment with tween 80 showed potential antibacterial activity against Ralstonia *solanacearum* by disrupting cellular proteins and cell membrane in vitro analysis and in vivo analysis like pot experiment of plant with the pathogen (Chen et al., 2016). In some studies, silver nanoparticles are stabilized by biological methods like using matrix of bovine submaxillary mucin and bile salt capping showed better antibacterial activity against *Acidovorax, Xanthomonas, Clavibacter,* and anthracnose, causing phytopathogen (Makarovsky et al., 2018; Shanmugam et al., 2015). Ag nanoparticles prepared on graphene oxide and double-stranded DNA inhibited the Xanthomonas *perforans* both in vitro completely and in plants (Ocsoy et al., 2013).

6.4.1.3 TITANIUM OXIDE AND ZINC NPs

Lemon fruit extract was used as a reducing agent for producing zinc oxide and titanium dioxide (TiO_2) nanoparticles. Such nanoparticles showed promising antibacterial activity against *Dickeya dadantii,* which causes stem and root rot disease in sweet potato. Macro zinc oxide and titanium dioxide did not exhibit any antibacterial activity, whereas nanoparticles synthesized from them showed antibacterial activity in a concentration-dependent manner (Hossain et al., 2019). TiO_2/Zn nanocomposite was evaluated for their antibacterial activity against *Xanthomonas sp,* which caused leaf spot disease and reduced the disease by 62–71% in rose plants in the pot experiment (Graham et al., 2016). Similarly, plate-like and particulates-like zinc formulations, namely Zinkicide SG4 and Zinkicide SG6, significantly suppressed *Xanthomonas citri,* which causes citrus canker lesions in grape fruit (Adisa et al., 2019).

6.4.1.4 OTHER NPs

Chemically synthesized Magnesium oxide nanoparticles exhibited a similar mechanism of action like foresaid nanoparticles against *Ralstonia solanacearum,* which causes bacterial wilt, and in addition to the above mechanisms, magnesium oxide nanoparticles also interfere with the reactive oxygen species level in the pathogen causing DNA damage and death (Cai et al., 2018). Surface modification of nanoparticles using suitable chemicals has enhanced antimicrobial activity. Halloysite nanotube is a clay material nanoparticle, and on the enhancement with surfactants like cetyl trimethyl-ammonium bromide (CTAB), sodium dodecyl sulfate, and Tween 80, the physicochemical properties of the nanotube were improved. CTAB modified nanotube showed better antibacterial activity against *Xanthomonas oryzae, Agrobacterium tumefaciens,* and *Ralstonia solanacearum* by disrupting the cell membrane integrity biofilm formation by the pathogens and induced high level of reactive oxygen species (Abhinayaa et al., 2019). Starch capped sulfur nanoparticles produced by spectral and microscopic methods showed significant antibacterial activity against the potato ring rot pathogen *Clavibacter michiganensis subsp. sepedonicus* (Lesnichaya et al., 2021). Copper-resistant phytopathogens like *Xanthomonas oryzae, F. oxysporum, P. syringae,* and *A. niger* were inhibited by graphene oxide nanoparticles (Rajwade et al., 2020).

6.4.2 NANOPARTICLES AGAINST FUNGAL PHYTOPATHOGENS

Various studies have been carried out to evaluate the antifungal activity of the nanoparticles produced by both biological and chemical methods that act against different phytopathogenic fungi.

6.4.2.1 ANTIFUNGAL PATHOGENIC NANOPARTICLES PRODUCED BY BIOLOGICAL SYNTHESIS METHOD

Many nanoparticles synthesized using plant-based extracts are studied for their antifungal activity against plant pathogenic fungi like *Fusarium, Curvularia sp.*, etc.

6.4.2.1.1 Sliver NPs

In comparison with 10 different species of *Cassia, Cassia roxburghii* produced stable silver nanoparticles of size 35 nm, and it showed significant antifungal activity against plant pathogens like *Rhizoctonia solani, Fusarium oxysporum,* and *Curvularia sp.* which was significantly better than amphotericin B (Balashanmugam et al., 2016). Silver nanoparticles of 13 nm size synthesized using leaf extract of *Ligustrum lucidum* showed reduction in conidia germination, colony growth, and also significant antifungal activity was observed by disk diffusion and in vitro inoculation analysis against *Setosphaeria turcica*, which causes northern corn leaf blight in maize (Huang et al., 2020). Silver nanoparticles synthesized using *Justicia peploides* and *Withania coagulans* showed effective antifungal activity against *Fusarium oxysporum, Fusarium solani,* and *Fusarium lateritium,* respectively. Silver nanoparticles when entering the cell, they may dissociate into silver ion, and they can bind with phosphate backbone of DNA as they have high affinity for phosphorus, which may cause cell death or they may increase the reactive oxygen species level in fungi-causing death (Hashmi et al., 2019).

Silver (Ag) nanoparticles synthesized using the leaf extract of *Acalypha indica* as reducing agents showed a significant antifungal mechanism by disrupting membrane integrity and DNA damage against *Alternaria alternata, Sclerotinia sclerotiorum, Macrophomina phaseolina, Rhizoctonia solani, Botrytis cinerea,* and *Curvularia lunata* with the highest effect against *Alternaria alternata* (Krishnaraj et al., 2012). Ag nanomaterials synthesized using aqueous extract of Artemisia absinthium inhibited both *Phythopthora parasitica* and *P. capsica* in tobacco with 100 mgL^{-1} concentration had similar inhibitory activity as of mefenoxam, a commercial fungicide (Adisa et al., 2019). Ag nanoparticles synthesized from potato steroidal alkaloids α-chaconine and α-solanine showed antifungal activity against *Alternaria alternate, Rhizoctonia solani, Botrytis cinerea,* and *Fusarium oxysporum f. sp. lycopersici* with highest and lowest activity against *R. solani* and *F. oxysporum,* respectively (Almadiy and Nenaah, 2018).

Antifungal activity of biosynthesized Ag nanomaterials against *Alternaria solani* was evidenced by the reduction in 100% spore count after 3 days and 73% reduction in biomass after 7 days (Kumari et al., 2017). Increasing concentration of Ag nanoparticles against *Bipolaris sorokiniana* was evidenced by inhibition of fungal conidial growth (Mishra et al., 2014), and against *Sclerotium rolfsii* in chick pea was evidenced by sclerotial rind disruption and accumulation inside the cells (Mishra et al., 2017). These antifungal mechanisms may be attributed to Ag^+ ions released from the Ag nanoparticles capable of reacting with functional groups like carboxyl, thiol, imidazole, phosphate, and amino on the cell wall inactivation of the various cellular process leading to the death of the pathogen (Adisa et al., 2019).

6.4.2.1.2 Zinc NPs

Zinc oxide nanoparticles synthesized using the aqueous extract of *Melia azedarach* by the solution combustion method showed significant antifungal activity in a dose-dependent manner against *Cladosporium cladosporioides* and *Fusarium oxysporum,* which are the soybean seed-borne pathogens. It kills the fungi by reducing ergosterol and cell membrane integrity by increasing reactive oxygen species and cellular damage in the pathogen (Lakshmeesha et al., 2020). Zinc oxide nanoparticles synthesized chemically and by green synthesis showed antifungal activity against *A. alternata, A. niger, B. cinerea, F. oxysporum,* and *P. expansum* with green synthesized Zinc oxide (ZnO) nanoparticles showing higher efficacy (Jamdagni et al., 2018). The shape of the nanoparticles also plays a major role in antifungal activity against the phytopathogens. Zinc oxide nanoparticles of three different shapes like nanoparticles, lamellar platelets, and hexagonal rods were studied for their antifungal activity against fungal pathogens of plants. Platelet-shaped ZnO nanoparticles showed antifungal activity against them *Fusarium solani*, reducing growth up to 65% (Pariona et al., 2020). ZnO nanoparticles also inhibited the growth of *Fusarium graminearum* after 7 days of exposure with 75% and 63% inhibition in agar and sand, respectively (Dimkpa et al., 2013). The major mechanism associated was reactive oxygen species (ROS) which damages the cell wall of the pathogens and causes death. Zinc-based nanoparticles are generally eco-friendly as it is less toxic to the plants (Adisa et al., 2019).

6.4.2.1.3 Other NPs

Copper nanoparticles synthesized using the leaf extract of *Azadirachta indica* showed potential activity against *Alternaria mali, Diplodia seriata,* and *Botryosphaeria dothidea,* which are observed in the common pathogens of apple in both the nanoparticles and in synergy with *A. indica* extract (Ahmad et al., 2020). Spherical gold nanoparticles of size 60 nm synthesized using *Mentha piperita* essential oil showed significant antifungal activity against *Aspergillus niger* (Islam, 2020). Selenium nanoparticles of size 15–40 nm synthesized using water extract of *Emblica officinalis* fruits exhibited significant antifungal activity against Aspergillus *brasiliensis, Rhizopus stolonifera, A. flavus, A. ochraceus, Fusarium anthophilum,* and *A. oryzae* (Gunti et al., 2019).

Many fungi and yeast have also been used as the reducing agent in synthesizing various nanoparticles with antifungal activity against phyto-pathogens. Mycosilver nanoparticles produced using endophytic fungi *Setosphaeria rostrata* isolated from *Solanum nigrum* showed mild and high antifungal activity against *Aspergillus niger* and *Fusarium udum,* respectively (Akther and Hemalatha, 2019). Silver nanoparticles produced using the supernatant of yeasts *Cryptococcus laurentii* and *Rhodotorula glutinis* were evaluated against phytopathogens like *Botrytis cinerea, Penicillium expansum, Aspergillus niger, Alternaria sp.,* and *Rhizopus sp* that limits the shelf life of fruits and vegetables in which both showed better activity than chemically synthesized nanoparticles and similar activity as commercial fungicide iprodione (Fernández et al., 2016). Silver nanoparticles produced using *Alternaria sp.* fungi isolated from banana cultivated soil showed significant antifungal activity against plant pathogen *Fusarium oxysporum* and *Alternaria sp* by destructing membrane integrity of pathogens (Win et al., 2020). Similarly, silver nanoparticles synthesized using *Aspergillus versicolor* were effective against *Sclerotinia sclerotiorum* and *Botrytis cinerea* in strawberry plants, and concentration-dependent activity was observed with higher activity against *B. cinerea* and silver nanoparticles synthesized using *Guignardia mangiferae* were effective against *Colletotrichum sp., Rhizoctonia solani*, and *Curvularia lunata* (Guilger-Casagrande and Lima, 2019). Many beneficial microbes like *Serratia sp., Bacillus, Trichoderma sp.,* and *Spirulina platensis* have been utilized to produce silver nanoparticles with antimicrobial activity against a variety of phytopathogens like *B. sorokiniana* (Mishra and Singh, 2015). Silver nanoparticles synthesized using *Trichoderma longibrachiatum* cell filtrate showed 90% activity against *Fusarium verticillioides, Fusarium moniliforme, Penicillium brevicompactum, Helminthosporium oryzae,* and *Pyricularia grisea*. The authors have suggested that the antifungal activity may be due to disruption of membrane-bound enzymes and lipids, inactivation of the sulfhydryl group in the cell wall, pore formation in the cell wall, DNA damage, and disturbing replication, elevation of hydrogen peroxide, hydroxyl radicals, and superoxide anions (Elamawi et al., 2018).

Copper nanoparticles produced using actinomycetes *Streptomyces capillispiralisCa-1* isolated from a medicinal plant *Convolvulus arvensis* evaluated against *Alternaria spp., Aspergillus niger,* and *Pythium spp.* showed antifungal activity by causing structural and functional changes of the fungal cell causing cell death (Hassan et al., 2018). Ag and Copper oxide nanoparticles produced using cell filtrate of *Trichoderma harzianum* showed effective activity against *Alternaria alternata, Pyricularia oryzae,* and *Sclerotinia sclerotiorum* in a dose-dependent manner (Consolo et al., 2020). Chitosan nanoparticles prepared using anionic proteins from *Penicillium oxalicum* were tested against *Pyricularia grisea, Alternaria solani, Fusarium oxysporum*. Improved morphology of chickpea seeds was observed on the application of nanoparticles (Sathiyabama and Parthasarathy, 2016).

6.4.2.2 ANTIFUNGAL PATHOGENIC NANOPARTICLES PRODUCED BY CHEMICAL SYNTHESIS METHOD

6.4.2.2.1 Chitosan NPs

Chitosan, chitosan saponin, and copper chitosan nanoparticles were reported for antifungal activity against phytopathogens like *Alternaria alternata, Macrophomina phaseolina,* and *Rhizoctonia solani.* Copper chitosan nanoparticles were more effective followed by chitosan nanoparticles (Saharan et al., 2013). Chitosan nanomaterials showed significant antifungal activity against *Ceratocystis fimbriata* and caused nearly 70% mortality in 3 h of exposure. Antifungal activity was evidenced by studying the mechanisms like reduction of mycelial expansion, irreversible damage of membrane, hyphal morphology alteration, induction of membrane permeability, causing cellular components like potassium ions to leak out of cells leading to necrotic death (Xing et al., 2018). Chitosan nanomaterials reduced the blast disease in finger millet by around 64% by possible ROS production mechanisms and delayed the disease onset (Adisa et al., 2019). Similarly, chitosan nanomaterials decreased nearly 82% of downy mildew disease caused by *Sclerospora graminicola* in millet seeds, and genomic analysis revealed the upregulation of genes related to ROS enzymes like catalase, peroxidase, superoxide dismutase, ammonia-lyase, which may be the possible mechanism for the suppression of the disease (Siddaiah et al., 2018). Membrane structure damage was also observed as a possible antifungal mechanism in chitosan nanocomposite with silver in its activity against *F. oxysporum* (Dananjaya et al., 2017).

6.4.2.2.2 Copper NPs

Copper oxide nanoparticles showed effective activity against *Phytophthora infestans* which causes damage in tomato, and several copper-based nanoparticles have been reported with sustained release of $Cu2^{+}$ions than copper sulfate (Adisa et al., 2019). Nanoparticles made of copper, copper oxide, silver, zinc oxide (ZnO) on evaluation against seven common fungal pathogens causing foliar and seed-borne diseases showed that copper nanoparticles showed significant reduction in mycelial growth. ZnO nanoparticles and copper nanoparticles showed better antifungal activity against zinc sulfate, and copper sulfate and nanoparticles showed 100-fold higher activity against spores than hyphae (Malandrakis et al., 2019). Copper nanoparticles,

Zinc nanoparticles, Chitosan, zinc oxide, and copper nanocomposites were evaluated against *A. alternata, R. solani,* and *B. cinerea* for their antifungal activity where the nanocomposite was effective (Al-Dhabaan et al., 2017).

6.4.2.2.3 Magnesium Oxide NPs

Magnesium oxide nanoparticles and macroscale magnesium oxide were compared for their antifungal activity against *Phytophthora nicotianae* and *Thielaviopsis basicola,* which causes black shank and black root rot in many plants like tobacco. Magnesium oxide nanoparticles showed better antifungal activity against these phytopathogens by causing morphological changes and elevation of ROS level in pathogen leading to death, suggesting that magnesium oxide in its nano form develops potential antifungal activity (Chen et al., 2020). Comparing magnesium oxide nanoparticles and sepiolite magnesium oxide nanocomposites against *Fusarium verticilliodes, Bipolaris oryzae,* and *Fusarium fujikuroi* showed that magnesium oxide in sepiolite-based nanocomposite was effective in protecting the seeds from the fungal pathogen. The antifungal activity was due to the distortion of hyphae and the collapse of spores after nanocomposite treatment (Sidhu et al., 2020).

6.4.2.2.4 Silver NPs

Silver nanoparticles synthesized chemically showed antifungal activity against *Colletotrichum gloesporioides* which caused anthracnose and its associated lesions in a wide variety of fruits. Antifungal activity increased in a dose-dependent manner, evidenced by reduction of mycelial growth (Aguilar-Méndez et al., 2011). Silver nanoparticles caused hyphae damage, causing separation of hyphal layers and collapse in *Rhizoctonia solani, Sclerotinia sclerotiorum,* and *S. minor*, which forms sclerotiorum in plant diseases. Higher activity was reported against *R. solani* followed by *S. sclerotiorum* and *S. minor* (Min et al., 2009). Silver nanoparticles inhibited radial fungal growth and reduced mycelial growth of phytopathogens like *Pythium aphanidermatum, Sclerotinia sclerotiorum,* and *Macrophomina phaseolina.* Greenhouse experiment also showed positive antifungal activity of silver nanoparticles (Mahdizadeh et al., 2015).

Silica silver nanoparticles combined with water-soluble polymer produced by the radiation method effectively eliminated powdery mildew of pumpkin caused by pathogenic fungi. Even the low concentration of 0.3 ppm was effective in its antifungal activity (Park et al., 2006). Silver chitosan nanocomposites showed antifungal activity against phytopathogens in the following order: *Aspergillus*>*Alternaria*>*Rhizoctonia*. Silver chitosan nanocomposite antifungal activity was higher in comparison with silver and chitosan nanoparticles (Kaur et al., 2012). Nano colloidal silver exhibited 95% activity in curing powdery mildew disease caused by *Sphaerotheca pannosa Var rosae* and recurrence was inhibited in rose plants (Patel et al., 2014). Silver-doped hollow titanium oxide nanoparticles showed effective antifungal activity against causative agents of fusarium wilt in potato and apple scab disease, namely *Fusarium solani* and *Venturia inaequalis,* respectively. Since Ag-doped TiO_2 nanoparticles have two different mechanisms of antifungal activity, that is, $TiO_{2,}$ which is positively charged, are attracted to negatively charged fungal cell wall, causing oxidation, cell wall perforation, and fungal death, whereas silver nanoparticles produce silver ions that damage DNA. Due to this dual antifungal mechanism, Ag-doped TiO_2 nanoparticles showed enhanced activity than individual nanoparticles (Boxi et al., 2016). Several other studies have also demonstrated nanoparticles' disruption of the transport system as a possible mechanism of antifungal activity (Mishra and Singh, 2015).

6.4.2.2.5 Other NPs

Sulfur nanoparticles have been studied for effective antifungal activity against *Erysiphe cichoracearum* at 1000 mg/L and against *Fusarium solani, Venturia inaequalis* by damaging cell wall result of deposition (Khan et al., 2019; Rajwade et al., 2020). β-D-glycan nanoparticles at 0.1%, w/v concentration on foliar treatment on turmeric plants, rhizome rot disease were suppressed up to 77% by increasing peroxidase activity polyphenol oxidase and protease inhibitors (Anusuya and Sathiyabama, 2015). Cell membrane disruption and hyphal malformation were observed as the possible mechanism of antifungal activity of mesoporous alumina sphere nanoparticles against *Fusarium oxysporum,* and the antifungal activity was better than tolclofosmethyl, a commercial fungicide (Kaur et al., 2012). Silica nanoparticles are also studied for their effective action against *Aspergillus niger* and *F. oxysporum* in maize (Suriyaprabha et al., 2014).

6.4.3 NANOPARTICLES AGAINST VIRAL PHYTOPATHOGENS

Research on the application of nanoparticles for combating plant viruses is still in its initial step (Vargas-Hernandez et al., 2020). Silver nanoparticles synthesized using *Bacillus pumilus, Bacillus licheniformis,* and *B. persicus* showed significant suppression of Bean *Yellow Mosaic Virus* infection in fava beans in pre, post, and simultaneous inoculation conditions of nanoparticles with the virus. Silver nanoparticles of size nm synthesized using *B. licheniformis* showed complete antiviral activity by completely suppressing the viral infection (Elbeshehy et al., 2015).

Fe_3O_4, Zinc oxide, and SiO_2 NPs showed antiviral activity against the *Tobacco Mosaic virus* by causing damage to TMV shell protein due to the interaction of nanoparticles with glycoprotein, which eventually led to aggregation (Cai et al., 2020, p. 4). Post-inoculation and pre-inoculation of SiO_2 nanoparticles were able to suppress the viral concentration of *tomato curly virus* and TMV and complete inhibition of disease, respectively (Hao et al., 2018). Pre-inoculation of nanoparticles has been shown to suppress various viruses in the plant. Pre-inoculation of Zinc oxide and TiO_2 nanoparticles have suppressed the *turnip mosaic virus* (21 days pre-inoculation) and *Tobacco mosaic virus* (12 days pre-inoculation) by limiting the replication of the virus. Gold nanoparticles against *Barley yellow Dwarf virus* in *Hordeum vulgare* (Aref et al., 2012), silver nanoparticles against *tomato mosaic virus* and *potato virus* by binding to coat proteins and *sunhemp rosette virus* in *Cymopsis tetragonaloba* plant by inactivation of the virus were also observed (Jain and Kothari, 2014). Application of graphene oxide-based silver nanoparticles against in *Solanum tuberosum* conferred disease resistance of plants against *potato virus Y* (El-Shazly et al., 2017).

6.4.4 MECHANISM OF ANTI-PHYTOPATHOGENIC ACTIVITY OF NANOPARTICLES

Generally, nanomaterials have the large surface, which is a major reason for enhanced activity and function in the biological system like plants, and due to their nano-size, it is easier for the plants to take them up than their bulk or ionic form (Adisa et al., 2019). In fungal cells nanomaterials internalization may be nonspecific direct internalization due to small size, specific receptor-mediated adsorption or by membrane-spanning ion transport proteins (Rajwade et al., 2020). The major mechanism of action includes ROS production and the production of ions that react with various cellular components causing damages.

Many nanoparticles are associated with elevated ROS production, which disturbs microbial cellular homeostasis, creates oxidative burst, or may cause mutations, single-stranded breaks, deletion, and protein cross-linking, eventually leading to death (Adisa et al., 2019; Ali et al., 2020). TiO_2-based nanoparticles have been studied to reduce the white and brown rot of fungi, and this antifungal activity is associated with ROS generation like superoxide anions, hydrogen peroxide, and hydroxyl radicals (Pho et al., 2020). Other mechanisms including protein dysfunction like oxidation of cysteine in the iron-binding site, membrane function disruption, interfering nutrient uptake by the pathogens are causing genotoxicity (Lemire et al., 2013). Generally, metallic nanoparticles release ions like Ag^+, $Cu^{2+}Zn^{2+}$, and $Ti^{4+,}$ which bind to functional groups and disrupt the cell membrane structure by altering the function of the membrane. They can also cause genotoxicity and disturb the electron transport chain, which causes the death of the pathogen (El-Argawy et al., 2017).

The major factor involved in the antimicrobial activity of chitosan nanoparticles is the presence of positive charge due to the amino functional group on its surface. Since microbial membrane has an anionic surface, the chitosan nanoparticles can bind to it by electrostatic interaction, cause cell membrane permeability, intracellular component leakage, and cause the death of the pathogen by cell lysis. They can also chelate with various metals in the cell, electrostatically interact with DNA, inhibit mRNA and protein synthesis, and interrupt many essential enzymes' activity (Adisa et al., 2019). Zinc oxide nanoparticles are also studied for their antibacterial activity by ROS production due to increased surface area. It causes lipid peroxidation, enzyme deactivation, and membrane destruction by accumulation, diffusion, interaction with the cell membrane, and intracellular components electrostatic interactions (Yi et al., 2019). TiO_2 nanoparticles doped with nitrogen modified with palladium oxide showed effective antifungal activity against *Fusarium graminearum,* and these nanoparticles can be activated by visible light, which produces ROS and disrupts the cell membrane and cellular components (Pho et al., 2020).

Many nanoparticles like Ag, Al_2O_3, CdO, Cu, TiO_2, and Zn are toxic toward pathogens, increase ROS production, disrupt cell membrane, and cause intracellular leakage (Hernández-Díaz et al., 2021). These mechanisms are common in most nanoparticles. But few other mechanisms are associated with the antimicrobial activity of nanoparticles. Al_2O_3 nanoparticles cause depolarization of cell membranes (Sadiq et al., 2009); Gold nanoparticles interfere with uncoiling and transcription of DNA and cause porosity in the cell membrane, causing cellular leakage (Rai et al., 2010). Cerium oxide nanoparticles inhibit electron and ion transport pumps and interfere with cellular respiration. It interacts with the bacterial membrane by electrostatic and redox interaction (Hernández-Díaz et al., 2021). Selenium nanoparticles cause depletion of ATP in cells and inhibit fungal spore germination (Huang et al., 2019). Silica nanoparticles have been shown to cause mechanical damage to the bacterial membrane in a dose-dependent manner (Hernández-Díaz et al., 2021). Iron and iron oxide-based nanoparticles cause cellular damage by causing confirmational DNA changes, protein, and cellular membrane. Another mechanism of action is the Fenton reaction, in which more potent ROS like OH• and HO^2 • are produced, which damages the cell and eventually causes the death of the pathogen (Arakha et al., 2015).

6.5 MECHANISM OF ANTIVIRAL ACTIVITY

Many studies have suggested that the major mechanism of nanoparticles to exhibit antiviral activity is by the production of ROS. Plant cells produce various antioxidant enzymes such as catalase, guaiacol peroxidase, and ascorbate peroxidase (Vargas-Hernandez et al., 2020). Several studies suggest that various antioxidant enzymes are elevated in the presence of nanoparticles which are reported in Table 6.1.

Several plant hormones are also suggested as possible mechanisms for antiviral activity. Many studies have reported the increase in various phytohormones after exposure to nanoparticles. Salicylic acid and jasmonic acid levels in *Arabidopsis thaliana* on exposure to CuO nanoparticles, jasmonic acid level in tomato plants on exposure to chitosan polyvinyl alcohol hydrogels, abscisic acid and salicylic acid level in *Nicotiana benthamiana* on exposure to Fe_3O_4NPs, cytokinin in *Capsicum annuum* L on exposure to CeO_2 nanoparticles were elevated (Vargas-Hernandez et al., 2020). Antiviral activity of nanoparticles against various viral plant pathogens is also associated with the nanoparticles to bind to envelope glycoprotein (Elbeshehy et al., 2015), interfere with the fusion of viral membrane and penetration of the virus into host cells (Mehrbod et al., 2009). Silver nanoparticles have been reported to inhibit viral RNA and virion production (Lara et al., 2010). Antiviral activity against the yellow mosaic virus was observed due to the binding of nanoparticles to the disulphide bond regions of CD4-binding domain and inhibiting the virus (Mohamed and Abd–Elsalam, 2018).

TABLE 6. 1 Antioxidant Enzymes Elevation in the Presence of Nanoparticles.

Nanoparticles	Enzyme/gene	Impacts	Reference
ZnO	Catalase, peroxide	Foliar application of ZnO nanoparticles inhibited *Tobacco Mosaic Virus* replication	(Cai et al., 2019)
Fe_3O_4	Catalase, peroxide	Promoted plant growth and elicited defence response against Tobacco Mosaic virus	(Cai et al., 2020, p 4)
SiO_2	Pox gene expression (Peroxidase)	Increased defence mechanism against papaya ring spot virus in cucumber plant	(Elsharkawy and Mousa, 2015)
Ag	Polyphenol oxidase, peroxide	Reduced the disease severity and relative concentration of *Tomato Mosaic virus* and *Potato virus Y* in tomato plants by inducing a systemic acquired resistance	(Noha et al., 2018)
Nickel oxide	Pod gene expression (peroxidase)	Reduced disease severity, accumulation of *Cucumber Mosaic Virus* accumulation and induced systemic resistance to the virus in cucumber plants	(Derbalah and Elsharkawy, 2019)

Thus, by various mechanisms like intracellular accumulation, ROS production, membrane disruption, deactivation of enzymes, interfering important cellular processes, and cell lysis, nanoparticles play a major role in acting against plant pathogens.

6.6 ISSUES AND CHALLENGES

Various challenges must be solved and overcome to achieve sustainable use of nanoparticles in agriculture for pesticidal and anti-phytopathogenic activity (Gahukar and Das, 2020). Even though many studies have been carried out to understand the efficacy of nanoparticles as pesticides and against plant pathogens, most of the studies are carried out in laboratory environment. Soil and field-based studies are required to confirm the reproducibility of nanoparticles in agricultural field application. Many experiments are required to estimate the concentration, type, and mode of application to get better results. Concentration must be fixed such that metallic nanoparticles do not cause any metal contamination or pollution in the soil (Adisa et al., 2019).

Bioefficacy should be confirmed to facilitate the required concentration of nanoparticles to be available at its site of activity. Stability and site-specific release by application of hydrophilic carriers can enhance the bioavailability of nanoparticles. Some nanoparticles may affect the plant in attacking pests and pathogens, causing cytotoxic and genotoxic effects in plants. During the experimental studies, it should be ensured that nanoparticles have no or low toxicity against the plants. Since target organisms are not the only living organism in the field, the nanoparticles used should be specific only to the target pest or pathogen and should not affect the beneficial microbes and other organisms like earthworms. Studies have shown that nanocapsules of 400 nm prepared using *Azadirachta indica* were specific in their activity and did not cause any adverse effect on soil microbiota even after 300 days. Apart from toxicity to soil organisms, even humans and animals are also vulnerable to adverse health effects due to the ingestion, inhalation, and dermal contact of nanoparticles, especially NPs based on metals. So, to reduce the risk posed mainly by field workers and consumers, metal nanoparticles should be modified to reduce their toxicity (Gahukar and Das, 2020).

In brief, the major challenges and issues associated with applying nanoparticles in agriculture include bioefficacy, toxicity to plants, animals, humans, and specificity. Nanoparticle-based pesticides and anti-phytopathogens should be further studied for enhancing their easy and safe application, cost-effectiveness, and easy availability with specificity, eco-friendly, and without toxicity.

6.7 CONCLUSIONS

These studies show that nanoparticle-based pesticides and anti-phytopathogenic agents hold great potential for nano-enabled agriculture. Till now, few nanoparticles like silica oxide-based nanopesticides have been approved for field use in Europe (Adisa et al., 2019). Nanoparticles offer various benefits like reducing the quantity to be used, specific activity, and cost-effective and eco-friendly agents. But further studies are required to modify and develop nanopesticides and anti-phytopathogenic agents that possess bioefficacy, stability, and toxicity to humans, animals, and the environment. Thus, nanoparticles can be developed as promising plant protective agents for eliminating the pest and pathogen in agriculture, thereby preventing economic loss and supporting the economic backbone of our country.

KEYWORDS

- **antibacterial**
- **antifungal**
- **crop protection**
- **nanoparticles**
- **pesticides**

REFERENCES

Abhinayaa, R.; Jeevitha, G.; Mangalaraj, D.; Ponpandian, N.; Meena, P. Toxic Influence of Pristine and Surfactant Modified Halloysite Nanotubes on Phytopathogenic Bacteria. *Appl. Clay Sci.* **2019,** *174*, 57–68.

Adisa, I. O.; Pullagurala, V. L. R.; Peralta-Videa, J. R.; Dimkpa, C. O.; Elmer, W. H.; Gardea-Torresdey, J. L.; White, J. C. Recent Advances in Nano-Enabled Fertilizers and Pesticides: A Critical Review of Mechanisms of Action. *Environ. Sci. Nano* **2019,** *6*(7), 2002–2030.

Aguilar-Méndez, M. A.; San Martín-Martínez, E.; Ortega-Arroyo, L.; Cobián-Portillo, G.; Sánchez-Espíndola, E. Synthesis and Characterization of Silver Nanoparticles: Effect on Phytopathogen *Colletotrichum Gloesporioides*. *J. Nanoparticle Res.* **2011,** *13*(6), 2525–2532.

Ahmad, H.; Venugopal, K.; Bhat, A. H.; Kavitha, K.; Ramanan, A.; Rajagopal, K.; Srinivasan, R.; Manikandan, E. Enhanced Biosynthesis Synthesis of Copper Oxide Nanoparticles (CuO-NPs) for Their Antifungal Activity Toxicity Against Major Phyto-Pathogens of Apple Orchards. *Pharm. Res.* **2020,** *37*(12), 1–12.

Ahmed, A. I.; Lee, Y. S. Nanoparticles as Alternative Pesticides: Concept, Manufacturing and Activities. *Korean J. Mycol.* **2015,** *43*(4), 207–215.

Akther, T.; Hemalatha, S. Mycosilver Nanoparticles: Synthesis, Characterization and Its Efficacy Against Plant Pathogenic Fungi. *BioNanoScience* **2019,** *9*(2), 296–301.

Al-Dhabaan, F. A.; Shoala, T.; Ali, A. A.; Alaa, M.; Abd-Elsalam, K.; Abd-Elsalam, K. Chemically-Produced Copper, Zinc Nanoparticles and Chitosan-Bimetallic Nanocomposites and Their Antifungal Activity Against Three Phytopathogenic Fungi. *Int. J. Agric. Technol.* **2017,** *13*(5), 753–769.

Ali, M.; Ahmed, T.; Wu, W.; Hossain, A.; Hafeez, R.; Islam Masum, M.; Wang, Y.; An, Q.; Sun, G.; Li, B. Advancements in Plant and Microbe-Based Synthesis of Metallic Nanoparticles and Their Antimicrobial Activity against Plant Pathogens. *Nanomaterials* **2020,** *10*(6), 1146.

Almadiy, A. A.; Nenaah, G. E. Ecofriendly Synthesis of Silver Nanoparticles Using Potato Steroidal Alkaloids and Their Activity against Phytopathogenic Fungi. *Braz. Arch. Biol. Technol.* **2018,** *61*.

Anusuya, S.; Sathiyabama, M. Foliar Application of β-D-Glucan Nanoparticles to Control Rhizome Rot Disease of Turmeric. *Int. J. Biol. Macromol.* **2015,** *72*, 1205–1212.

Arakha, M.; Pal, S.; Samantarrai, D.; Panigrahi, T. K.; Mallick, B. C.; Pramanik, K.; Mallick, B.; Jha, S. Antimicrobial Activity of Iron Oxide Nanoparticle Upon Modulation of Nanoparticle-Bacteria Interface. *Sci. Rep.* **2015,** *5*(1), 1–12.

Aref, N.; Alkubaisi, N.; Marraiki, N.; Hindi, A. In *Multi-Functional Effects of Gold Nano-Particles Inducing Plant Virus Resistance Crops*, Proceedings of the 5th Annual World Congress of Industrial Biotechnology-2012; Xi'an: China, 2012; vol 25.

Alif Alisha, A. S.; Thangapandiyan, S. Comparative Bioassay of Silver Nanoparticles and Malathion on Infestation of Red Flour Beetle, Tribolium Castaneum. *J. Basic Appl. Zool.* **2019,** *80*(1), 1–10.

Athanassiou, C. G.; Kavallieratos, N. G.; Benelli, G.; Losic, D.; Rani, P. U.; Desneux, N. Nanoparticles for Pest Control: Current Status and Future Perspectives. *J. Pest Sci.* **2018,** *91*(1), 1–15.

Ayoub, H. A.; Khairy, M.; Elsaid, S.; Rashwan, F. A.; Abdel-Hafez, H. F. Pesticidal Activity of Nanostructured Metal Oxides for Generation of Alternative Pesticide Formulations. *J. Agric. Food Chem.* **2018,** *66*(22), 5491–5498.

Balashanmugam, P.; Balakumaran, M. D.; Murugan, R.; Dhanapal, K.; Kalaichelvan, P. T. Phytogenic Synthesis of Silver Nanoparticles, Optimization and Evaluation of in Vitro Antifungal Activity against Human and Plant Pathogens. *Microbiol. Res.* **2016,** *192*, 52–64.

Bharani, R. A.; Namasivayam, S. K. R.; Shankar, S. S. Biocompatible Chitosan Nanoparticles Incorporated Pesticidal Protein Beauvericin (Csnp-Bv) Preparation for the Improved Pesticidal Activity Against Major Groundnut Defoliator *Spodopteralitura (Fab.) (Lepidoptera; Noctuidae). Int. Chem. Tech. Res.* **2014,** *6*(12), 5007–5012.

Boxi, S. S.; Mukherjee, K.; Paria, S. Ag Doped Hollow TiO2 Nanoparticles as an Effective Green Fungicide Against Fusarium Solani and Venturia Inaequalis Phytopathogens. *Nanotechnology* **2016,** *27*(8), 085103.

Cai, L.; Cai, L.; Jia, H.; Liu, C.; Wang, D.; Sun, X. Foliar Exposure of Fe3O4 Nanoparticles on *Nicotiana Benthamiana*: Evidence for Nanoparticles Uptake, Plant Growth Promoter and Defense Response Elicitor Against Plant Virus. *J. Hazard. Mater.* **2020,** *393*, 122415.

Cai, L.; Chen, J.; Liu, Z.; Wang, H.; Yang, H.; Ding, W. Magnesium Oxide Nanoparticles: Effective Agricultural Antibacterial Agent against *Ralstonia Solanacearum. Front. Microbiol.* **2018,** *9*, 790.

Cai, L.; Liu, C.; Fan, G.; Liu, C.; Sun, X. Preventing Viral Disease by ZnONPs through Directly Deactivating TMV and Activating Plant Immunity in *Nicotiana Benthamiana. Environ. Sci. Nano* **2019,** *6*(12), 3653–3669.

Chandra, J. H.; Raj, L. A.; Namasivayam, S. K. R.; Bharani, R. A. In *Improved Pesticidal Activity of Fungal Metabolite From Nomureae Rileyi With Chitosan Nanoparticles,* International Conference on Advanced Nanomaterials & Emerging Engineering Technologies; IEEE, 2013; pp 387–390.

Chandrashekharaiah, M.; Kandakoor, S. B.; Gowda, G. B.; Kammar, V.; Chakravarthy, A. K. Nanomaterials: A Review of Their Action and Application in Pest Management and Evaluation of DNA-Tagged Particles. In *New Horizons in Insect Science: Towards Sustainable Pest Management*; 2015; pp 113–126.

Chen, J.; Li, S.; Luo, J.; Wang, R.; Ding, W. Enhancement of the Antibacterial Activity of Silver Nanoparticles Against Phytopathogenic Bacterium *Ralstonia Solanacearum* by Stabilization. *J. Nanomater.* **2016,** *2016,* 1–15.

Chen, J.; Mao, S.; Xu, Z.; Ding, W. Various Antibacterial Mechanisms of Biosynthesized Copper Oxide Nanoparticles against Soilborne *Ralstonia Solanacearum*. *RSC Adv.* **2019,** *9*(7), 3788–3799.

Chen, J.; Wu, L.; Lu, M.; Lu, S.; Li, Z.; Ding, W. Comparative Study on the Fungicidal Activity of Metallic MgO Nanoparticles and Macroscale MgO against Soilborne Fungal Phytopathogens. *Front. Microbiol.* **2020,** *11*, 365.

Chinnaperumal, K.; Govindasamy, B.; Paramasivam, D.; Dilipkumar, A.; Dhayalan, A.; Vadivel, A.; Sengodan, K.; Pachiappan, P. Bio-Pesticidal Effects of *Trichoderma Viride* Formulated Titanium Dioxide Nanoparticle and Their Physiological and Biochemical Changes on Helicoverpa Armigera (Hub.). *Pestic. Biochem. Physiol.* **2018,** *149*, 26–36.

Clemente, Z.; Grillo, R.; Jonsson, M.; Santos, N. Z. P.; Feitosa, L. O.; Lima, R.; Fraceto, L. F. Ecotoxicological Evaluation of Poly (ε-Caprolactone) Nanocapsules Containing Triazine Herbicides. *J. Nanosci. Nanotechnol.* **2014,** *14*(7), 4911–4917.

Consolo, V. F.; Torres-Nicolini, A.; Alvarez, V. A. Mycosinthetized Ag, CuO and ZnO Nanoparticles From a Promising *Trichoderma Harzianum* Strain and Their Antifungal Potential Against Important Phytopathogens. *Sci. Rep.* **2020,** *10*(1), 1–9.

Dananjaya, S. H. S.; Erandani, W.; Kim, C.-H.; Nikapitiya, C.; Lee, J.; De Zoysa, M. Comparative Study on Antifungal Activities of Chitosan Nanoparticles and Chitosan Silver Nano Composites Against *Fusarium Oxysporum* Species Complex. *Int. J. Biol. Macromol.* **2017,** *105*, 478–488.

Dantas, J.; Motta, I.; Vidal, L.; Bílio, J.; Pupe, J. M.; Veiga, A.; Carvalho, C. H.; Lopes, R. B.; Rocha, T. L.; Silva, L. P. A Comprehensive Review of the Coffee Leaf Miner *Leucoptera Coffeella* (Lepidoptera: Lyonetiidae), With Special Regard to Neotropical Impacts, Pest Management and Control. 2020.

Derbalah, A. S. H.; Elsharkawy, M. M. A New Strategy to Control *Cucumber Mosaic Virus* Using Fabricated NiO-Nanostructures. *J. Biotechnol.* **2019,** *306*, 134–141.

Dimkpa, C. O.; McLean, J. E.; Britt, D. W.; Anderson, A. J. Antifungal Activity of ZnO Nanoparticles and Their Interactive Effect With a Biocontrol Bacterium on Growth Antagonism of the Plant Pathogen *Fusarium Graminearum*. *Biometals* **2013,** *26*(6), 913–924.

Du, L.; Miao, X.; Jiang, Y.; Jia, H.; Tian, Q.; Shen, J.; Liu, Y. An Effective Strategy for the Synthesis of Biocompatible Gold Nanoparticles Using Danshensu Antioxidant: Prevention of Cytotoxicity via Attenuation of Free Radical Formation. *Nanotoxicology* **2012,** *7*(3), 294–300.

Elabasy, A.; Shoaib, A.; Waqas, M.; Jiang, M.; Shi, Z. Synthesis, Characterization, and Pesticidal Activity of Emamectin Benzoate Nanoformulations Against *Phenacoccus Solenopsis* Tinsley (Hemiptera: Pseudococcidae). *Molecules* **2019,** *24*(15), 2801.

Elamawi, R. M.; Al-Harbi, R. E.; Hendi, A. A. Biosynthesis and Characterization of Silver Nanoparticles Using *Trichoderma Longibrachiatum* and Their Effect on Phytopathogenic Fungi. *Egypt. J. Biol. Pest Control* **2018,** *28*(1), 1–11.

Elango, G.; Roopan, S. M.; Dhamodaran, K. I.; Elumalai, K.; Al-Dhabi, N. A.; Arasu, M. V. Spectroscopic Investigation of Biosynthesized Nickel Nanoparticles and Its Larvicidal, Pesticidal Activities. *J. Photochem. Photobiol. B Biol.* **2016,** *162*, 162–167.

El-Argawy, E.; Rahhal, M. M. H.; El-Korany, A.; Elshabrawy, E. M.; Eltahan, R. M. Efficacy of Some Nanoparticles to Control Damping-off and Root Rot of Sugar Beet in El-Behiera Governorate. *Asian J. Plant Pathol.* **2017,** *11*, 35–47.

Elbeshehy, E. K.; Elazzazy, A. M.; Aggelis, G. Silver Nanoparticles Synthesis Mediated by New Isolates of *Bacillus* Spp., Nanoparticle Characterization and Their Activity Against *Bean Yellow Mosaic Virus* and Human Pathogens. *Front. Microbiol.* **2015,** *6*, 453.

El-Moslamy, S.; Kabeil, S.; Hafez, E. J. O. Bioprocess Development for *Chlorella Vulgaris* Cultivation and Biosynthesis of Anti-Phytopathogens Silver Nanoparticles. *J. Nanomater. Mol. Nanotechnol.* **2016,** *5*(1) ..

El-Saadony, M. T.; El-Hack, A.; Mohamed, E.; Taha, A. E.; Fouda, M. M.; Ajarem, J. S.; N Maodaa, S.; Allam, A. A.; Elshaer, N. Ecofriendly Synthesis and Insecticidal Application of Copper Nanoparticles against the Storage Pest *Tribolium Castaneum*. *Nanomaterials* **2020,** *10*(3), 587.

Elsharkawy, M. M.; Mousa, K. M. Induction of Systemic Resistance against *Papaya Ring Spot Virus* (PRSV) and Its Vector *Myzus Persicae* by *Penicillium Simplicissimum* GP17-2 and Silica (SiO2) Nanopowder. *Int. J. Pest Manag.* **2015,** *61*(4), 353–358.

El-Shazly, M. A.; Attia, Y. A.; Kabil, F. F.; Anis, E.; Hazman, M. Inhibitory Effects of Salicylic Acid and Silver Nanoparticles on Potato Virus Y-Infected Potato Plants in Egypt. *Middle East J. Agric. Res.* **2017,** *6*(3), 835–848.

Fernández, J. G.; Fernández-Baldo, M. A.; Berni, E.; Camí, G.; Durán, N.; Raba, J.; Sanz, M. I. Production of Silver Nanoparticles Using Yeasts and Evaluation of Their Antifungal Activity against Phytopathogenic Fungi. *Process Biochem.* **2016,** *51*(9), 1306–1313.

Gahukar, R. T.; Das, R. K. Plant-Derived Nanopesticides for Agricultural Pest Control: Challenges and Prospects. *Nanotechnol. Environ. Eng.* **2020,** *5*(1), 1–9.

Gao, S.; Zhao, Y.; Gou, P.; Chen, N.; Xie, Y. Preparation of CuAlO2 Nanocrystalline Transparent Thin Films With High Conductivity. *Nanotechnology* **2003,** *14*(5), 538.

Graham, J. H.; Johnson, E. G.; Myers, M. E.; Young, M.; Rajasekaran, P.; Das, S.; Santra, S. Potential of Nano-Formulated Zinc Oxide for Control of Citrus Canker on Grapefruit Trees. *Plant Dis.* **2016,** *100*(12), 2442–2447.

Guilger-Casagrande, M.; Lima, R. de. Synthesis of Silver Nanoparticles Mediated by Fungi: A Review. *Front. Bioeng. Biotechnol.* **2019,** *7*, 287.

Gunti, L.; Dass, R. S.; Kalagatur, N. K. Phytofabrication of Selenium Nanoparticles From Emblica Officinalis Fruit Extract and Exploring Its Biopotential Applications: Antioxidant, Antimicrobial, and Biocompatibility. *Front. Microbiol.* **2019,** *10*, 931.

Hamza, R. Z. M. M. Larvicidal, Antioxidant Activities and Perturbation of Transaminases Activities of Titanium Dioxide Nanoparticles Synthesized Using Moringa Oleifera Leaves Extract Against the Red Palm WEEVIL (Rhynchophorus Ferrugineus; Innovare Academic Sciences, 2015; pp. 49–54.

Hao, Y.; Yuan, W.; Ma, C.; White, J. C.; Zhang, Z.; Adeel, M.; Zhou, T.; Rui, Y.; Xing, B. Engineered Nanomaterials Suppress *Turnip Mosaic Virus* Infection in Tobacco (*Nicotiana Benthamiana*). *Environ. Sci. Nano* **2018,** *5*(7), 1685–1693.

Hashmi, S. S.; Abbasi, B. H.; Rahman, L.; Zaka, M.; Zahir, A. Phytosynthesis of Organo-Metallic Silver Nanoparticles and Their Anti-Phytopathogenic Potency Against Soil Borne *Fusarium* Spp. *Mater. Res. Express* **2019,** *6*(11), 1150a9.

Hassan, S. E.-D.; Salem, S. S.; Fouda, A.; Awad, M. A.; El-Gamal, M. S.; Abdo, A. M. New Approach for Antimicrobial Activity and Bio-Control of Various Pathogens by Biosynthesized Copper Nanoparticles Using Endophytic Actinomycetes. *J. Radiat. Res. Appl. Sci.* **2018,** *11*(3), 262–270.

Hernández-Díaz, J. A.; Garza-García, J. J.; Zamudio-Ojeda, A.; León-Morales, J. M.; López-Velázquez, J. C.; García-Morales, S. Plant-Mediated Synthesis of Nanoparticles and

Their Antimicrobial Activity Against Phytopathogens. *J. Sci. Food Agric.* **2021,** *101*(4), 1270–1287.

Hossain, A.; Abdallah, Y.; Ali, M.; Masum, M.; Islam, M.; Li, B.; Sun, G.; Meng, Y.; Wang, Y.; An, Q. Lemon-Fruit-Based Green Synthesis of Zinc Oxide Nanoparticles and Titanium Dioxide Nanoparticles Against Soft Rot Bacterial Pathogen *Dickeya Dadantii*. *Biomolecules* **2019,** *9*(12), 863.

Huang, H.; Yang, X. Synthesis of Chitosan-Stabilized Gold Nanoparticles in the Absence/ Presence of Tripolyphosphate. *Biomacromolecules* **2004,** *5*(6), 2340–2346.

Huang, T.; Holden, J. A.; Heath, D. E.; O'Brien-Simpson, N. M.; O'Connor, A. J. Engineering Highly Effective Antimicrobial Selenium Nanoparticles Through Control of Particle Size. *Nanoscale* **2019,** *11*(31), 14937–14951.

Huang, W.; Yan, M.; Duan, H.; Bi, Y.; Cheng, X.; Yu, H. Synergistic Antifungal Activity of Green Synthesized Silver Nanoparticles and Epoxiconazole against Setosphaeria Turcica. *J. Nanomater.* **2020,** *2020*.

Islam, M. T. Application of Nanomaterials in Plant Protection. Ph.D. Thesis, Bangabandhu Sheikh Mujibur Rahman Agricultural University, 2020.

Jain, D.; Kothari, S. L. Green Synthesis of Silver Nanoparticles and Their Application in Plant Virus Inhibition. *J. Mycol. Plant Pathol.* **2014,** *44*(1), 21–24.

Jamdagni, P.; Rana, J. S.; Khatri, P.; Nehra, K. Comparative Account of Antifungal Activity of Green and Chemically Synthesized Zinc Oxide Nanoparticles in Combination With Agricultural Fungicides. *Int. J. Nano Dimens.* **2018,** *9*(2), 198–208.

Jameel, M.; Shoeb, M.; Khan, M. T.; Ullah, R.; Mobin, M.; Farooqi, M. K.; Adnan, S. M. Enhanced Insecticidal Activity of Thiamethoxam by Zinc Oxide Nanoparticles: A Novel Nanotechnology Approach for Pest Control. *ACS Omega* **2020,** *5*(3), 1607–1615.

Javed, B.; Ikram, M.; Farooq, F.; Sultana, T.; Mashwani, Z.-R.; Raja, N. I. Biogenesis of Silver Nanoparticles to Treat Cancer, Diabetes, and Microbial Infections: A Mechanistic Overview. *Appl. Microbiol. Biotechnol.* **2021,** *105*(6), 2261–2275.

Kah, M.; Hofmann, T. Nanopesticide Research: Current Trends and Future Priorities. *Environ. Int.* **2014,** *63*, 224–235.

Kamaraj, C.; Gandhi, P. R.; Elango, G.; Karthi, S.; Chung, I.-M.; Rajakumar, G. Novel and Environmental Friendly Approach; Impact of Neem (*Azadirachta Indica*) Gum Nano Formulation (NGNF) on *Helicoverpa Armigera* (Hub.) and *Spodoptera Litura* (Fab.). *Int. J. Biol. Macromol.* **2018,** *107,* 59–69.

Kaur, P.; Thakur, R.; Choudhary, A. An in Vitro Study of the Antifungal Activity of Silver/ Chitosan Nanoformulations Against Important Seed Borne Pathogens. *Int. J. Sci. Technol. Res.* **2012,** *1*(6), 83–86.

Khan, M. R.; Ahamad, F.; Rizvi, T. F. Effect of Nanoparticles on Plant Pathogens. In *Advances in Phytonanotechnology*; Elsevier, 2019; pp 215–240.

Ki, H. Y.; Kim, J. H.; Kwon, S. C.; Jeong, S. H. A Study on Multifunctional Wool Textiles Treated With Nano-Sized Silver. *J. Mater. Sci.* **2007,** *42*(19), 8020–8024.

Kitherian, S. Nano and Bio-Nanoparticles for Insect Control. *Res. J. Nanosci. Nanotechnol.* **2017,** *7*(1), 1–9.

Krishnaraj, C.; Ramachandran, R.; Mohan, K.; Kalaichelvan, P. T. Optimization for Rapid Synthesis of Silver Nanoparticles and Its Effect on Phytopathogenic Fungi. *Spectrochim. Acta A Mol. Biomol. Spectrosc.* **2012,** *93*, 95–99.

Kumari, M.; Pandey, S.; Bhattacharya, A.; Mishra, A.; Nautiyal, C. S. Protective Role of Biosynthesized Silver Nanoparticles Against Early Blight Disease in Solanum Lycopersicum. *Plant Physiol. Biochem.* **2017,** *121*, 216–225.

Lakshmeesha, T. R.; Murali, M.; Ansari, M. A.; Udayashankar, A. C.; Alzohairy, M. A.; Almatroudi, A.; Alomary, M. N.; Asiri, S. M. M.; Ashwini, B. S.; Kalagatur, N. K. Biofabrication of Zinc Oxide Nanoparticles From *Melia Azedarach* and Its Potential in Controlling Soybean Seed-Borne Phytopathogenic Fungi. *Saudi J. Biol. Sci.* **2020,** *27*(8), 1923–1930.

Lara, H. H.; Ayala-Núnez, N. V.; Turrent, L. del C. I.; Padilla, C. R. Bactericidal Effect of Silver Nanoparticles Against Multidrug-Resistant Bacteria. *World J. Microbiol. Biotechnol.* **2010,** *26*(4), 615–621.

Lemire, J. A.; Harrison, J. J.; Turner, R. J. Antimicrobial Activity of Metals: Mechanisms, Molecular Targets and Applications. *Nature Rev. Microbiol.* **2013,** *11*(6), 371–384.

Lesnichaya, M.; Gazizova, A.; Perfileva, A.; Nozhkina, O.; Graskova, I.; Sukhov, B. Starch-Capped Sulphur Nanoparticles Synthesised From Bulk Powder Sulphur and Their Anti-Phytopathogenic Activity Against *Clavibacter Sepedonicus*. *IET Nanobiotechnol.* **2021,** *15*(7), 585–593.

Lydia, N. J.; Atchayadana, U.; Mubina, J. Marine Endophytic Fungi Isolated From Gulf of Mannar—A Source for New Generation of Pharmaceutical Drugs and Biosynthesis of Silver Nanoparticles and Its Antibacterial Efficacy. In *Fungi Bio-Prospects in Sustainable Agriculture, Environment and Nano-technology*; Elsevier, 2021; pp 175–185.

Mahdizadeh, V.; Safaie, N.; Khelghatibana, F. Evaluation of Antifungal Activity of Silver Nanoparticles against Some Phytopathogenic Fungi and *Trichoderma Harzianum*. *J. Crop Protect.* **2015,** *4*(3), 291–300.

Majumdar, T. D.; Singh, M.; Thapa, M.; Dutta, M.; Mukherjee, A.; Ghosh, C. K. Size-Dependent Antibacterial Activity of Copper Nanoparticles Against *Xanthomonas Oryzae* Pv. *Oryzae*–A Synthetic and Mechanistic Approach. *Colloid Interface Sci. Commun.* **2019,** *32*, 100190.

Makarovsky, D.; Fadeev, L.; Salam, B. B.; Zelinger, E.; Matan, O.; Inbar, J.; Jurkevitch, E.; Gozin, M.; Burdman, S. Silver Nanoparticles Complexed With Bovine Submaxillary Mucin Possess Strong Antibacterial Activity and Protect Against Seedling Infection. *Appl. Environ. Microbiol.* **2018,** *84*(4), e02212–e02217.

Mala, R.; Arunachalam, P.; Sivasankari, M. Synergistic Bactericidal Activity of Silver Nanoparticles and Ciprofloxacin Against Phytopathogens. *J. Cell Tissue Res.* **2012,** *12*(2), 3249.

Malaikozhundan, B.; Vinodhini, J. Nanopesticidal Effects of Pongamia Pinnata Leaf Extract Coated Zinc Oxide Nanoparticle Against the Pulse Beetle, *Callosobruchus Maculatus*. *Mater. Today Commun.* **2018,** *14*, 106–115.

Malaj, E.; Peter, C.; Grote, M.; Kühne, R.; Mondy, C. P.; Usseglio-Polatera, P.; Brack, W.; Schäfer, R. B. Organic Chemicals Jeopardize the Health of Freshwater Ecosystems on the Continental Scale. *Proc. Natl. Acad. Sci.* **2014,** *111*(26), 9549–9554.

Malandrakis, A. A.; Kavroulakis, N.; Chrysikopoulos, C. V. Use of Copper, Silver and Zinc Nanoparticles Against Foliar and Soil-Borne Plant Pathogens. *Sci. Total Environ.* **2019,** *670*, 292–299.

Mehrbod, P.; Motamed, N.; Tabatabaeian, M.; Soleymani, E. R.; Amini, E.; Shahidi, M.; Kheyri, M. T. In Vitro Antiviral Effect of "Nanosilver" on Influenza Virus. *DARU J. Pharm. Sci.* **2009**, *17*(2),88–93.

Min, J.-S.; Kim, K.-S.; Kim, S.-W.; Jung, J.-H.; Lamsal, K.; Kim, S.-B.; Jung, M.-Y.; Lee, Y.-S. Effects of Colloidal Silver Nanoparticles on Sclerotium-Forming Phytopathogenic Fungi. *Plant Pathol. J.* **2009,** *25*(4), 376–380.

Misra, A. N.; Misra, M.; Singh, R. Nanotechnology in Agriculture and Food Industry. *Int. J. Pure Appl. Sci. Technol.* **2013,** *16*(2), 1.

Mishra, A.; Saini, R. K.; Bajpai, A. K. Polymer Formulations for Pesticide Release. In *Controlled Release of Pesticides for Sustainable Agriculture*; Springer, 2020; pp 185–206.

Mishra, S.; Singh, H. B. Biosynthesized Silver Nanoparticles as a Nano-weapon against Phytopathogens: Exploring Their Scope and Potential in Agriculture. *Appl. Microbiol. Biotechnol.* **2015,** *99*(3), 1097–1107.

Mishra, S.; Singh, B. R.; Naqvi, A. H.; Singh, H. B. Potential of Biosynthesized Silver Nanoparticles Using *Stenotrophomonas* Sp. BHU-S7 (MTCC 5978) for Management of Soil-Borne and Foliar Phytopathogens. *Sci. Rep.* **2017,** *7*(1), 1–15.

Mishra, S.; Singh, B. R.; Singh, A.; Keswani, C.; Naqvi, A. H.; Singh, H. B. Biofabricated Silver Nanoparticles Act as a Strong Fungicide against *Bipolaris Sorokiniana* Causing Spot Blotch Disease in Wheat. *Plos One* **2014,** *9*(5), e97881.

Mohamed, M. A.; Abd–Elsalam, K. A. Nanoantimicrobials for Plant Pathogens Control: Potential Applications and Mechanistic Aspects. In *Nanobiotechnology Applications in Plant Protection*; Springer, 2018; pp 87–109.

Namasivayam, S. K. R.; Bharani, R. A. Biocompatible Silver Nanoparticles-Loaded Fungal Metabolites Nanoconjugate (AgNp–FM) Preparation for the Noteworthy Pesticidal Activity. *National Academy Science Letters* 2021; pp 1–7.

Narendrakumar, G.; Karthick Raja Namasivayam, S. Surface-Modified Nanosilica–Chitinase (SiNp-Chs)-Doped Nano Enzyme Conjugate and Its Synergistic Pesticidal Activity with Plant Extracts against Armyworm *Spodoptera Litura (Fab.)(Lepidoptera: Noctuidae)*. *IET Nanobiotechnol.* **2021,** *15*(1), 117–134.

Noha, K.; Bondok, A. M.; El-Dougdoug, K. A. Evaluation of Silver Nanoparticles as Antiviral Agent against ToMV and PVY in Tomato Plants. *Sciences* **2018,** *8*(01), 100–111.

Nuruzzaman, M. D.; Rahman, M. M.; Liu, Y.; Naidu, R. Nanoencapsulation, Nano-Guard for Pesticides: A New Window for Safe Application. *J. Agric. Food Chem.* **2016,** *64*(7), 1447–1483.

Ocsoy, I.; Paret, M. L.; Ocsoy, M. A.; Kunwar, S.; Chen, T.; You, M.; Tan, W. Nanotechnology in Plant Disease Management: DNA-Directed Silver Nanoparticles on Graphene Oxide as an Antibacterial against *Xanthomonas Perforans*. *Acs Nano* **2013,** *7*(10), 8972–8980.

Oh, J.-W.; Chun, S. C.; Chandrasekaran, M. Preparation and in Vitro Characterization of Chitosan Nanoparticles and Their Broad-Spectrum Antifungal Action Compared to Antibacterial Activities Against Phytopathogens of Tomato. *Agronomy* **2019,** *9*(1), 21.

Pariona, N.; Paraguay-Delgado, F.; Basurto-Cereceda, S.; Morales-Mendoza, J. E.; Hermida-Montero, L. A.; Mtz-Enriquez, A. I. Shape-Dependent Antifungal Activity of ZnO Particles against Phytopathogenic Fungi. *Appl. Nanosci.* **2020,** *10*(2), 435–443.

Park, H.-J.; Kim, S.-H.; Kim, H.-J.; Choi, S.-H. A New Composition of Nanosized Silica-Silver for Control of Various Plant Diseases. *Plant Pathol. J.* **2006,** *22*(3), 295–302.

Pascual-Villalobos, M. J.; Cantó-Tejero, M.; Vallejo, R.; Guirao, P.; Rodríguez-Rojo, S.; Cocero, M. J. Use of Nanoemulsions of Plant Essential Oils as Aphid Repellents. *Ind. Crops Products* **2017,** *110*, 45–57.

Patel, N.; Desai, P.; Patel, N.; Jha, A.; Gautam, H. K. Agronanotechnology for Plant Fungal Disease Management: A Review. *Int. J. Curr. Microbiol. App. Sci.* **2014,** *3*(10), 71–84.

Pho, Q. H.; Losic, D.; Ostrikov, K. K.; Tran, N. N.; Hessel, V. Perspectives on Plasma-Assisted Synthesis of N-Doped Nanoparticles as Nanopesticides for Pest Control in Crops. *React. Chem. Eng.* **2020**, *5*(8), 1374–1396.

Prasad, R.; Kumar, V.; Prasad, K. S. Nanotechnology in Sustainable Agriculture: Present Concerns and Future Aspects. *Afr. J Biotechnol.* **2014,** *13*(6), 705–713.

Rai, A.; Prabhune, A.; Perry, C. C. Antibiotic Mediated Synthesis of Gold Nanoparticles with Potent Antimicrobial Activity and Their Application in Antimicrobial Coatings. *J. Mater. Chem.* **2010,** *20*(32), 6789–6798.

Rajput, V., Singh, A., Minkina, T., Shende, S., Kumar, P., Verma, K., Bauer, T., Gorobtsova, O., Deneva, S. and Sindireva, A. Potential Applications of Nanobiotechnology in Plant Nutrition and Protection for Sustainable Agriculture. *Nanotechnology in Plant Growth Promotion and Protection*, 2021; pp 79–92

Rajwade, J. M.; Chikte, R. G.; Paknikar, K. M. Nanomaterials: New Weapons in a Crusade against Phytopathogens. *Appl. Microbiol. Biotechnol.* **2020,** *104*(4), 1437–1461.

Ramkumar, G.; Asokan, R.; Ramya, S.; Gayathri, G. Characterization of Trigonella Foenum-Graecum Derived Iron Nanoparticles and Its Potential Pesticidal Activity Against Tuta Absoluta (Lepidoptera). *J. Cluster Sci.* **2020,** 1–6.

Sabbour, M. M.; Abd El-Aziz, S. E.; Shadia, E. Efficacy of Nano-Diatomaceous Earth against Red Flour Beetle, *Tribolium Castaneum* and Confused Flour Beetle, *Tribolium Confusum (Coleoptera: Tenebrionidae)* Under Laboratory and Storage Conditions. *Bull. Environ. Pharmacol. Life Sci.* **2015,** *4*(7), 54–59.

Sadiq, I. M.; Chowdhury, B.; Chandrasekaran, N.; Mukherjee, A. Antimicrobial Sensitivity of *Escherichia Coli* to Alumina Nanoparticles. *Nanomed. Nanotechnol. Biol. Med.* **2009,** *5*(3), 282–286.

Sahab, A. F.; Waly, A. I.; Sabbour, M. M.; Nawar, L. S. Synthesis, Antifungal and Insecticidal Potential of Chitosan (CS)-g-Poly (Acrylic Acid)(PAA) Nanoparticles Against Some Seed Borne Fungi and Insects of Soybean. *Int. J. Chem. Tech. Res.* **2015,** *8*(2), 589–598.

Saharan, V.; Mehrotra, A.; Khatik, R.; Rawal, P.; Sharma, S. S.; Pal, A. Synthesis of Chitosan Based Nanoparticles and Their in Vitro Evaluation against Phytopathogenic Fungi. *Int. J. Biol. Macromol.* **2013,** *62*, 677–683.

Sahayaraj, K.; Madasamy, M.; Radhika, S. A. Insecticidal Activity of Bio-Silver and Gold Nanoparticles Against *Pericallia Ricini* Fab.(Lepidaptera: Archidae). *J. Biopestic.* **2016,** *9*(1), 63.

Samih, M. A.; Rouhani, M.; Aslani, A.; Beiki, K. In *Insecticidal Properties of Amitraz, Nano-Amitraz, Nano-ZnO and Nano-ZnO-Al2O3 Nanoparticles on Agonoscena Pistaciae (Hem.: Aphelaridae)*, Proceedings Symposium: Third International Symposium on Insect Physiology, Biochemistry and Molecular Biology; 2011; pp 2–5.

Sankar, M. V.; Abideen, S. Pesticidal Effect of Green Synthesized Silver and Lead Nanoparticles Using *Avicennia Marina* Against Grain Storage Pest *Sitophilus Oryzae*. *Int. J. Nanomater. Biostruct.* **2015,** *5*(3), 32–39.

Saratale, R. G.; Benelli, G.; Kumar, G.; Kim, D. S.; Saratale, G. D. Bio-Fabrication of Silver Nanoparticles Using the Leaf Extract of an Ancient Herbal Medicine, Dandelion (*Taraxacum Officinale*), Evaluation of Their Antioxidant, Anticancer Potential, and Antimicrobial Activity against Phytopathogens. *Environ. Sci. Pollut. Res.* **2018,** *25*(11), 10392–10406.

Sasson, Y.; Levy-Ruso, G.; Toledano, O.; Ishaaya, I. Nanosuspensions: Emerging Novel Agrochemical Formulations. In *Insecticides design using advanced technologies*; Springer, 2007; pp 1–39.

Sathiyabama, M.; Muthukumar, S. Chitosan Guar Nanoparticle Preparation and Its in Vitro Antimicrobial Activity Towards Phytopathogens of Rice. *Int. J. Biol. Macromol.* **2020,** *153*, 297–304.

Sathiyabama, M.; Parthasarathy, R. Biological Preparation of Chitosan Nanoparticles and Its in Vitro Antifungal Efficacy Against Some Phytopathogenic Fungi. *Carbohydr. Polym.* **2016,** *151*, 321–325.

Schnoor, B.; Elhendawy, A.; Joseph, S.; Putman, M.; Chacón-Cerdas, R.; Flores-Mora, D.; Bravo-Moraga, F.; Gonzalez-Nilo, F.; Salvador-Morales, C. Engineering Atrazine Loaded Poly (Lactic-Co-Glycolic Acid) Nanoparticles to Ameliorate Environmental Challenges. *J. Agric. Food Chem.* **2018,** *66*(30), 7889–7898.

Servin, A.; Elmer, W.; Mukherjee, A.; De la Torre-Roche, R.; Hamdi, H.; White, J. C.; Bindraban, P.; Dimkpa, C. A Review of the Use of Engineered Nanomaterials to Suppress Plant Disease and Enhance Crop Yield. *J. Nanoparticle Res.***2015,** *17*(2), 1–21.

Shanmugam, C.; Gunasekaran, D.; Duraisamy, N.; Nagappan, R.; Krishnan, K. Bioactive Bile Salt-Capped Silver Nanoparticles Activity Against Destructive Plant Pathogenic Fungi Through in Vitro System. *RSC Adv.* **2015,** *5*(87), 71174–71182.

Siddaiah, C. N.; Prasanth, K. V. H.; Satyanarayana, N. R.; Mudili, V.; Gupta, V. K.; Kalagatur, N. K.; Satyavati, T.; Dai, X.-F.; Chen, J.-Y.; Mocan, A. Chitosan Nanoparticles Having Higher Degree of Acetylation Induce Resistance against Pearl Millet Downy Mildew through Nitric Oxide Generation. *Sci. Rep.* **2018,** *8*(1), 1–14.

Sidhu, A.; Bala, A.; Singh, H.; Ahuja, R.; Kumar, A. Development of MgO-Sepoilite Nanocomposites Against Phytopathogenic Fungi of Rice *(Oryzae Sativa*): A Green Approach. *ACS Omega* **2020,** *5*(23), 13557–13565.

Singh, A.; Rajput, V. D., Mehrotra, R.; Pal, N.; Singh, V. K.; Minkina, T.; Chokheli, V. A.; Singh, R. K. In *Sustainable Soil Fertility Management*; Nova Science Publishers Inc., 2020a; vol 1, pp 73–100.

Singh, A., Rajput, V., Rawat, S., Kumar Singh, A., Bind, A., Kumar Singh, A., Chernikova, N.; Voloshina, M., Lobzenko, I. Monitoring Soil Salinity and Recent Advances in Mechanism of Salinity Tolerance in Plants. *Biogeosyst. Tech.* **2020b,**7(2). https://doi.org/10.13187/bgt.2020.2.66

Singh, A., Rajput, V., Singh, A., Sengar, R., Singh, R., Minkina, T. Transformation Techniques and Their Role in Crop Improvements: A Global Scenario of GM Crops. *Policy Issues Genet. Modified Crops* **2021,** *1,* 515–542.

Siva, C.; Kumar, M. S. Pesticidal Activity of Eco-Friendly Synthesized Silver Nanoparticles Using *Aristolochia Indica* Extract against *Helicoverpa Armigera Hubner (Lepidoptera: Noctuidae)*. *Int. J. Adv. Sci. Tech. Res.* **2015,** *2*, 197–226.

Sivapriya, V.; Azeez, A. N.; Deepa, S. V. Phyto Synthesis of Iron Oxide Nano Particles Using the Agro Waste of *Anthocephalus Cadamba* for Pesticidal Activity against *Sitophilus Granaries*. *J. Entomol. Zool. Stud.* **2018,** *6*, 1050–1057.

Song, M.-R.; Cui, S.-M.; Gao, F.; Liu, Y.-R.; Fan, C.-L.; Lei, T.-Q.; Liu, D.-C. Dispersible Silica Nanoparticles as Carrier for Enhanced Bioactivity of Chlorfenapyr. *J. Pest. Sci.* **2012,** *37*(3), 258–260.

Stephenson, G. R. Pesticide Use and World Food Production: Risks and Benefits, Environmental Fate and Effects of Pesticides. Am. Chem. Soc. **2003,** *853,* 261–270.

Suriyaprabha, R.; Karunakaran, G.; Kavitha, K.; Yuvakkumar, R.; Rajendran, V.; Kannan, N. Application of Silica Nanoparticles in Maize to Enhance Fungal Resistance. *IET Nanobiotechnol.* **2014,** *8*(3), 133–137.

Vanti, G. L.; Nargund, V. B.; Vanarchi, R.; Kurjogi, M.; Mulla, S. I.; Tubaki, S.; Patil, R. R. Synthesis of Gossypium Hirsutum-Derived Silver Nanoparticles and Their Antibacterial Efficacy against Plant Pathogens. *Appl. Organomet. Chem.* **2019,** *33*(1), e4630.

Vargas-Hernandez, M.; Macias-Bobadilla, I.; Guevara-Gonzalez, R. G.; Rico-Garcia, E.; Ocampo-Velazquez, R. V.; Avila-Juarez, L.; Torres-Pacheco, I. Nanoparticles as Potential Antivirals in Agriculture. *Agriculture* **2020,** *10*(10), 444.

Win, T. T.; Khan, S.; Fu, P. Fungus-(*Alternaria* Sp.) Mediated Silver Nanoparticles Synthesis, Characterization, and Screening of Antifungal Activity Against Some Phytopathogens. *J. Nanotechnol.* **2020,** *2020*.

Worrall, E. A.; Hamid, A.; Mody, K. T.; Mitter, N.; Pappu, H. R. Nanotechnology for Plant Disease Management. *Agronomy* **2018,** *8*(12), 285.

Wuithschick, M.; Birnbaum, A.; Witte, S.; Sztucki, M.; Vainio, U.; Pinna, N.; Rademann, K.; Emmerling, F.; Kraehnert, R.; Polte, J. Turkevich in New Robes: Key Questions Answered for the Most Common Gold Nanoparticle Synthesis. *ACS Nano* **2015,** *9*(7), 7052–7071.

Xing, K.; Xing, Y.; Liu, Y.; Zhang, Y.; Shen, X.; Li, X.; Miao, X.; Feng, Z.; Peng, X.; Qin, S. Fungicidal Effect of Chitosan via Inducing Membrane Disturbance against *Ceratocystis Fimbriata. Carbohydr. Polym.* **2018,** *192*, 95–103.

Yang, F.-L.; Li, X.-G.; Zhu, F.; Lei, C.-L. Structural Characterization of Nanoparticles Loaded With Garlic Essential Oil and Their Insecticidal Activity Against *Tribolium Castaneum (Herbst)(Coleoptera: Tenebrionidae). J. Agric. Food Chem.* **2009,** *57*(21), 10156–10162.

Yi, G.; Li, X.; Yuan, Y.; Zhang, Y. Redox Active Zn/ZnO Duo Generating Superoxide (˙ O 2-) and H 2 O 2 Under All Conditions for Environmental Sanitation. *Environ. Sci. Nano* **2019,** *6*(1), 68–74.

Zahir, A. A.; Bagavan, A.; Kamaraj, C.; Elango, G.; Rahuman, A. A. Efficacy of Plant-Mediated Synthesized Silver Nanoparticles Against *Sitophilus Oryzae. J. Biopestic.* **2012,** *5*, 95.

Zhao, X.; Cui, H.; Wang, Y.; Sun, C.; Cui, B.; Zeng, Z. Development Strategies and Prospects of Nano-Based Smart Pesticide Formulation. *J. Agric. Food Chem.* **2017,** *66*(26), 6504–6512.

Zong, R.-L.; Zhou, J.; Li, B.; Fu, M.; Shi, S.-K.; Li, L.-T. Optical Properties of Transparent Copper Nanorod and Nanowire Arrays Embedded in Anodic Alumina Oxide. *J. Chem. Phys.* **2005,** *123*(9), 094710.

CHAPTER 7

Role of Nanomaterials in Plants Under Abiotic Stress

FATHY ELBEHIRY[1], HASSAN EL-RAMADY[2], and HEBA ELBASIOUNY[3]

[1]*Basic and Applied Sciences Department, Higher Institute for Agricultural Co-Operation, Shubra El-Kheima, Egypt*

[2]*Soil and Water Department, Faculty of Agriculture, Kafrelsheikh University, Kafr El-Sheikh, Egypt*

[3]*Department of Environmental and Biological Sciences, Home Economics Faculty, Al-Azhar University, Tanta, Egypt*

ABSTRACT

Plant production faces many challenges including biotic and abiotic stresses. Consequently, the accumulation of reactive oxygen species (ROS) is activated and resulted in oxidative stress and may cause death in plants; however, in some cases, the plant has a strong defense system that may face these circumstances. Using nanomaterials is an attractive way for such circumstances; it can boost the defense system of plants, improve plant growth and development, and enhance the quality and the quantity of yields. Also, nanomaterials are applied as fertilizers to effectively supply nutrients (macro- and micronutrients) because of many of their advantages such as high surface area, ability to penetrate plants, higher reactivity with other chemicals, and others. Furthermore, nanomaterials are widely used to remediate toxic compounds in the environment. However, the use of nanomaterials in the agriculture sector needs to be precise because the high concentration of these materials may make them stressors for plants.

Nanotechnology for Sustainable Agriculture: An Innovative and Eco-Friendly Approach. Vishnu D. Rajput, Abhishek Singh, Tatiana M. Minkina, Krishan K. Verma, and Awani Kumar Singh (Eds.)

7.1 INTRODUCTION

Global agricultural production faces several challenges including the incredibly increasing world population, changes in climate, and other biotic/abiotic stresses (Sreelakshmi et al. 2021). These stresses cause losses in crop productivity especially drought, salinity, waterlogging, and heat stress as well as phytopathogens (Rhaman et al., 2021). Hence, cultivated plants under stress have a great defense system, which supports plants against these stresses through endogenous materials like nonenzymatic antioxidants (e.g., ascorbic acid, glutathione, proline) or enzymatic antioxidants (e.g., catalase, peroxidases, and superoxide dismutase). Several exogenous materials have been applied to mitigate these abiotic stresses like melatonin (Tiwari et al., 2020; Menhas et al., 2021), nitric oxide (Zhou et al., 2021), 5-aminolevulinic acid as plant growth regulators (Rhaman et al., 2021), and nanofertilizers (NFs) (Sharma et al., 2021). Nanofertilizers or nano-encapsulated nutrients are considered effective slow-release nutrients for plant demand that further regulate plant growth (Madzokere et al., 2021). It could be encapsulated the conventional fertilizers using nanoparticles for the slow and sustainable release of nutrients over an extended time (Madzokere et al., 2021; Sharma et al., 2021).

Recently, many engineered NFs have been applied for supporting crops under abiotic stresses such as nano copper (González-García et al., 2021; Nguyen et al., 2021; Noman et al., 2020, 2021; Rai et al., 2021; Xu et al., 2021), nano iron (Mohammadi et al., 2020; Ahmed et al., 2021; Bidi et al., 2021; Manzoor et al., 2021; Sreelakshmi et al., 2021), nano selenium (Zahedi et al., 2020; Shalaby et al., 2021; Wang et al., 2020, 2021), and nano zinc (Hassanpour-aghdam et al., 2020; Faizan et al., 2021a, 2021b; Priyanka et al., 2021; Semida et al., 2021; Shah et al., 2021; Yasmin et al., 2021). The NFs can supply cultivated plants with nutrients, improving nutrient efficiency and enhancing plant growth under stress (Fatima et al., 2021). On the other hand, several nanomaterials (NMs) have many applications under stressful conditions like nano-phytoremediation (Romeh and Saber, 2020), although sometimes the over-application of NMs may cause a serious problem as nano-pollutants, which could be found in water (Gui et al., 2021) and soil (Naghdi et al., 2017). Recently, these nano-pollutants have an increasing global concern because of their possible harmful impacts on human health and the entire environment.

Therefore, this chapter will focus on different NMs on cultivated plants under different abiotic stresses. The role of nanofertilizers for stressful plants, nano-phytoremediation, and nano-pollutants as stressors on cultivated plants will be also discussed.

7.2 NANOMATERIALS FOR CULTIVATED PLANTS UNDER ABIOTIC STRESS

The relationship between plants and NMs is long-standing, and plants are regularly exposed to natural NMs. Plants have powerful defensive systems to combat diverse stressors since they have survived and developed genetically and phenotypically under severe environments throughout their lives. Synthetic NMs have recently put these mechanisms to the challenge (Dev et al., 2018), although advanced NT offers appealing advantages such as improved efficiency of chemicals through enhanced nanostructure that decreases their environmental loads in the chemical priming applications. Despite related lack of knowledge about the interactions between NPs and plants, their applications could be a promising approach in agriculture (Gohari et al., 2021). Abiotic stresses can be counted as the most severe negative circumstances that plants face. The severity of any of them can be increased in plants as they are sessile organisms and hence cannot move out avoiding the disturbance of whole plant metabolism and performance. As a result of confronting abiotic stresses, too much accumulation of ROS is activated which leads to a circumstance of oxidative stress and ultimately causes cell death. Normally, chloroplasts, mitochondrial, and peroxisomes processes create ROS; however, environmental conditions that boost ROS accumulation cause oxidative stress conditions. The uncontrolled oxidation encouraged or motivated by any environmental stress can be overcome by cellular defense mechanisms, which include variable antioxidant enzymes and plant secondary metabolites such as flavonoids, phenolics, and proline (Souza et al., 2017; Hussain et al., 2021).

The applications of NPs include mitigation of abiotic stresses among other applications. In different crop species, the application of NMs reduces reactive oxygen species (ROS), malondialdehyde content, and chlorophyll degradation accelerating the activities of antioxidative enzymes, photosynthetic parameters, Rubisco expression, and genes of chlorophyll-binding protein. Moreover, NPs can protect the photosynthesis process, a cellular process susceptible to abiotic stresses, by reducing osmotic and oxidative stresses. Generally, NMs have shown enhanced crop growth in adverse conditions. In contrast, the transformation and fate of NMs in agro-ecosystems also raise significant issues since these NMs in certain conditions may generate ROS and N and demonstrate toxic effects (Singh and Husen, 2020; Gohari et al., 2021). For example, in comparison to control and other treatments, foliar application of nano-Se on cucumber under combined stressors (salinity and heat) enhanced growth metrics, growth yield, and yield quality (H_2O_2 and Si) (Shalaby et al., 2021). The nano-Fe oxide is also stated to assist plants to cope with salt. Photosystem II (PSII) performance and photosynthetic rate were enhanced in seeds primed with sorghum iron oxide NP (Kukde et al., 2019). It has also been found that plant sensitivity to NPs correlated with an increase in lettuce P abundance and absorption, as well as enhanced plant development. Stress is thought to have played a role in enhanced root exudation in lettuce following exposure to TiO_2 NPs in the soil. As the TiO_2 NP concentration in the soil elevated, shoot and root length, and dry weight enhanced in comparison to regulation (Montes et al., 2017). The usage of Ti NPs is becoming increasingly popular. It is described that seeds treated with Ti NPs suspensions were found to have higher germination rates, longer root lengths, and better seedling development in a variety of plant species. Furthermore, Ti NPs improved plant resistance to abiotic and biotic stresses such as cold, drought, Cd toxicity, and bacterial spot disease by *Xanthomonas perforans* (Gómez-Merino and Trejo-Téllez, 2018).

7.3 NANOFERTILIZERS FOR STRESSFUL PLANTS

After drought and salt, plant nutrient deficiency is a major limiting factor for plant growth and yield. Plants require nutrients for growth and development, and their roots absorb the majority of mineral nutrients from the soil. Fertilizers, on the other hand, are given to the soil to compensate for nutritional deficiencies, allowing plants to develop normally. For their growth and development, plants require 17 elements, 14 of which are essential nutrients. Six of these key nutrients, Ca, K, Mg, N, P, and S, are macronutrients that must be absorbed in substantial amounts. Cl, Cu, Mn, Fe, Zn, Co, Mo, and Ni are micronutrients that are needed in tiny amounts. The nutrient shortage shown in commercially accessible economic crops has an impact on human health, particularly among rural residents, but NT's long-term strategy is alleviating these issues (Amjad and Serajuddin, 2021). In addition, high demands for agricultural products put enormous pressure on the new agricultural strategies including the use of high-yielding chemically manufactured fertilizer, farmyard manure, bio-manure, etc., because it has been shown that fertilizer use boosts the production and its products in the proper proportions. Thus, chemical fertilizers have become an important element in meeting the growing demands. Because of the ever-increasing population with plentiful food supply, agricultural resources need additional development. The expensive cost of traditional fertilizers is a challenge for all producers at the net root level since they are utilized in large amounts (Fatima et al., 2021). Recently, NT-based products are developed for the utilization in agriculture. Agri-NT products include many materials such as NFs, and nano-pesticides which are modernized agriculture and associated fields. They are created in parallel with the progress of the emerging agri-NT. The novel NFs can be made by many NPs for increasing crop production. For example, ZnO NPs can stimulate the lateral roots that can modify the root architecture and enhance the whole uptake of nutrients in plants (Thirugnanasambandan, 2021).

At the nanoscale, the matter has changed characteristics that are unique and distinct from those found at the macroscopic level. These nano-formulations might be used in the agricultural and food sectors because of the unique features of nano-agrochemicals, such as high surface-to-volume ratio, high reactivity, effectiveness, and efficiency (Pirzadah et al., 2020). Thus, recently, NT has been widely utilized for producing NFs and for agricultural purposes. Thus, NFs are modified or synthesized materials that are applied to improve the soil quality for enhanced crop quality and quantity of different crops. Nanofertilizers are a customized version of traditional fertilizers that are created from bulk materials or vegetative or reproductive portions of plants utilizing nanotechnology by a variety of physical, chemical, and biological techniques. Nanoparticles have entirely different molecular characteristics than bulk materials due to their tiny size and high surface-to-volume ratio. Nanofertilizing can be a significant tool for achieving sustainable agriculture by replacing traditional fertilizers (El-Gamry et al., 2018; Shukla et al., 2019). Nanofertilizers are small enough to penetrate the root, epidermis, and stomata of leaves, and also move rapidly through the xylem and phloem vessels of plants. The NFs research and uses to benefit crops have been described (Wang and Nguyen, 2018). The NPs affect plants at very low concentrations, and their effects vary depending on the type and dose; they have been demonstrated to be an appealing alternative for the production of NFs that are more efficient and effective than standard fertilizers (Khan and Upadhyaya). Surprisingly, most of the NMs examined on crops as "nanofertilizers" were either created or marketed by chemical companies for commercial purposes. Furthermore, the likelihood of applying NFs in large-scale agriculture systems remains a challenge. Many countries, on the other hand, are going forward with their plans to use NFs in their agriculture systems to improve crop productivity (Pirzadah et al., 2020).

Some of NF researches have focused on micronutrients such as Zn, Cu, Mn, and Fe; however, plants require large amounts of macronutrients (N, P, K, Ca, S, and Mg) to increase crop productivity, so macronutrients have received more attention than micronutrients (Vázquez-Núñez et al., 2018). Sharma et al. (2021) reported that C nanotubes, Cu, Mn, Mo, Zn, Fe, Si, their oxides, and NMs of commonly used agricultural inputs like urea, P, and S are among the NMs accessible. The ingredients described above can be administered in a variety of ways, including soil application, plant injection, in vitro, and foliar treatment. Nutrients applied foliarly have been shown to quickly correct nutrient deficits and improve crop production (Sharma et al., 2021).

Many authors applied NFs to improve crop quality and quantity in different ways. For example, Wang and Nguyen (2018) stated that ZnO and TiO NFs boost nutrient absorption and the content of chlorophyll, as well as enzymatic activity, resulting in better crop growth, yields, and quality. Furthermore, Zn, Ti, Mn, and Fe oxide NPs are not only micronutrients for plants, but they also limit the growth of plant disease fungi. Micronutrient NFs have been used to some crops such as barley, soybeans (*Glycin max L.*), spinach (*Spinacia oleracea*), cluster beans (*Cyamopsis tetragonoloba L.*), peanuts (*Arachis hypogae*), pearl millet (*Pennisetum americanum*), and chickpeas (*Cicer arietinum L.*) (Wang and Nguyen, 2018). Kandil et al. (2020) conducted a field experiment to study the effects of combined foliar application treatments (fulvic acid and NPK NPs) and boron (B) fertilizer on sugar beet plants for 120 and 150 days after planting. Fulvic acid + NPK NPs significantly increased the root length, diameter, and weight, and the root to shoot ratio of plants. The interaction between fulvic acid or NPK NPs and B significantly affected the parameters of plants' yield and quality. However, the conjunction of foliar applications of fulvic acid + NPK NPs with B resulted in the greatest mean values of these parameters in the environmental and soil conditions at Alexandria, Egypt. Singh et al. (2021) applied and evaluated synthesized ZnO NP as an alternative NF to conventional Zn sulfate in Zn-deficient soils varied with their pH from 7.2 to 8.7. They applied several doses ranging from 0 to 500 mg L^{-1} with different sizes (30–100 nm). It has shown that the maximum soil microbial biomass C and bacterial population were recorded at 100 mg L^{-1} of ZnONPs, while a sharp decline was recorded at higher concentrations. Furthermore, soil application of 5 mg kg^{-1} ZnO

NPs produced higher root elongation, shoot elongation, total chlorophyll, grain yield, and grain Zn-content compared to the conventional Zn sulfate. They concluded the application of spherically synthesized ZnO NPs with an average of 36.7 nm at 5 mg kg^{-1} in the soil and 100 mg L^{-1} as a foliar application to maintain soil microbial biomass C and bacterial population improves total chlorophyll, and grain Zn-contents, thus sustaining wheat production in Zn-deficient soils (neutral and alkaline).

Salem et al. (2021) reported some advantages of NFs that make them more effective and beneficial than traditional fertilizers:

1) Increased surface area which increases nutrient absorption and efficiency and improved NF reactivity with other chemicals.
2) Higher solubility: In a range of solvents, particularly water, the NF is highly soluble. This NF characteristic helps to solubilize and disperse soil insoluble nutrients increasing nutrient bioavailability.
3) Small particle size: NFs have a particle size of less than 100 nm, which enhances their capacity to enter plants from applied surfaces like soil or leaves, increasing plant nutrient uptake.
4) Fertilizer encapsulating at NPs: This causes increased nutrient availability and absorption in agricultural plants. Zeolite-based NFs enhance fertilizer availability in crops in the growth cycle, limit nutrient losses by denitrification, volatilization, and leaching, and fix nutrients in the soil.
5) Rapid penetration and regulated fertilizer release: Because of this, NFs contribute to enhancing plant nutrient availability, healthy seedling development, and reduced fertilizer toxicity. When compared to bulk zinc sulfate, nano-ZnO delivers a greater percentage of germination and root development in peanut seeds.
6) Nutrient absorption efficiency of NFs: This increases the proportion of soil nutrients absorbed by crops. Furthermore, NFs help to prevent fertilizer loss through leaching.
7) NFs release period: Bulk fertilizers with activated nutrient release periods are beneficial in the short term; however, NFs can lengthen the nutrient release period as in Figure 7.1.

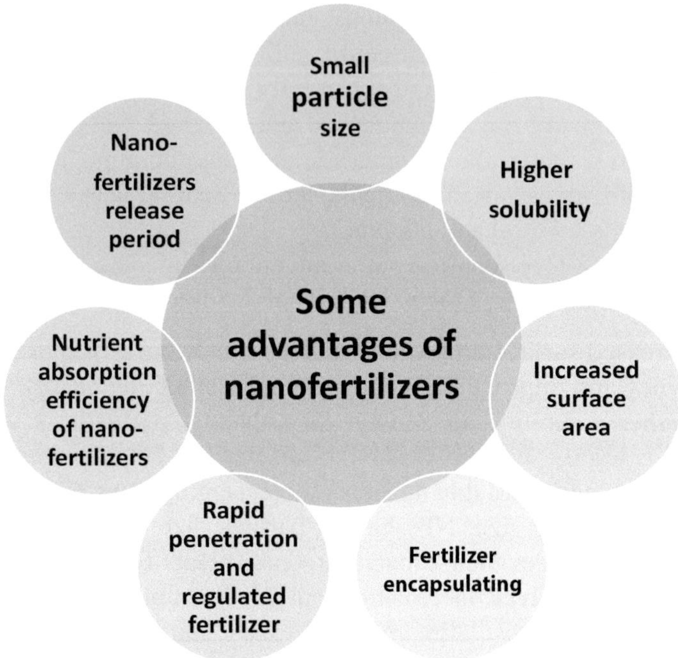

FIGURE 7.1 Some advantages of nanofertilizers.

Although more attention is being paid to the role of nanofertilizers in increasing nutrient usage efficiency and water purification, their positive effects on decreasing the negative impact of various abiotic stresses on plants, resulting in increased agricultural yields, have been reported also. As a result, determining the precise beneficial concentration of NPs is critical to minimize plant toxicity, which varies depending on the kind of NPs, plant stage during the application, and NPs concentration (Salem et al., 2021). Table 7.1 includes the effect of using some nanofertilizers under abiotic stresses on plants.

7.4 NANOMATERIALS AND REMEDIATION OF TOXIC CONTAMINATES

Environmental crises, fossil fuel combustion, urbanization, and industrialization have all contributed to soil contamination over time, creating toxic metals, which is a recurring concern. Since a result, toxic contaminants in soil remediation are constantly a hot subject, as polluted soil may have

TABLE 7.1 A Survey on Some Engineered Nanofertilizers on Some Crops Under Abiotic Stresses

Crop (scientific name)	Stress type and its details	Nanofertilizer and its concentration	Roles of nanofertilizer	Reference
Nano-copper				
Bell pepper (*Capsicum annuum* L.)	Salt stress (25 and 50 mM NaCl)	Chemical Cu-NPs at 100 and 500 mg l^{-1} (50 nm)	Increased the bioactive compounds in fruits like flavonoids, carotene, and carotenoids	González-García et al. (2021)
Maize (*Zea mays* L.)	Drought stress (drying soil for 7 days)	Chemical zero-valent copper (Cu^{o}-NPs; 30–40 nm; 69.4 μM)	Maintained water status in leaves; photosynthetic pigments, and improved detoxification of ROS	Nguyen et al. (2021)
Maize (*Zea mays* L.)	Soil salinity (0.5 mM NaCl = 4 dS m^{-1})	Biological Cu-NPs (28.55 nm; 25, 50, 100 mg kg^{-1})	Dose 100 mg kg^{-1} prevented oxidative damage and lipid peroxidation by activating the antioxidants and demoting ROS cellular levels	Noman et al. (2021)
Rice (*Oryza sativa* L.)	CuO-NPs stress (100 μM)	Chemical CuO-NPs (50 nm)	Applied Si and H_2S supported rice oxidative stress by CuO-NPs, via stimulated enzymes of ascorbate-glutathione cycle	Rai et al. (2021)
NO plants	Soil stress by different levels of pH (4.89–8.05), texture, DOC (8.39–107 mg l^{-1})	CuO-NPs (100 mg l^{-1}) at three forms: CuO-NPs, CuO-NTs, and CuO-MPs (50, 10–12 × 75–100 nm, 10 μm resp.)	Dissolution of CuO-NPs in soil totally controlled by soil DOC and pH; greater Cu^{2+} release by nano-CuO than CuO-NTs and CuO-MPs	Xu et al. (2021)
Wheat (*Triticum aestivum* L.)	Cadmium stress (soil mixed with 4.75 mg kg^{-1})	Biological Cu-NPs (17–38 nm; 25, 50, 100 mg kg^{-1})	Cu-NPs (100 mg kg^{-1}) decreased in wheat Cd translocation by 49.62 %	Noman et al. (2020)
Nano- iron				
Rice (*Oryza sativa* L.)	Cd stress (16.31 mg kg^{-1} soil) and drought stress	Biological iron oxide-NPs (25, 50, 100 mg kg^{-1})	Increased biomass, enzymatic antioxidants, decrease in ROS	Ahmed et al. (2021)

TABLE 7.1 *(Continued)*

Crop (scientific name)	Stress type and its details	Nanofertilizer and its concentration	Roles of nanofertilizer	Reference
Rice (*Oryza sativa* L.)	As stress (50 μM) in Hoagland solution for 21 days	Chemical iron oxide-NPs (20–30 nm; 50 mg L^{-1})	Reduced As accumulation in rice roots and leaves and phytotoxicity	Bidi et al. (2021)
Wheat (*Triticum aestivum* L.)	Cd (3.78 mg kg^{-1}) and salinity stress (EC 8.07 dS m^{-1})	Biological FeO-NPs (19–40 nm; 25 to 100 mg kg^{-1})	FeO-NPs (100 mg kg^{-1}) increased antioxidants, biomass, photosynthetic pigments under stress	Manzoor et al. (2021)
Foxtail millet (*Setaria italica* L.)	Drought stress by employing 10% PEG	Biological Fe_3O_4-NPs (10–18 nm; 5, 10, 15, 20, 50, 90, and 120 mg l^{-1})	Increased the uptake of iron, H_2O_2, and proline contents in plants, which ameliorated drought stress	Sreelakshmi et al. (2021)
Sunflower (*Helianthus annuus*)	Chromium stress (Cr IV 75 and 150 mg kg^{-1})	Chemical Fe^0-NPs (35–45 nm; 1.0 and 2.0%)	Reduced Cr uptake and enhanced activity of de-toxification plant enzymes (SOD, CAT, and POX)	Mohammadi et al. (2020)
Nano- selenium				
Cucumber (*Cucumis sativus* L.)	Soil salinity (EC 4.49 dS m^{-1}) and heat stress (41°C)	Biological nano-Se (50–200 nm; 25 mg L^{-1})	Improved productivity and growth under combined heat and salinity stress	Shalaby et al. (2021)
Paddy rice (*Oryza sativa* L.)	Cd and Pb stress (3.0 and 300 mg kg^{-1}, respectively.)	Chemical nano-Se (160 nm; 25–100 μmol L^{-1})	Dose of 50 μmol L^{-1} Se-NPs is the best; decreased Cd accumulation, and improved photosynthesis	Wang et al. (2021)
Paddy rice (*Oryza sativa* L.)	Cd and Pb stress (3.0 and 300 mg kg^{-1}, respectively)	Chemical nano-Se (50 nm; 4, 6, 12 mg L^{-1})	Combined nano-silica and nano-Se reduced Cd and Pb accumulation; mitigated oxidative stress damage	Wang et al. (2020)
Strawberry (*Fragaria* × *ananassa* Duch.)	Drought stress (30, 60, and 100% field capacity)	Chemical Se-NPs (25 mg L^{-1}, 60 nm)	Applied Se/SiO_2-NPs at 100 mg L^{-1} managed drought stress by higher level of osmolytes like carbohydrate and proline	Zahedi et al. (2020)

TABLE 7.1 *(Continued)*

Crop (scientific name)	Stress type and its details	Nanofertilizer and its concentration	Roles of nanofertilizer	Reference
Nano- zinc				
Tomato (*Lycopersicon esculentum* Mill.)	Cu stress (100 mg kg^{-1} soil)	Chemical ZnO-NPs (20–100 nm; 50 mg L^{-1})	Improved photosynthetic activity, growth, proline, stomatal aperture, and enzymatic antioxidants	Faizan et al. (2021a)
Tomato (*Lycopersicon esculentum* Mill.)	Salt stress (150 mM NaCl)	Chemical ZnO-NPs (20–100 nm; 10, 50, 100 mg L^{-1})	Alleviated salt toxicity by boosting the growth and mitigated the adverse effects on plants	Faizan et al. (2021b)
Cotton (*Gossypium hirsutum* L.)	In hydroponic system (Cd 12.5 and 150 mg L^{-1} Pb) for 21 days	Phycomolecules coated zinc oxide-NPs (75 mg L^{-1})	Protected cotton seedlings by alleviating Cd and Pb phytotoxicity; promoted antioxidant systems	Priyanka et al. (2021)
Eggplant (*Solanum melongena* L.)	Drought stress (60% of ET_c) and saline soil (7.33 dS m^{-1})	Chemical ZnO-NPs (200 nm; 50 and 100 mg kg l^{-1})	Alleviated damage of cell membrane and increased leaf and stem anatomical parameters	Semida et al. (2021)
Muskmelon (*Cucumis melo* L.)	Cd stress (75 mg kg^{-1})	Chemical ZnO-NPs (20 mg kg^{-1})	Alleviated phytotoxicity of Cd by reducing uptake of Cd and protein/non protein bound thiols under Cd-stress	Shah et al. (2021)
Safflower (*Carthamus tinctorius* L.)	Salt stress (250 mM NaCl)	Biological ZnO-NPs (17 mg L^{-1})	Protected plants to stress by regulating ion homeostasis and antioxidant systems	Yasmin et al. (2021)
Rosemary (*Rosmarinus officinalis* L.)	Salt stress (75, 150, and 225 mM NaCl)	Chemical nano-zinc (10–30 nm; at 3 mg L^{-1})	Alleviated impacts of salinity by decreasing both H2O2 and MDA	Hassanpour-aghdam et al. (2020)

CAT, catalase; CuO-MPs, spherical microsized CuO; CuO-NTs, tubular nano-CuO; DOC, dissolved organic carbon; ET_c, crop evapotranspiration; Fe^0, nano-zerovalent iron; H_2O_2, hydrogen peroxide; MDA, malondialdehyde; PEG, poly ethylene glycol; POX, peroxidase; ROS, reactive oxygen species; SOD, super oxide dismutase.

negative impacts on the environment, agricultural safety, and human health. Numerous remediation techniques have been created; nevertheless, it is critical to ensure their safety and to consider each methodology's limitations that include high-energy inputs and residues generation (Souza et al., 2020). One of the most environmental applications of NT is the remediation of contaminated environments and hazardous substances. Some of NPs are used to remove contaminants, while others sequester them. For example, C nanotubes have been noted for their significantly stronger capacity to adsorb dioxin than conventional activated carbon. Furthermore, the use of microorganisms for intracellular/extracellular creation of different NPs (either in chemical composition, in size/shape, or regulated mono-dispersity) can be a modern, commercially proper, and eco-friendly approach that can decrease toxic chemicals in the traditional protocol (Juwarkar et al., 2010).

Much research in recent years have demonstrated that NT combined with phytoremediation can be a promising remediation approach for contaminated soil. This advanced technology has gained prominence in the remediation of potentially toxic contaminants from soil, and a variety of nanomaterials, nano-TiO_2, nano-Ni, nano-Ag, nano-SnO_2, nano-Zn, nZVI (nanozero valent Fe), biochar-supported nanohydroxyapatite, and nano-C black can be used to remove potentially toxic metals such as Pb, Cd, and other metals. When heavy metal-contaminated soils are treated with a sophisticated nanophytoremediation technology, different NMs have varying remediation effects. Furthermore, the effect of potentially harmful heavy metal immobilization changes in contaminated soils is influenced by varied particle sizes of NMs, which affects the heavy metal restoration process (Shikha and Singh, 2021). Souza et al. (2020) stated also that Fe oxide NPs have higher 10 times sorption capacity of microscale particles. The low cost, low energy consumption, and suitability to in situ treatment are all advantages of this technique. However, it is critical to examine the environmental impacts of NPs (Souza et al., 2020). Souza et al. (2020) added also in this context that many NPs have been investigated for As adsorption (and hence remediation). The most promising materials for this include Fe oxide NPs like hematite and magnetite, as well as zerovalent iron NPs. Depending on the degree of oxidation of As, the mechanism of interaction between As and Fe NPs complexes such as FeS differs. For example, at $pH > 6$, the As(V) interacts with FeS particles by creating an outer sphere complexation and chemical complexation, whereas the As(III) interacts with FeS particles via surface sorption. At pH 5, the interaction happens by the next equation

$$3FeS+H_3AsO_3+3H^+ \rightleftharpoons 1/2\ Fe_3S_4+AsS+3/2\ Fe_2+ 3H_2O$$

Zand et al. (2020) reported that the use of NMs to promote phytoremediation of contaminated soils is gaining traction around the world. Because of its large specific surface area, great reducibility, cheap cost, and low toxicity, nanoscale zero-valent iron (nZVI) have been utilized in the remediation of soils contaminated with organic and inorganic pollutants. Immobilization of heavy metals in soil by the use of nZVI has been discovered to be a key mechanism for achieving remediation goals in metal-contaminated locations. The efficacy of nZVI to immobilize metals in soils is influenced by many factors, including soil characteristics, nZVI dosage, and the presence of nontarget contaminants. Anionic heavy metals (As and Cr) were shown to be more immobilized in acidic soils, while cationic heavy metals (Pb, Cd, and Zn) were found to be more retained in calcareous soils. The P Pb, As, and Cr immobilization increased by more than 82% when % nZVI was applied; however, nZVI had little effect on Cd availability/immobilization (only 13–42%), independent of soil pH. NPs' effect in promoting plant development and tolerance to metal stress has been widely researched (El-Moneim et al., 2021). Si NPs may improve the nutrient status, photosynthesis, morphology, and physiology of metal-stressed plants, allowing them to develop faster. In wheat grains, ZnO NPs increased Zn^{2+} concentration, photosynthetic pigments, antioxidant enzymes, and decreased Cd^{2+} concentration. It is stated that ZnO NPs coupled with biochar reduced Cd^{2+} levels in rice (*Oryza sativa*) and maize (*Zea mays*) in a short time of growth. It is reported also that applying zero-valent Fe to calcareous or acidic soils reduced the availability of heavy metals (El-Moneim et al., 2021).

7.5 NANOMATERIALS AS STRESSORS ON CULTIVATED PLANTS

Metal and metalloid NPs have the potential to help crops cope with the detrimental effects of abiotic stressors. On the other side, it has been discovered that at a certain concentration, certain NPs have detrimental impacts on crop yield (El-Moneim et al., 2021). Also, the explosive growth of NPs resulting from human activities contributes to environmental pollution and may have an adverse effect on the growth and yield of economically important crops, while accumulation of toxic components such as heavy metals in eatable parts of plants poses a health risk (Jampílek and Kráľová, 2019).

Toxic reactions and aberrant cell division are possible side effects. In onion root tips, silver nanoparticles (Ag NPs) and silver ions (Ag+) were shown to lower the mitotic index and produce many chromosomal abnormalities (*Allium cepa*) (El-Moneim et al., 2021). Also, Au NPs, for example, promoted vegetative development and fruit/seed production at lower concentrations but inhibited them at greater concentrations (Liu et al., 2020).

The plant promotes antioxidant defense function to minimize oxidative damage and improve plant susceptibility to NPs toxicity (Rajput et al., 2019). Thus, different plants interact with the NMs through physical or chemical interaction. These interactions induce chemical and physical signaling, resulting in the production of unknown new chemicals that may be detrimental to plants, such as ROS and lipid peroxidation, as well as changes in ion transport across cell membranes. The major players that determine the interaction between NMs with plants include nanoparticle characteristics, plant species, and environmental variables (Dev et al., 2018).

7.6 CONCLUSIONS

Plants face a variety of challenges, including a growing global population and biotic and abiotic stresses. Agricultural yield losses are caused by drought, salt, waterlogging, and heat stress, as well as phytopathogens. As a result, stressed grown plants have a robust defense mechanism that uses endogenous resources like nonenzymatic antioxidants to defend them from these stresses. The use of nanotechnology in many fields including the agriculture field is promising in many aspects including plant growth and development under stress, the efficacy of fertilizing, remediation, and many other aspects. However, the explosive use of NPs contributes to environmental pollution and some negative impacts on plant growth and development, as well as on the accumulation of toxic components in edible parts of plants posing a health risk.

KEYWORDS

- **nanomaterials**
- **abiotic stress**
- **reactive oxygen species**
- **nanofertilizers**
- **nano-pesticides**

REFERENCES

Ahmed, T.; Noman, M.; Manzoor, N.; Shahid, M.; Abdullah, M.; Ali, L.; Wang, G.; Hashem, A.; Al-Arjani, A. F.; Alqarawi, A. A.; Abd-Allah, E. F.; Li, B. Nanoparticle-Based Amelioration of Drought Stress and Cadmium Toxicity in Rice Via Triggering the Stress Responsive Genetic Mechanisms and Nutrient Acquisition. *Ecotoxicol. Environ. Saf.* **2021,** *209,* 111829. https://doi.org/10.1016/j.ecoenv.2020.111829

Amjad, S.; Serajuddin, M. Applications of Nanobiotechnology to Mitigate Mineral Nutrients Deficiency Stress in Crop Plants. In *Nanobiotechnology*; Al-Khayri, J. M., Ansari, M. I., Singh, A. K., Eds.; Springer: Cham, 2021. https://doi.org/10.1007/978-3-030-73606-4_19

Bidi, H.; Fallah, H.; Niknejad, Y.; Tari, D. B. Iron Oxide Nanoparticles Alleviate Arsenic Phytotoxicity in Rice by Improving Iron Uptake, Oxidative Stress Tolerance and Diminishing Arsenic Accumulation. *Plant Physiol. Biochem.* **2021,** *163,* 348–357. https://doi.org/10.1016/j.plaphy.2021.04.020

Dev, A.; Srivastava, A. K.; Karmakar, S. Nanomaterial Toxicity for Plants. *Environ. Chem. Lett.* **2018,** *16,* 85–100. https://doi.org/10.1007/s10311-017-0667-6

El-Ghamry, A.; Mosa, A. A.; Alshaal, T.; El-Ramady, H. Nanofertilizers Vs. Biofertilizers: New Insights. *Environ. Biodivers. Soil Secur.* 2018, 2(2018), 51–72. DOI: 10.21608/jenvbs.2018.3880.1029

El-Moneim, D. A.; Dawood, M. F. A.; Moursi, Y. S.; et al. Positive and Negative Effects of Nanoparticles on Agricultural Crops. *Nanotechnol. Environ. Eng.* **2021,** *6*(21). https://doi.org/10.1007/s41204-021-00117-0

Faizan, M.; Bhat, J. A.; Chen, C.; Alyemeni, M. N.; Wijaya, L.; Ahmad, P.; Yu, F.. Zinc Oxide Nanoparticles (ZnO-NPs) Induce Salt Tolerance by Improving the Antioxidant System and Photosynthetic Machinery in Tomato. *Plant Physiol. Biochem.* **2021b,** *161,* 122–130. https://doi.org/10.1016/j.plaphy.2021.02.002

Faizan, M.; Bhat, J. A.; Noureldeen, A.; Ahmad, P.; Yu, F. Zinc Oxide Nanoparticles and 24-epibrassinolide Alleviates Cu Toxicity in Tomato by Regulating ROS Scavenging, Stomatal Movement and Photosynthesis. *Ecotoxicol. Environ. Saf.* **2021a,** *218,* 112293. https://doi.org/10.1016/j.ecoenv.2021.112293

Fatima, F.; Hashim, A.; Anees, S. Efficacy of Nanoparticles as Nanofertilizer Production: A Review. Environ. *Sci. Pollut. Res.* **2021,** *28,* 1292–1303. https://doi.org/10.1007/s11356-020-11218-9

Gohari, G.; Panahirad, S.; Sepehri, N.; Akbari, A.; Zahedi, S. M., Jafari, H.; Dadpour, M. R.; Fotopoulos, V. Enhanced Tolerance to Salinity Stress in Grapevine Plants Through Application of Carbon Quantum Dots Functionalized by Proline. *Environ. Sci. Pollut. Res.* **2021,** *28*(31), 42877–42890. https://doi.org/10.1007/s11356-021-13794-w

Gómez-Merino, F. C.; Trejo-Téllez, L. I. The Role of Beneficial Elements in Triggering Adaptive Responses to Environmental Stressors and Improving Plant Performance. In *Biotic and Abiotic Stress Tolerance in Plants*; Vats, S., Eds.; Springer: Singapore, 2018. https://doi.org/10.1007/978-981-10-9029-5_6

González-García, Y.; Cárdenas-Álvarez, C.; Cadenas-Pliego, G.; Benavides-Mendoza, A.; Cabrera-de-la-Fuente, M.; Sandoval-Rangel, A.; Valdés-Reyna, J.; Juárez-Maldonado, A. Effect of Three Nanoparticles (Se, Si and Cu) on the Bioactive Compounds of Bell Pepper Fruits under Saline Stress. *Plants (Basel)* **2021,** *10*(2), 217. DOI: 10.3390/plants10020217

Gui, W.; Liu, J.; Song, X.; Zhang, H.; Lin, J.; Luan, B. A New Microfiltration Membrane With Three-Dimensional Reticular Architecture for Nano-pollutants Removal From

Wastewater. Progress Natl. Sci. Mater. Int. **2021,** *31*(3), 414–419. https://doi.org/10.1016/j.pnsc.2021.04.002

Hassanpouraghdam, M. B.; Mehrabani, L. V.; Tzortzakis, N. Foliar Application of Nano-zinc and Iron Affects Physiological Attributes of Rosmarinus officinalis and Quietens NaCl Salinity Depression. *J. Soil Sci. Plant Nutr.* **2020,** *20,* 335–345. https://doi.org/10.1007/s42729-019-00111-1

Hussain, F.; Hadi, F.; Rongliang, Q. Effects of Zinc Oxide Nanoparticles on Antioxidants, Chlorophyll Contents, and Proline in Persicaria Hydropiper L. and its Potential for Pb Phytoremediation. *Environ. Sci. Pollut. Res.* **2021,** *28,* 34697–34713. https://doi.org/10.1007/s11356-021-13132-0

Jampílek, J.; Kráľová, K. Impact of Nanoparticles on Photosynthesizing Organisms and Their Use in Hybrid Structures With Some Components of Photosynthetic Apparatus. In *Plant Nanobionics*, Nanotechnology in the Life Sciences; Prasad, R., Eds.; Springer: Cham, 2019; pp 255–332. https://doi.org/10.1007/978-3-030-12496-0_11

Juwarkar, A. A.; Singh, S. K.; Mudhoo, A. A Comprehensive Overview of Elements in Bioremediation. *Rev. Environ. Sci. Biotechnol.* **2010,** *9,* 215–288. https://doi.org/10.1007/s11157-010-9215-6

Kandil, E. E.; Abdelsalam, N. R.; Abd EL Aziz, A. A.; Siddiqui, M. H. Efficacy of Nanofertilizer, Fulvic Acid and Boron Fertilizer on Sugar Beet (Beta vulgaris L.) Yield and Quality. *Sugar Tech.* **2020,** *22,* 782–791. https://doi.org/10.1007/s12355-020-00837-8

Kukde, S.; Sarangi, B. K.; Purohit, H. Antioxidant Role of Nanoparticles for Enhancing Ecological Performance of Plant System. *Compr. Anal. Chem.* **2019,** *87,* 159–187.

Liu, Y.; Pan, B.; Li, H.; Lang, D.; Zhao, Q.; Zhang, D.; Wu, M.; Steinberg, C. E.; Xing, B. Can the Properties of Engineered Nanoparticles be Indicative of Their Functions and Effects in Plants? *Ecotoxicol. Environ. Saf.* **2020,** *205,* 111128.

Madzokere, T. C.; Murombo, L. T.; Chiririwa, H. Nano-Based Slow Releasing Fertilizers for Enhanced Agricultural Productivity. In *Materials Today: Proceedings*, 2021; vol 45(3), pp 3709–3715. https://doi.org/10.1016/j.matpr.2020.12.674

Manzoor, N.; Ahmed, T.; Noman, M.; Shahid, M.; Nazir, M. M.; Ali, L.; Alnusaire, T. S.; Li, B.; Schulin, R.; Wang, G. Iron Oxide Nanoparticles Ameliorated the Cadmium and Salinity Stresses in Wheat Plants, Facilitating Photosynthetic Pigments and Restricting Cadmium Uptake. *Sci. Total Environ.* **2021,** *769,* 145221. https://doi.org/10.1016/j.scitotenv.2021.145221

Menhas, S.; Yang, X.; Hayat, K.; Aftab, T.; Bundschuh, J.; Arnao, M. B.; Zhou, Y.; Zhou, P. Exogenous Melatonin Enhances Cd Tolerance and Phytoremediation Efficiency by Ameliorating Cd-Induced Stress in Oilseed Crops: A Review. *J. Plant Growth Regulat.* **2021,** *41,* 922–935. https://doi.org/10.1007/s00344-021-10349-8

Mohammadi, H.; Amani-Ghadim, A. R.; Matin, A. A.; Ghorbanpour, M. Fe0 Nanoparticles Improve Physiological and Antioxidative Attributes of Sunflower (*Helianthus annuus*) Plants Grown in Soil Spiked With Hexavalent Chromium. *3 Biotech* **2020,** *10,* 19. https://doi.org/10.1007/s13205-019-2002-3

Montes, A.; Bisson, M. A.; Gardella, J. A.; Aga, D. S. Uptake and Transformations of Engineered Nanomaterials: Critical Responses Observed in Terrestrial Plants and the Model Plant Arabidopsis Thaliana. *Sci. Total Environ.* **2017,** *607–608,* 1497–1516.

Naghdi, M.; Metahni, S.; Ouarda, Y.; Brar, S. K.; Das, R. K.; Cledon, M. Instrumental Approach Toward Understanding Nano-Pollutants. *Nanotechnol. Environ. Eng.* **2017,** *2,* 3. https://doi.org/10.1007/s41204-017-0015-x

Nguyen, D. V.; Nguyen, H. M.; Le, N. T.; Nguyen, K. H.; Nguyen, H. T.; Le, H. M.; Nguyen, A. T.; Dinh, N. T. T.; Hoang, S. A.; Ha, C. V. Copper Nanoparticle Application Enhances Plant Growth and Grain Yield in Maize Under Drought Stress Conditions. *J. Plant Growth Regul.* **2021,** *41,* 364–375. https://doi.org/10.1007/s00344-021-10301-w

Noman, M.; Ahmed, T.; Hussain, S.; Niazi, M. B. K.; Shahid, M.; Song, F. Biogenic Copper Nanoparticles Synthesized by Using a Copper-Resistant Strain *Shigella flexneri* SNT22 Reduced the Translocation of Cadmium From Soil to Wheat Plants. *J. Hazard. Mater.* **2020,** *398,* 123175. https://doi.org/10.1016/j.jhazmat.2020.123175

Noman, M.; Ahmed, T.; Shahid, M.; Niazi, M. B. K.; Qasim, M.; Kouadri, F.; Abdulmajeed, A. M.; Alghanem, S. M.; Ahmad, N.; Zafar, M.; Ali, S. Biogenic Copper Nanoparticles Produced by Using the Klebsiella Pneumoniae Strain NST2 Curtailed Salt Stress Effects in Maize by Modulating the Cellular Oxidative Repair Mechanisms. *Ecotoxicol. Environ. Saf.* **2021,** *217,* 112264. https://doi.org/10.1016/j.ecoenv.2021.112264

Pirzadah, B.; Pirzadah, T. B.; Jan, A.; Hakeem, K. R. Nanofertilizers: A Way Forward for Green Economy. In *Nanotechnology in the Life Sciences*; Hakeem, K., Pirzadah, T., Eds.; Springer: Cham, 2020. https://doi.org/10.1007/978-3-030-39978-8_5

Priyanka, N.; Geetha, N.; Manish, T.; Sahi, S. V.; Venkatachalam, P. Zinc Oxide Nanocatalyst Mediates Cadmium and Lead Toxicity Tolerance Mechanism by Differential Regulation of Photosynthetic Machinery and Antioxidant Enzymes Level in Cotton Seedlings. *Toxicol. Rep.* **2021,** *8,* 295–302. https://doi.org/10.1016/j.toxrep.2021.01.016

Rajput, V. D.; Minkina, T.; Sushkova, S.; Chokheli, V.; Soldatov, M. Toxicity Assessment of Metal Oxide Nanoparticles on Terrestrial Plants. Compr. Anal. Chem. **2019,** *87,* 189–207.

Rhaman, M. S.; Imran, S.; Karim, M. M.; Chakrobortty, J.; Mahamud, M. A.; Sarker, P.; Tahjib-Ul-Arif, M.; Robin, A. H. K.; Ye, W.; Murata, Y.; Hasanuzzaman, M. 5-aminolevulinic Acid-Mediated Plant Adaptive Responses to Abiotic Stress. *Plant Cell Rep.* **2021,** *40,* 1451–1469. https://doi.org/10.1007/s00299-021-02690-9

Rizwan, M.; Ali, S.; Ali, B.; Adrees, M.; Arshad, M.; Hussain, A.; Zia ur Rehman, M.; Waris, A. A. Zinc and Iron Oxide Nanoparticles Improved the Plant Growth and Reduced the Oxidative Stress and Cadmium Concentration in Wheat. *Chemosphere* **2019,** *214,* 269–277.

Romeh, A. A.; Saber, R. A. I. Green Nano-Phytoremediation and Solubility Improving Agents for the Remediation of Chlorfenapyr Contaminated Soil and Water. *J. Environ. Manag.* **2020,** *260,* 110104. https://doi.org/10.1016/j.jenvman.2020.110104

Salem, K. F. M.; Alloosh, M. T.; Saleh, M. M.; Alnaddaf, L. M.; Almuhammady, A. K.; Al-Khayri, J. M. Utilization of Nanofertilizers in Crop Tolerance to Abiotic Stress. In *Nanobiotechnology*; Al-Khayri, J. M., Ansari, M. I., Singh, A. K., Eds.; Springer: Cham, 2021. https://doi.org/10.1007/978-3-030-73606-4_11

Shalaby, T. A.; Abd-Alkarim, E.; El-Aidy, F.; Hamed, E.; Sharaf-Eldin, M.; Taha, N.; El-Ramady, H.; Bayoumi, Y.; Reis, A. R. D. Nano-Selenium, Silicon and H2O2 Boost Growth and Productivity of Cucumber Under Combined Salinity and Heat Stress. *Ecotoxicol. Environ. Saf.* **2021,** *212,* 111962.

Semida, W. M.; Abdelkhalik, A.; Mohamed, G. F.; Abd El-Mageed, T. A.; Abd El-Mageed, S. A.; Rady, M. M.; Ali, E. F. Foliar Application of Zinc Oxide Nanoparticles Promotes Drought Stress Tolerance in Eggplant (*Solanum melongena* L.). *Plants (Basel)* **2021,** *10*(2), 421. DOI: 10.3390/plants10020421

Shah, A. A.; Aslam, S.; Akbar, M.; Ahmad, A.; Khan, W. U.; Yasin, N. A.; Ali, B.; Rizwan, M.; Ali, S. Combined Effect of *Bacillus fortis* IAGS 223 and Zinc Oxide Nanoparticles to

Alleviate Cadmium Phytotoxicity in *Cucumis melo*. *Plant Physiol. Biochem.* **2021,** *158,* 1–12. https://doi.org/10.1016/j.plaphy.2020.11.011

Shalaby, T. A.; Abd-Alkarim, E.; El-Aidy, F.; Hamed, E.; Sharaf-Eldin, M.; Taha, N.; El-Ramady, H.; Bayoumi, Y.; dos Reis, A. R. Nano-selenium, Silicon and H_2O_2 Boost Growth and Productivity of Cucumber Under Combined Salinity and Heat Stress. *Ecotoxicol. Environ. Saf.* **2021,** *212,* 111962. https://doi.org/10.1016/j.ecoenv.2021.111962

Sharma, S.; Rana, V. S.; Pawar, R.; Lakra, J.; Racchapannavar, V. K. Nanofertilizers for Sustainable Fruit Production: A Review. *Environ. Chem. Lett.* **2021,** *19,* 1693–1714. https://doi.org/10.1007/s10311-020-01125-3

Shikha, D.; Singh, P. K. In Situ Phytoremediation of Heavy Metal–Contaminated Soil and Groundwater: A Green Inventive Approach. *Environ. Sci. Pollut. Res.* **2021,** *28,* 4104–4124. https://doi.org/10.1007/s11356-020-11600-7

Shukla, Y. M. Nanofertilizers: A Recent Approach in Crop Production. In *Nanotechnology for Agriculture: Crop Production & Protection*; Springer: Singapore, 2019; pp 25–58.

Singh, S.; Husen, A. Behavior of Agricultural Crops in Relation to Nanomaterials Under Adverse Environmental Conditions. In *Nanomaterials for Agriculture and Forestry Applications*; Husen, A., Jawaid, M.; pp. 219–256. https://doi.org/10.1016/B978-0-12-817852-2.00009-3.

Singh, K.; Madhusudanan, M.; Verma, A. K.; et al. Engineered Zinc Oxide Nanoparticles: An Alternative to Conventional Zinc Sulphate in Neutral and Alkaline Soils for Sustainable Wheat Production. *3 Biotech* **2021,** *11,* 322. https://doi.org/10.1007/s13205-021-02861-1

Sreelakshmi, B.; Induja, S.; Adarsh, P. P.; Rahul, H. L.; Arya, S. M.; Aswana, S.; Haripriya, R.; Aswathy, B. R.; Manoj, P. K.; Vishnudasan, D. Drought Stress Amelioration in Plants Using Green Synthesised Iron Oxide Nanoparticles. Mater. Today Proc. **2021,** *41,* 723–727. https://doi.org/10.1016/j.matpr.2020.05.801

Souza, L. A.; Monteiro, C. C.; Carvalho, R. F.; Gratão, P. L.; Azevedo, R. A. Dealing With Abiotic Stresses: An Integrative View of How Phytohormones Control Abiotic Stress-Induced Oxidative Stress. *Theor. Exp. Plant Physiol.* **2017,** *29,* 109–127. https://doi.org/10.1007/s40626-017-0088-8

Souza, L. R. R.; Pomarolli, L. C.; da Veiga, M. A. M. S. From Classic Methodologies to Application of Nanomaterials for Soil Remediation: An Integrated View of Methods for Decontamination of Toxic Metal(oid)s. *Environ. Sci. Pollut. Res.* **2020,** *27,* 10205–10227. https://doi.org/10.1007/s11356-020-08032-8

Thirugnanasambandan, T. Advances of Engineered Nanofertilizers for Modern Agriculture. In *Plant-Microbes-Engineered Nano-particles (PM-ENPs) Nexus in Agro-Ecosystems, Advances in Science, Technology & Innovation*; Singh, P., et al., Eds.; Springer Nature: Switzerland, AG, 2021; pp 131–152. https://doi.org/10.1007/978-3-030-66956-0_9

Tiwari, R. K.; Lal, M. K.; Naga, K. C.; Kumar, R.; Chourasia, K. N.; Subhash, S.; Kumar, D.; Sharma, S. Emerging Roles of Melatonin in Mitigating Abiotic and Biotic Stresses of Horticultural Crops. *Sci. Hortic.* **2020,** *272,* 109592. https://doi.org/10.1016/j.scienta.2020.109592

Vázquez-Núñez, E.; López-Moreno, M. L.; de la Rosa Álvarez, G.; Fernández-Luqueño, F. Incorporation of Nanoparticles Into Plant Nutrients: The Real Benefits. In *Agricultural Nanobiotechnology*; Springer: Cham, 2018; pp 49–76.

Wang, C.; Cheng, T.; Liu, H.; Zhou, F.; Zhang, J.; Zhang, M.; Liu, X.; Shi, W.; Cao, T. Nano-Selenium Controlled Cadmium Accumulation and Improved Photosynthesis in Indica

Rice Cultivated in Lead and Cadmium Combined Paddy Soils. *J. Environ. Sci.* **2021,** *103,* 336–346. https://doi.org/10.1016/j.jes.2020.11.005

Wang, S. L.; Nguyen, A. D. Effects of Zn/B Nanofertilizer on Biophysical Characteristics and Growth of Coffee Seedlings in a Greenhouse. *Res. Chem. Intermed.* **2018,** *44,* 4889–4901. https://doi.org/10.1007/s11164-018-3342-z

Wang, C.R.; Rong, H.; Zhang, X. B.; Shi, W. J.; Hong, X.; Liu, W. C.; Cao, T.; Yu, X.; Yu, Q. Effects and Mechanisms of Foliar Application of Silicon and Selenium Composite Sols on Diminishing Cadmium and Lead Translocation and Affiliated Physiological and Biochemical Responses in Hybrid Rice (*Oryza sativa* L.) Exposed to Cadmium and Lead. *Chemosphere* **2020,** *251,* 126347. https://doi.org/10.1016/j.chemosphere.2020.126347

Xu, M.; Wang, Y.; Mu, Z.; Li, S.; Li, H. Dissolution of Copper Oxide Nanoparticles is Controlled by Soil Solution pH, Dissolved Organic Matter, and Particle Specific Surface Area. *Sci. Total Environ.* **2021,** *772,* 145477. https://doi.org/10.1016/j.scitotenv.2021.145477

Yasmin, H.; Mazher, J.; Azmat, A.; Nosheen, A.; Naz, R.; Hassan, M. N.; Noureldeen, A.; Ahmad, P. Combined Application of Zinc Oxide Nanoparticles and Biofertilizer to Induce Salt Resistance in Safflower by Regulating Ion Homeostasis and Antioxidant Defence Responses. *Ecotoxicol. Environ. Saf.* **2021,** *218,* 112262. https://doi.org/10.1016/j.ecoenv.2021.112262

Zahedi, S. M.; Moharrami, F.; Sarikhani, S.; Padervand, M. Selenium and Silica Nanostructure-Based Recovery of Strawberry Plants Subjected to Drought Stress. *Sci. Rep.* **2020,** *10,* 17672. DOI: 10.1038/s41598-020-74273-9

Zand, A. D.; Tabrizi, A. M.; Heir, A. V. The Influence of Association of Plant Growth-Promoting Rhizobacteria and Zero-Valent Iron Nanoparticles on Removal of Antimony From Soil by Trifolium Repens. *Environ. Sci. Pollut. Res.* **2020,** *27,* 42815–42829. https://doi.org/10.1007/s11356-020-10252-x

Zhou, X.; Joshi, S.; Khare, T.; Patil, S.; Shang, J.; Kumar, V. Nitric Oxide, Crosstalk With Stress Regulators and Plant Abiotic Stress Tolerance. *Plant Cell Rep.* **2021,** *40*(8), 1395–1414. https://doi.org/10.1007/s00299-021-02705-5

CHAPTER 8

The Behavior of Nanomaterials in Soil and Interaction with Soil Biota

MONIKA MAHAJAN, ANUCHAYA DEVI, BHAVISHA SHARMA, and RAJEEV PRATAP SINGH

Institute of Environment and Sustainable Development, Banaras Hindu University, Varanasi, Uttar Pradesh, India

ABSTRACT

Human civilization cannot exist without the food and agriculture domains, which are inextricably linked to human life. As a result, now that the regulatory authorities have legalized nanotechnology, it has a lot of potential in this industry. Traditional agricultural technologies' limitations have limited the effective use of available farmland to meet demand and ensure food security. Nanotechnology has emerged as one of the most promising alternatives for addressing the shortcomings of traditional agriculture approaches. Nanotechnology has opened a new chapter in agriculture's long-term viability. It reduces the amount of synthetic chemicals in the field by a significant amount. Engineered nanoparticles have a promising future in agriculture; it can promote soil amelioration from toxic pollutants and improve plant growth and productivity. Adequate amount and duration of exposure of nanoparticles such as $nSiO_2$, $nTiO_2$, AgNP, single- and multi-walled carbon nanotubes have several advantages such as stimulating the synthesis of the bioactive compound, increasing the efficacy of germination rate, and enhancing the vegetative biomass. But the inappropriate concentration of nanoparticles diminishes the metabolism, enzymatic activities of soil and plants. This chapter is an approach to point out the possible prospects and

Nanotechnology for Sustainable Agriculture: An Innovative and Eco-Friendly Approach. Vishnu D. Rajput, Abhishek Singh, Tatiana M. Minkina, Krishan K. Verma, and Awani Kumar Singh (Eds.)

implications of nanoparticles with the current guidelines to their use in agriculture, food safety, and security of the nation.

8.1 INTRODUCTION

It is projected that the exponential increase in the demography will lead the world to reach 9 billion by the year 2050 (Silva, 2017). The major challenge here is to accomplish food security and feed the huge mass within the restrictions of inadequate resources (Singh et al., 2021a; Rajput et al., 2021). Current agriculture practices exhaust around 4 million tons of pesticide, 187 million tons of fertilizer, 2700 billion cubic meters of freshwater, and tremendous energy (Kah et al., 2019). Moreover, post induction of green revolution, ampule use of chemical fertilizer and pesticide are declining the health of the environment (Mukhopadhyay, 2014). Therefore, bio-friendly growth boosters came into existence as an alternative to ensure the environmental worry. But this pro-environment method also has significant weaknesses, such as large dose requirements, field stability, and inadequate performance under various environmental conditions (Pandey, 2018). Here, to overcome the inadequate yield nano-based formulation is introduced in agriculture (Iavicoli et al., 2017; Prasad et al., 2017; Lowry et al., 2019). Nanotechnology (NT) is a very advanced and revolutionary option in the agriculture sector. In this modern era, nanotechnology emerges with intensive benefits such as reduced input cost and advancement in current agriculture practices to improve yield with proper support of the "sustainable intensification" (Perez-de-Luque and Hermosin, 2013; Sekhon, 2014). Nanotechnology is very innovative; it converts the macro-sized element into nanoparticles (1–100 nm) and small-sized nanoparticles that behave differently from parent material (Bhattacharya et al., 2011). This technology can be used for multiple purposes (1) to truncate the plant pathology via restricted transport of functional molecules, (2) to diagnose the disease through nanosensors (Boom, 2011).

The use of nanotechnology has a wide spectrum in the agricultural field because of the nano-enhanced solution that enables seed germination, plant protection, and growth. It can curtail the excessive use of agrochemical and nutrient loss in fertilization. Along with these, it revamps ecological factors such as water and soil health (Sonkaria et al., 2012; Sekhon, 2014). Nanotechnology having various nano-devices, nano ingredients, nanoformulation to enhance agricultural practices, such as nano-pesticide, nano herbicides for pest and weed management, nano fertilizer for proficient nutrient supply, nano biosensors diagnose appropriate moisture content and help in maintaining nutrient budget (Iavicoli et al., 2017). An engineered nanoparticle has enough potency to positively or negatively affect soil physicochemical, soil microflora, and biological properties; for instance, cerium oxide nanoparticle is one of them (Keller et al., 2013; Li et al., 2017). According to Li et al.'s (2017) experiment performed at Clemson, United States, cerium oxide nanoparticles exhibited an immense effect on the improvement of phosphatase enzymatic activity with a significant increase from 97.46 to 131.37 to 181.45% at a concentration of 100, 500, and 1000 mg/kg, respectively.

Nanomaterials have concomitant characteristics such as tiny size and comparatively high specific surface area, reactivity that invokes recalcitrant contaminants from complex environments. A repercussion of nanotechnologies (NTs) on plant physiology, morphology, biological properties are mixed type; it varied from the plant to plant. In some plants, it promotes growth and seed germination, but in some plants such as wheat, it inhibits 60% root growth via stimulating the lateral roots increasing the oxidative stress (Sang et al., 2019; Shukla et al., 2019). So, instead of numerous benefits it has several implications that resist the widespread of nanotechnological use in agriculture. This chapter aims to elucidate the clear scope of guidelines for evaluating nano-based agri-input and food products in India. Article compiled the importance of nanotechnology and it is exhibited as a better substitute than conventional chemicals. But every technology has a darker side, so we discuss its loopholes and their future outlook in agriculture.

8.2 NEED FOR NANOTECHNOLOGY IN AGRICULTURE

Conventional techniques involve direct diffuse of concentrated fertilizer to the crop, but in reality, very less amount will reach to its targeted site due to leaching, drift, runoff, oxidation, evaporation, hydrolysis through microbial degradation, photolysis, and soil moisture. According to the previous studies of Ombodi and Saigusa (2000), it has been evaluated that out of the total percentage of sprayed fertilizer, 40–70% nitrogen, 80–90% phosphorus, and 50–90% of potassium were lost in the environment and could not reach to the plants that cause huge economic loss. According to the International Fertilizer Industry Association report illustration, fertilizer use rate sharply increased 5–6% in the year 2009–2011.

In the aftermath of the green revolution, fertilizer and pesticide practices amplify many times from 0.5 tons to 23 million tons in just 48 years (1960–2008) (Nelson et al., 2019). This paradigm shift has promoted the grain yield up to four times (Huke et al., 2007). But after lots of success, productivity becomes constant in certain crops, and the percentage of organic content in soil continuously decreases because of nonjudicious use of chemical fertilizer. Irrational use of an array of chemicals on fertile land disturbs the NPK proportion 10:2.7:1, which is much different from the standard ratio of 4:2:1 (Meghana et al., 2021). Excessive use of synthetic chemicals will be toxic for water reservoirs and nektons, ultimately incur health risks in humans. Excessive use of agrochemicals promotes pathogen, arthropods of pest resistance, soil microflora, enhances bioaccumulation of heavy metals, pesticides, and affects the tree biome (Manjunatha et al., 2016). Inappropriate agricultural practices can deteriorate soil health and exceed the fertilizer expenses and energy to regulate the productivity in demolishing soil (Tillman et al., 2002; Mukhopadhyay, 2014). The challenges agriculture faces has wide ambit such as continuous water availability, nutrient deficiency, productivity, climate change, land-use change of cultivable field, desertification, resistance to pesticide, and genetically modified organisms (Oliver, 2014; Raman, 2017). They create a hurdle in attaining 300 million tons food demand for feeding 1.5 billion projected population by 2025 in India (Kumar et al., 2012). It is imperative to maintain the equilibrium between organic and inorganic fertilizers for the conservation of soil health attributes, but it is very problematic to achieve. Conventional techniques in agriculture, such as organic farming, have shown a negative impression in the case of productivity and quality (Shennan et al., 2017). Altogether, it can be favorable that alternate cultivation practices be also tested and supply proportionate, adequate nutrients to promote plant growth, and combat environmental pollution. The limitation of conventional technology triggers the research in the field of nanotechnology.

An efficient, innovative nanotechnology study should be consisting of cost-effective nanoscale or nanostructured materials as fertilizers and a properly controlled efficient way to reduce environmental pollution; these aspects are required for smart fertilizer (Cinnamuthu and Bhopati, 2009). Their nanometer size makes them more efficient and useful in the easy delivery of nutrients to plants such as fertilizer. They are encapsulated with nanomaterial from inside and externally narrow coating of the protective polymer film in the form of a tiny particle (De Rosa et al., 2010; Manjunath et al., 2016). This specific packaging of nano-fertilizer particles increases the adherence with material more intensely because greater surface tension is created than the macro-fertilizer. The size of nano-based fertilizer and its higher surface tension makes it suitable to mobilize inside the plant easily and diminish the unnecessary input cost of farmers and upgrade their quality of living, which ultimately curtail environmental pollution (Kalia et al., 2020). In this way, nanotechnology in agriculture is becoming the foreseeable part of agriculture and is the need of the hour.

8.3 APPLICATION OF INNOVATIVE NANOTECHNOLOGY IN MODERN AGRICULTURE

8.3.1 IMPACT OF NANOMATERIAL ON SOIL

Engineered nanomaterial plays a crucial role in soil remediation by enhancing overall soil health, but in conventional ex-situ soil remediation techniques, it involves replacing contaminated soil with fresh one. Therefore, engineered nanomaterial becomes the silver lining for this field. There are various in-situ mechanisms available such as (1) Immobilization via carbon nanotubes and metal oxides nanomaterial, (2) photocatalytic degradation, and (3) oxidation/reduction (Wang et al., 2019).

Nowadays, immobilization/adsorption techniques are the biggest in remedial application because of cost-effectiveness and sustainable cleanup process for polluted soil (Barzegar et al., 2017). Nanoscale supplements such as carbon nanotube and carbon allotropes (fullerene, graphene) along with metal oxides such as (Fe_3O_4 and TiO_2), composites of nanosized material are used for the adsorption of organic and inorganic contaminants from soil ecosystem (Qian et al., 2020). The basic principle involves removing higher surface hydrophobicity, large adsorption capacity, and van der Waals forces (Qian et al., 2020). Towell et al.'s (2011) study illustrated that carbon nanotubes (CNTs) could combat from polyaromatic hydrocarbons in various soil conditions by inhibiting the process of mineralization and mobilization of aromatic compounds. After hampering the mechanism, it will reduce the bioavailability of PAHs to plants and microorganisms.

Despite CNTs, iron oxide and titanium dioxide have enough capacity to resist the mobilization of hazardous trace metals such as cadmium and arsenic (Sebastian et al., 2020). Liang et al. (2014) reported reduced metal stress in the terrestrial plant after the application of Fe_3O_4. Although, after the amendment of iron oxide for surface modification, stabilizing compounds such as the use of starch was necessary; otherwise, metal oxide tended to agglomerate and settled out as the soil particles in the aqueous phase (Liang and Zhao, 2014). TiO_2 has another important sorption benefit in the presence of humic acid and fluvic acid at a particular pH 4.0 (Qian et al., 2020). Tan et al. (2007) demonstrated with their result that titanium dioxide could remove Thorium (Th (IV) up to 94% efficacy. The combined potential for chemical oxidation via ENP and microbial biodegradation to alleviate Aroclor 1248 (polychlorineated biphenyl (PCB's)) was assessed in one experiment. At the start, 99%, 92% tetra, 84% penta, and 28% hexa-chlorinated biphenyls have been dechlorinated using bimetallic nanoparticles composed of palladium and iron. During the final step of PCB's amelioration, *Burkholderia xenovorans* were involved because there was no significant toxic effect of nanoparticles on *Burkholderia xenovorans* (Le et al., 2015).

8.3.2 IMPLICATION OF NANOMATERIAL OVER SOIL MICROORGANISM AND AOIL ENZYME

Nanomaterial has played a pivotal role in soil remediation with the help of carbon nanotubes and metal oxide, but this is a brighter part of the theme; however, every concept consists of a shady side also filled with implications. The use of nanomaterial in soil has some shortcomings over soil organisms and soil enzymes (Li et al., 2017; Achari and Kowshik, 2018; Asadishad et al., 2018). According to Asadishad et al.'s (2018) experiment, 1, 10, and 100 mg engineered nanoparticles (silver (nAg), zinc oxide (nZnO), copper oxide (nCuO), titanium dioxide ($nTiO_2$)) per kg amended in soil for 30-day period. At the observation time, zinc oxide and copper oxide had not shown any significant response on the enzymatic activity of soil. But the reaction was contrary to silver, it inhibited some enzymatic activity at a higher dose (100 mg/kg), and the result is expressed differently with $nTiO_2$. It affects nonsignificantly slightly diminishing enzymatic activity in the soil.

Nanoparticle shows the detrimental effect on soil microbiota along with its enzymatic equilibrium. Xu et al. (2015) illustrated with their study, nanoparticles of titanium dioxide and copper oxide deprive soil microbial biomass and their community flocs in paddy soil (Singh et al., 2020). Most of the metal oxide nano-particles ZnO, TiO_2, CeO_2, Fe_3O_4 reduce various soil enzyme invertase, urease, catalase, phosphatase, and deduce the bacterial community (Cao et al., 2016; You et al., 2017). Chai et al. (2015) reported ZnO and CeO_2 ENPs (engineered nanoparticles) restrict the bacterial count such as *P-solubilizing, K solubilizing,* and *Azotobactor* in a plate. The life of the subtle organism is very sensitive; the minute concentration of biogenic nanoparticles made up of gold creates a growth inhibition zone (Lakshmi et al., 2012; Maliszewska, 2016; Rajput et al., 2018). All nanoparticles are not toxic for the soil ecosystem, but few act as havoc for specific bacteria. Fajardo et al. (2014) demonstrated from their result that silver nanoparticles adversely affect soil bacteria such as *Bacillus cereus* and *Pseudomonas stutzeri*, but at the same time, Al_2O_3 was not shown any adverse effect. Some studies also support that Ag NPs affect organic matter (Grillo et al., 2015; Schlich and Hund- Rinke, 2015). In general, the previous report suggests low pH organic matter and cation exchange capacity obstruct the sorption of silver nanoparticles to colloidal structure in the soil. Low pH organic matter is an ideal condition for the mobilization of particles, toxicity, bioavailability, and higher pH organic content promotes the AgNPs sorption and restricts the toxicity. The nanoparticle of zinc oxide was found toxic for *Folsomia candida* in acidic soil (Waalewijin-Kool et al., 2013). Nanoparticles metal oxide affects soil properties and influences the soil biota, but it inserts inside the plant system and delivers both types of response, either positive or negative, as per the plant physiology and morphology accompanied with quality and quantity of nano-materials.

8.3.3 CONSEQUENCES OF NANOPARTICLES ON PLANT HEALTH

The application of adequate nanomaterial plays a pivotal role in improving the germination, growth, and productivity in some particular plants. NPs provoke stress-tolerant genes and stress protein that helps to overcome biotic and abiotic stresses in various plant species (Usman et al., 2020; Singh et al., 2021b).

TABLE 8.1 The Ramification of Nanomaterial on Soil Health and its Function.

Nanomaterial	Effects positive/negative	Function	References
nAg, $nTiO_2$	Negative	Soil enzyme	Asadishad et al., 2018
$nTiO_2$, nCuO	Negative	Soil enzyme	Xu et al., 2015
TiO_2	Positive	Remediation	Tan et al., 2007
CNT's, iron oxide, titanium dioxide	Positive	Remediation	Sebastian et al., 2019
CNT's	Positive	Remediation	Towell et al., 2011
Carbon allotropes, Fe_3O_4, and TiO_4	Positive	Remediation	Qian et al., 2020
TiO_2, Fe_3O_4, CeO_2, ZnO	Negative	Soil enzyme	Jiling et al., 2016; You et al., 2017

According to the study by Joshi et al. (2018), multi-walled carbon nanotubes (MWCNTs) posed augmentation in the process of seed germination in crops such as tomato, peanut, soybean, and garlic (Khodakovskaya et al., 2010; Lahiani et al., 2013; Srivastava and Rao, 2014).

There is some nanomaterial in nanoscience, such as silicon dioxide (SiO_2), titanium dioxide (TiO_2), and Zeolite, which have a beneficial response in yield enhancement, stimulating enzymatic activity in crops. Carbon fullerenes application in *Arabidopsis* had improved hypocotyl growth through cell division (Gao et al., 2011). Fullerol use for seed dressings can increase the quality and quantity of product up to 128% and promote the synthesis of bioactive compounds, such as lycopene, charantin in *Momordica charantia*. The use of nanoemulsions in seed priming can accelerate seed germination. Acharya et al.'s (2020) experiment result also follows the same conclusion. Turmeric oil nanoemulsions (TNE) and silver nanoparticles (AgNPs) have improved the germination of watermelon seeds, and the study demonstrates the significant increase of soluble sugar (glucose and fructose) contents in AgNP-treated seeds at 96 h. Nanoemulsion-treated seeds were expressed better growth and yield of watermelon.

Uptake and translocation of engineered nanomaterial differ from the type of plant, chemical composition, and size of nanoparticles (Aslani et al., 2014). The gene expression of plants is different after the confrontation with the nanoparticle. The expression of plant is associated with a change in the biological pathway (Ghormade et al., 2011; Bagheri et al., 2012); it might invoke the physiological, morphological, and phytotoxic responses (Siddiqui et al., 2015) for instant soaking of wheat seeds before germination in 40–60 μg/L MWCNTs solution for 4 h. It will have resulted in rapid root growth and vegetative biomass (Wang et al., 2012). Nanomaterial such as nZnO poses a more noticeable effect on tobacco growth (Nicotiana tabacum L.) calli because it accumulated more zinc when received than nanomaterials, and progression of calli growth also received higher protein contents (Mazaheri-Tirani and Dayani, 2020). In addition NPs are used as a foliar application; in an experiment performed by Elsheery et al. (2020), SiO_2 nanoparticles (5–15 nm; 300 μg/L) foliar spray was used on sugar cane (*Saccharum officinarum L.*) under stress condition and the result showed an improved effective quantum yield of cyclic electron flow. Despite all the benefits, many studies resist nanoparticle application in agriculture due to their ecotoxic response on various plants.

8.4 THE NEGATIVE REPERCUSSION OF NANOTECHNOLOGY ON PLANTS

Nanomaterial poses detrimental effects on germination, biomass, leaf number, root elongation (Doshi et al., 2008; Lee et al., 2010). However, the adverse effect depends on exposure time, size, nature of nanomaterial (Nouri et al., 2020). Fullerene is beneficial, but sometimes it disturbs the energy pathways and electron movement by suppressing the transcription gene (Hussain et al., 2016). Similarly, MWCNTs play an important role in the upregulation of gene that helps in channelizing water transport, formation of the cell wall as well as it further stimulates cell division in the plant (Khodakovskaya et al., 2012), but inappropriate concentration and exposure time of MWCNTs will be toxic (Usman et al., 2020). The plethora of studies reported on the adverse effect of nanoparticles because of the formation of antimicrobial metal ions from the surface of NPs. Induction of antibacterial effects on zinc oxide (ZnO) and Ag stimulates Zn ion and silver ions (Feng et al., 2000; Wang et al., 2016; Aziz et al., 2015, 2016). According to research on the phytotoxic effects of AgNPs on rice plants, NPs are taken up through the roots and cause intracellular damage. The effects of AgNPs with sizes ranging from 1

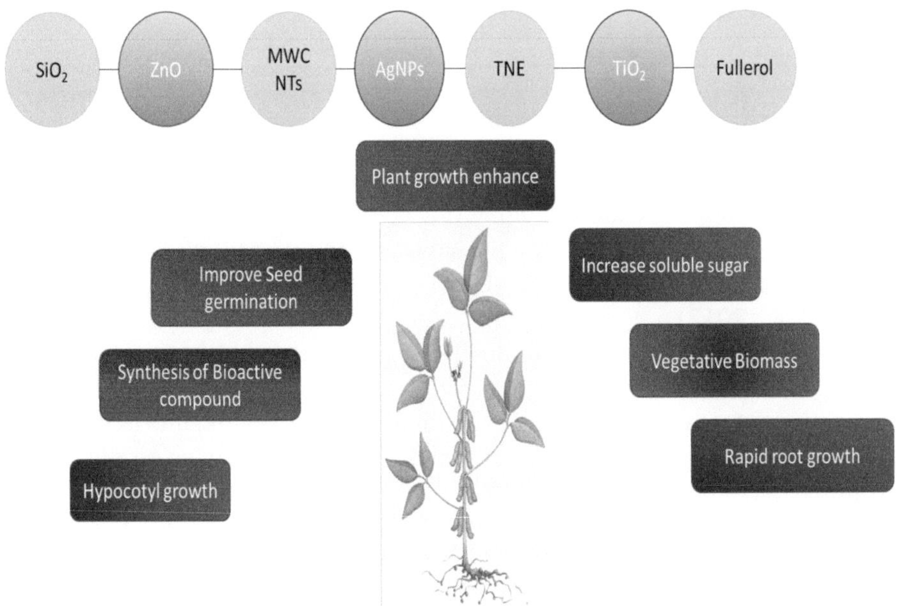

FIGURE 8.1 Beneficial impact of nanoparticle on plant growth and plant properties.

to 20 nm and concentrations ranging from 1 to 100 ppm on the germination of ryegrass, barley, and flax (*Linum usitatissimum*) were investigated, with different-sized NPs having distinct impacts on different plant species. The smallest sized particle inhibited ryegrass even at extremely low concentrations (El-Temash and Joner, 2012). Size-dependent toxicity experiments of AgNPs were also performed on Italian ryegrass (*Lolium multiflorum*), and it was revealed that smaller AgNPs significantly hindered growth, with shorter roots and shoots and less biomass, as compared to plants treated with larger NPs of equal concentrations (Yin et al., 2011). This indicated that the total NP surface area has a considerable effect on the AgNP toxicity. When seedlings were subjected to 40 ppm of gum arabic (GA)-coated AgNPs, they failed to develop root hairs with vacuolated and collapsed cortical cells and a damaged root cap, probably due to a loss of gravitropism in roots due to decreased auxin transport. AgNPs phytotoxicity experiments on mung bean and sorghum seedling growth found detrimental effects (Nair, 2016). The mechanisms of nanomaterial toxicity in plants vary, such as generation of reactive oxygen species, inhibition of cellular respiration, and cell damage (Fu et al., 2014). Metal ions (Chromium, Arsenic, Mercury, Cadmium, Iron, Copper) release intracellular ROS; the entire mechanism comes under the Fenton reaction (Birben et al., 2012). In this reaction, H_2O_2 oxidizes the reduced metal and generates hydroxide ion and hydroxyl radical ($OH^{\cdot}$).

$$H_2O_2 + Metal^{2+} \rightarrow Metal^{3+} + OH^- + OH^{\cdot}$$

Release of ROS causes oxidative stress that leads to free radical which induced protein and DNA structure damage. Copper is an essential element for regulating protein and enzymatic activity in plants. But the oxidative property of copper oxide nanomaterial having a considerable toxic effect such as CuO NPs has a detrimental effect on the photosynthesis of *Elodea densa* (waterweed) at (1 mg/L) higher concentration (Nekrasova et al., 2011). Larue et al. (2012) reported a similar sort of result with high amount of TiO_2 NPs. According to the study, 36 nm Titanium dioxide has been stored in the wheat root parenchyma. Several hydroponic phytotoxic experiments have been reported in previous studies, and their experimental results demonstrate the failed and detrimental effect of metal oxide nanoparticles on plants. For example, amendment of CeO_2 NPs over the cotton plants had hampered the function of the vascular bundle and reduced indole-3-acetic acid as well as abscisic acid (Nhan et al., 2015).

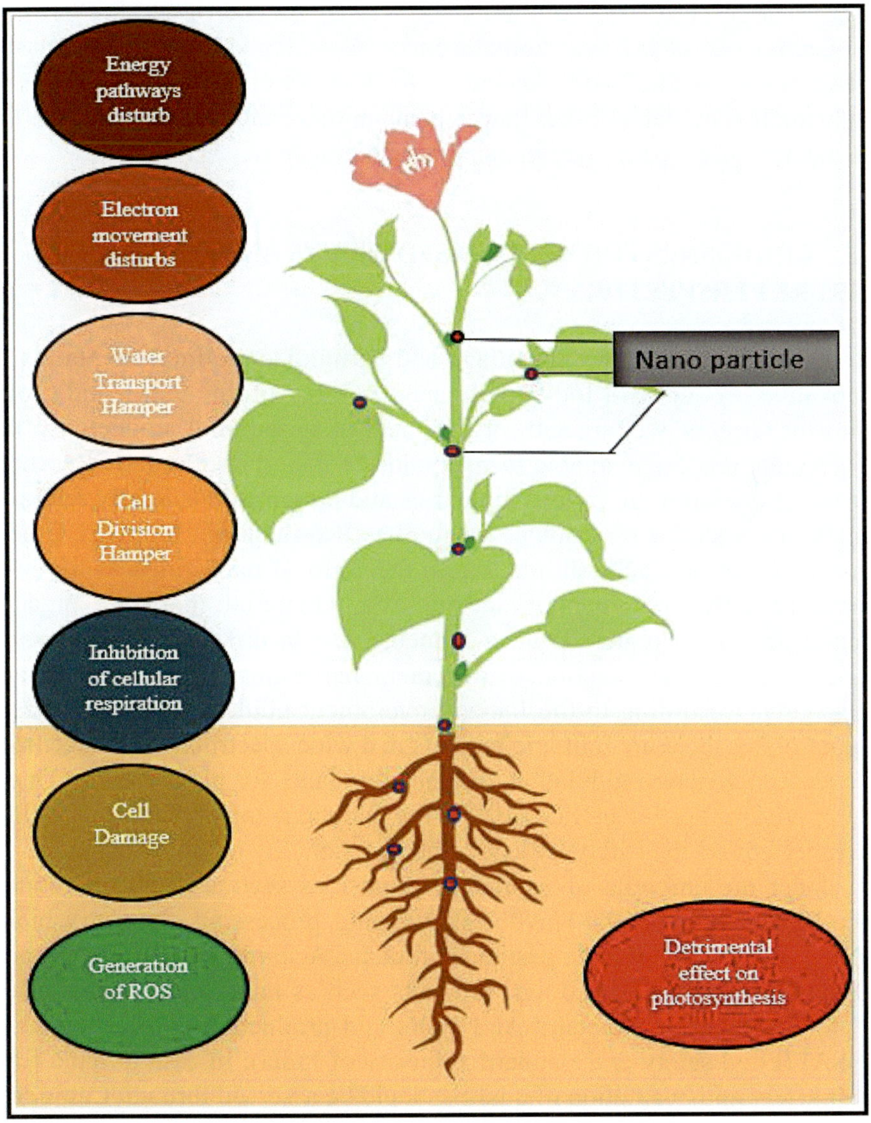

FIGURE 8.2 Implication of nanomaterials on plant health and metabolism.

NPs enter into the plant through root junction and its uptake through various physiological, chemical barriers. After translocation inside the plant, it first confronts with the cell wall. The cell wall permits the smaller particle and hinders the larger mobility (Dietz and Herth, 2011; Rastogi et al., 2017). But, sometimes, larger nanoparticles also succeed in entering cell walls through large pores and facilitate endocytosis (Etxeberria et al., 2016). If some remnants are left over from the endocytosis, it goes through symplastic transport (Ma et al., 2010). According to Wong et al. (2016), the proposed mathematical model specifies that size, magnitude, and zeta potential are the most determining parameter in nanoparticle transport.

8.5 GUIDELINES FOR NANO-BASED INPUTS AND FOCUS ON ITS FURURE PERSPECTIVE

Guidelines would always encourage and harmonize the implementation of innovative products. In India, there are a number of guidelines and provisions for farmers' welfare and enhancement of agricultural productivity, but few regulations are available to maintain the nano-based inputs. Recently, the conglomeration of various ministries and organizations jointly release a regulatory guideline for evaluating nano-based agricultural additives. Guidelines based on agricultural additive in the form of nano finished formulation as an active ingredient(s) made up of any material: inorganic, organic, composite. These materials have immense use in different dimensions of agriculture such as protection, management, harvesting, post-harvesting, and packaging. According to the Indian government guidelines on nano-based input (2020), there are multiple benefits in a wide spectrum of management, prevention, genetic modulation, and biostimulants for plant growth. Under the provision, Fertilizer (FCO,1985), the nano agri-input products (NAIPs) category covers the safety, efficacy, and functionality of nano fertilizers.

Under the umbrella of regulatory provisions section 9, the Insecticide Act, 1968 (Act 46 of 1968) involves packaging, processing, chemistry, bioefficacy, and toxicity of nano pesticide products. According to the government regulation, nano generated agri products such as nano food should follow all food safety and Standard Act, 2006, and products may be adopted by FSSAI (Food safety and standard authority of India). In case of nano feed, their safety and evaluation of quality should be assessed through Cattle feed (Regulation of Manufacture and Sale) Order, 2009, and with supplementary criteria for inclusion, FSSAI may employ it. But implementation of standards would always do under the regulation of BIS (Bureau of Indian Standard).

In the future prospective ENPs application in food safety and agriculture has increased due to their distinctive size and surface-related features when compared to their bulk or ionic analogues (Nuruzzaman et al., 2016; Dimkpa et al., 2019). ENPs could be used in agriculture as nanofertilizers, growth regulators, or nanopesticides to combat plant and zoonotic diseases. Furthermore, due to their increased surface reactivity, ENPs could be used as a sensor in the early detection of infections and/or the breakdown of pesticidal residues (Xiong et al., 2017). ZnONPs with significantly better physiochemical properties could be used as a new fertilizer to increase food quality and agricultural productivity (Tanha et al., 2020).

8.6 CONCLUSIONS

The application of nanoscience/nanotechnology in agriculture is very peculiar and novel. Flaws in conventional techniques stimulate the contemporary research in the field of nanoparticles. Amendment of engineered nanoparticles in the agricultural field led to improved germination, growth, and plant productivity. The list of advantages is so long, such as carbon nanotubes, oxides of metal used for amelioration soil from heavy metals, and polychlorineated biphenyls. The potential of nanotechnology will revolutionize the agricultural sector and allied fields such as food security, safety, and management. With lots of benefits, some inevitable implications are also associated with nanotechnology. It expresses consolidated toxicological response on soil and plant regime such as inappropriate amount, size, and exposure time suppresses the transcription gene, generating reactive oxygen species—Indian government launch recent guideline for evaluation and safety of nano-based food and fodders. But for better application worldwide, nanotechnology requires wide-ranging databank and international alliance for dissemination of policy, regulation, and innovations in this field. A strategic roadmap of implementation will reduce the downside of nanotechnology.

ACKNOWLEDGMENTS

The authors are grateful to Director and Dean, Institute of Environment and Sustainable Development (IESD), Banaras Hindu University (BHU), Varanasi, India. The authors express their sincere thanks to Head, Department of Environment and Sustainable Development, IESD, BHU for providing the necessary help. RPS is thankful to authorities of Banaras Hindu University (BHU), Varanasi, India for providing support under IOE (Institute of Excellence) scheme. AD thanks DBT for DBT-RA fellowship.

KEYWORDS

- **amelioration**
- **guidelines**
- **nanoparticle**
- **nanotechnology**
- **plant growth**
- **sustainable agriculture**

REFERENCES

Abdel-Aziz, M. S.; Abou-El-Sherbini, K. S.; Hamzawy, E. M.; Amr, M. H.; El-Dafrawy, S. Green Synthesis of Silver Nano-Particles by Macrococcus Bovicus and its Immobilization Onto Montmorillonite Clay for Antimicrobial Functionality. *Appl. Biochem. Biotechnol.* **2015,** *176*(8), 2225–2241.

Achari, G. A.; Kowshik, M. Recent Developments on Nanotechnology in Agriculture: Plant Mineral Nutrition, Health, and Interactions With Soil Microflora. *J. Agric. Food Chem.* **2018,** *66*(33), 8647–8661.

Acharya, P.; Jayaprakasha, G. K.; Crosby, K. M.; Jifon, J. L.; Patil, B. S. Nanoparticle-Mediated Seed Priming Improves Germination, Growth, Yield, and Quality of Watermelons (Citrullus lanatus) at Multi-Locations in Texas. *Sci. Rep.* **2020,** *10*(1), 1–16.

Asadishad, B.; Chahal, S.; Akbari, A.; Cianciarelli, V.; Azodi, M.; Ghoshal, S.; Tufenkji, N. Amendment of Agricultural Soil With Metal Nanoparticles: Effects on Soil Enzyme Activity and Microbial Community Composition. *Environ. Sci. Technol.* **2018,** *52*(4), 1908–1918.

Aslani, F.; Bagheri, S.; Muhd Julkapli, N.; Juraimi, A. S.; Hashemi, F. S. G.; Baghdadi, A. Effects of Engineered Nanomaterials on Plants Growth: An Overview. *Sci. World J.* **2014,** *2014,* 641759.

Aziz, H. M. A.; Hasaneen, M. N.; Omer, A. M. Nano Chitosan-NPK Fertilizer Enhances the Growth and Productivity of Wheat Plants Grown in Sandy Soil. *Spanish J. Agric. Res.* **2016,** *14*(1), 17.

Bagheri, H.; Afkhami, A.; Saber-Tehrani, M.; Khoshsafar, H. Preparation and Characterization of Magnetic Nanocomposite of Schiff Base/Silica/Magnetite as a Preconcentration Phase for the Trace Determination of Heavy Metal Ions in Water, Food and Biological Samples Using Atomic Absorption Spectrometry. *Talanta* **2012,** *97,* 87–95

Barzegar, G.; Jorfi, S.; Soltani, R. D. C.; Ahmadi, M.; Saeedi, R.; Abtahi, M.; Ramavandi, B.; Baboli, Z.. Enhanced Sono-Fenton-Like Oxidation of PAH-Contaminated Soil Using Nano-Sized Magnetite as Catalyst: Optimization With Response Surface Methodology. *Soil Sediment Contam. Int. J.* **2017,** *26*(5), 538–557.

Bhattacharya, S.; Shilpa, M. Mapping Nanotechnology Research and Innovation in India. *DESIDOC J. Libr. Inf. Technol.* **2011,** *31*(5).

Birben, E.; Sahiner, U. M.; Sackesen, C.; Erzurum, S.; Kalayci, O. Oxidative Stress and Antioxidant Defense. *World Allergy Organ. J.* **2012,** *5*(1), 9–19.

Boom, R. M. Nanotechnology in Food Production. Nanotechnology in the Agri-Food Sector, 2011; pp 37–57.

Cao, J.; Feng, Y.; Lin, X.; Wang, J. Arbuscular Mycorrhizal Fungi Alleviate the Negative Effects of Iron Oxide Nanoparticles on Bacterial Community in Rhizospheric Soils. *Front. Environ. Sci.* **2016,** *4,* 10.

Chai, H.; Yao, J.; Sun, J.; Zhang, C.; Liu, W.; Zhu, M.; Ceccanti, B. The Effect of Metal Oxide Nanoparticles on Functional Bacteria and Metabolic Profiles in Agricultural Soil. *Bull. Environ. Contam. Toxicol.* **2015,** *94*(4), 490–495.

Chinnamuthu, C. R.; Boopathi, P. M. Nanotechnology and Agroecosystem. *Madras Agric. J.* **2009,** *96*(1/6), 17–31.

DeRosa, M. C.; Monreal, C.; Schnitzer, M.; Walsh, R.; Sultan, Y. Nanotechnology in Fertilizers. *Nat. Nanotechnol.* **2010,** *5*(2), 91–91.

Dietz, K. J.; Herth, S. Plant Nanotoxicology. *Trends Plant Sci.* **2011,** *16*(11), 582–589.

Doshi, R.; Braida, W.; Christodoulatos, C.; Wazne, M.; O'Connor, G. Nano-aluminum: Transport Through Sand Columns and Environmental Effects on Plants and Soil Communities. *Environ. Res.* **2008,** *106*(3), 296–303.

Elsheery, N. I.; Sunoj, V. S. J.; Wen, Y.; Zhu, J. J.; Muralidharan, G.; Cao, K. F. Foliar Application of Nanoparticles Mitigates the Chilling Effect on Photosynthesis and Photoprotection in Sugarcane. *Plant Physiol. Biochem.* **2020,** *149,* 50–60.

El-Temsah, Y. S.; Joner, E. J. Impact of Fe and Ag Nanoparticles on Seed Germination and Differences in Bioavailability During Exposure in Aqueous Suspension and Soil. *Environ. Toxicol.* **2012,** *27*(1), 42–49.

Etxeberria, E.; Gonzalez, P.; Bhattacharya, P.; Sharma, P.; Ke, P. C. Determining the Size Exclusion for Nanoparticles in Citrus Leaves. *HortScience* **2016,** *51*(6), 732–737.

Fajardo, C.; Saccà, M. L.; Costa, G.; Nande, M.; Martin, M. Impact of Ag and Al2O3 Nanoparticles on Soil Organisms: In vitro and Soil Experiments. *Sci. Total Environ.* **2014,** *473,* 254–261.

Feng, Q. L.; Wu, J.; Chen, G. Q.; Cui, F. Z.; Kim, T. N.; Kim, J. O. A Mechanistic Study of the Antibacterial Effect of Silver Ions on Escherichia coli and Staphylococcus Aureus. *J. Biomed. Mater. Res.* **2000,** *52*(4), 662–668.

Fu, P. P.; Xia, Q.; Hwang, H. M.; Ray, P. C.; Yu, H. Mechanisms of Nanotoxicity: Generation of Reactive Oxygen Species. *J. Food and Drug Analy.* **2014,** *22*(1), 64–75.

Ganesh-Kumar, A., Mehta, R.; Pullabhotla, H.; Prasad, S. K.; Kavery, G.; Ashok, G. Demand and Supply of Cereals in India 2010–2025. *IFPRI-Discussion Papers*, 2012, (1158).

Gao, J., Wang, Y.; Folta, K. M.; Krishna, V.; Bai, W.; Indeglia, P.; Georgieva, A.; Nakamura, H.; Koopman, B.; Moudgil, B. Polyhydroxy Fullerenes (fullerols or fullerenols): Beneficial Effects on Growth and Lifespan in Diverse Biological Models. *PLoS One* **2011,** *6*(5), e19976.

Ghormade, V.; Deshpande, M. V.; Paknikar, K. M. Perspectives for Nano-biotechnology Enabled Protection and Nutrition of Plants. *Biotechnol. Adv.* **2011,** *29*(6), 792–803.

Grillo, R.; Rosa, A. H.; Fraceto, L. F. Engineered Nanoparticles and Organic Matter: A Review of the State-of-the-art. *Chemosphere* **2015**, 119, 608–619.

Huke, R. E. The Green Revolution. *J. Geog.* **2007,** *84*(6), 248–254.

Hussain, I.; Singh, N. B.; Singh, A.; Singh, H.; Singh, S. C. Green synthesis of Nanoparticles and its Potential Application. *Biotechnol. Lett.* **2016,** *38*(4), 545–560.

Iavicoli, I.; Leso, V.; Beezhold, D. H.; Shvedova, A. A. Nanotechnology in Agriculture: Opportunities, Toxicological Implications, and Occupational Risks. *Toxicol. Appl. Pharmacol.* **2017,** *329,* 96–111.

Joshi, M.; Adak, B.; Butola, B. S. Polyurethane Nanocomposite-Based Gas Barrier Films, Membranes and Coatings: A Review on Synthesis, Characterization and Potential Applications. *Progress Mater. Sci.* **2018,** *97,* 230–282.

Kah, M.; Tufenkji, N.; White, J. C. Nano-enabled Strategies to Enhance Crop Nutrition and Protection. *Nature Nanotechnol.* **2019,** *14*(6), 532–540.

Kalia, A.; Sharma, S. P.; Kaur, H.; Kaur, H. Novel Nanocomposite-Based Controlled-Release Fertilizer and Pesticide Formulations: Prospects and Challenges. In *Multifunctional Hybrid Nanomaterials for Sustainable Agri-Food and Ecosystems*, 2020; pp 99–134.

Keller, A. A.; McFerran, S.; Lazareva, A.; Suh, S. Global Life Cycle Releases of Engineered Nanomaterials. *J. Nanoparticle Res.* **2013,** *15*(6), 1–17.

Khodakovskaya, M. V.; de Silva, K.; Nedosekin, D. A.; Dervishi, E.; Biris, A. S.; Shashkov, E. V.; Galanzha, E. I.; Zharov, V. P. Complex genetic, photothermal, and photoacoustic analysis of nanoparticle-plant interactions. *Proc. Natl. Acad. Sci.* **2011,** *108*(3), 1028–1033.

Khodakovskaya, M.; Dervishi, E.; Mahmood, M.; Xu, Y.; Li, Z.; Watanabe, F.; Biris, A. S. Retraction Notice for Carbon Nanotubes are Able to Penetrate Plant Seed Coat and Dramatically Affect Seed Germination and Pant Growth. *ACS Nano* **2012,** *6*(8), 7541–7541.

Lahiani, M. H.; Dervishi, E.; Chen, J.; Nima, Z.; Gaume, A.; Biris, A. S.; Khodakovskaya, M. V. Impact of Carbon Nanotube Exposure to Seeds of Valuable Crops. *ACS Appl. Mater. Interfaces* **2013,** *5*(16), 7965–7973.

Lakshmi, R. V.; Bharathidasan, T.; Bera, P.; Basu, B. J. Fabrication of Superhydrophobic and Oleophobic Sol–gel Nanocomposite Coating. *Surf. Coat. Technol.* **2012,** *206*(19–20), 3888–3894.

Larue, C.; Laurette, J.; Herlin-Boime, N.; Khodja, H.; Fayard, B.; Flank, A. M.; Brisset, F.; Carriere, M. Accumulation, Translocation and Impact of TiO2 Nanoparticles in Wheat (Triticum aestivum spp.): Influence of Diameter and Crystal Phase. *Sci. Total Environ.* **2012,** *431,* 197–208.

Le, T. T.; Nguyen, K. H.; Jeon, J. R.; Francis, A. J.; Chang, Y. S. Nano/bio Treatment of Polychlorinated Biphenyls With Evaluation of Comparative Toxicity. *J. Hazard. Mater.* **2015,** *287,* 335–341.

Lee, W. M.; Kim, S. W.; Kwak, J. I.; Nam, S. H.; Shin, Y. J.; An, Y. J. Research Trends of Ecotoxicity of Nanoparticles in Soil Environment. *Toxicol. Res.* **2010,** *26*(4), 253–259.

Li, B.; Chen, Y.; Liang, W. Z.; Mu, L.; Bridges, W. C.; Jacobson, A. R.; Darnault, C. J. Influence of Cerium Oxide Nanoparticles on the Soil Enzyme Activities in a Soil-Grass Microcosm System. *Geoderma* **2017,** *299,* 54–62.

Liang, B.; Xie, Y.; Fang, Z.; Tsang, E. P. Assessment of the Transport of Polyvinylpyrrolidone-Stabilised Zero-valent Iron Nanoparticles in a Silica Sand Medium. *J. Nanoparticle Res.* **2014,** *16*(7), 1–11.

Liang, Q.; Zhao, D. Immobilization of Arsenate in a Sandy Loam Soil Using Starch-Stabilized magnetite Nanoparticles. *J. Hazard. Mater.* **2014,** *271,* 16–23.

Lowry, G. V.; Avellan, A.; Gilbertson, L. M. Opportunities and Challenges for Nanotechnology in the Agri-tech Revolution. *Nat. Nanotechnol.* **2019,** *14*(6), 517–522.

Ma, X.; Geiser-Lee, J.; Deng, Y.; Kolmakov, A. Interactions Between Engineered Nanoparticles (ENPs) and Plants: Phytotoxicity, Uptake and Accumulation. *Sci. Total Environ.* **2010,** *408*(16), 3053–3061.

Maliszewska, I. Effects of the Biogenic Gold Nanoparticles on Microbial Community Structure and Activities. *Ann. Microbiol.* **2016,** *66*(2), 785–794.

Manjunatha, S. B.; Biradar, D. P.; Aladakatti, Y. R. Nanotechnology and its Applications in Agriculture: A Review. *J. Farm Sci.* **2016,** *29*(1), 1–13.

Mazaheri-Tirani, M.; Dayani, S. In Vitro Effect of Zinc Oxide Nanoparticles on Nicotiana Tabacum Callus Compared to ZnO Micro Particles and Zinc Sulfate (ZnSO 4). *Plant Cell Tissue Organ Culture* (PCTOC) **2020,** *140*(2), 279–289.

Meghana, K. T.; Wahiduzzaman, M. D.; Vamsi, Golla. Nano Fertilizers in Agriculture. *Acta Sci. Agric.* **2021,** *5*(3) 35–46.

Mukhopadhyay, S. S. Nanotechnology in Agriculture: Prospects and Constraints. *Nanotechnol. Sci. Appl.* **2014**, *7,* 63.

Nair, R. Effects of Nanoparticles on Plant Growth and Development. In *Plant Nanotechnology*; Springer: Cham, 2016; pp 95–118.

National Center for Biotechnology Information, d. PubChem database. Zinc sulfate, CID= 24424 [Online]. https://pubchem.ncbi.nlm.nih.gov/compound/Zinc-sulfate.

Nekrasova, G. F.; Ushakova, O. S.; Ermakov, A. E.; Uimin, M. A.; Byzov, I. V. Effects of Copper (II) Ions and Copper Oxide Nanoparticles on Elodea Densa Planch. *Russ. J. Ecol.* **2011,** *42*(6), 458–463.

Nelson, A. R. L. E.; Ravichandran, K.; Antony, U. The Impact of the Green Revolution on Indigenous Crops of India. *J. Ethnic Foods* **2019,** *6*(1), 1–10.

Nhan, L. V.; Ma, C.; Rui, Y.; Liu, S.; Li, X.; Xing, B.; Liu, L. Phytotoxic Mechanism of Nanoparticles: Destruction of Chloroplasts and Vascular Bundles and Alteration of Nutrient Absorption. *Sci. Rep.* **2015,** *5,* 11618.

Ninawe, S.; Krishna, A V. Guidelines for Evaluation of Nano-Based Agri- Input and Food Products in India, 2020 [Online]. https://www.jatinverma.org.

Nouri, Z.; Hajialyani, M.; Izadi, Z.; Bahramsoltani, R.; Farzaei, M. H.; Abdollahi, M. Nanophytomedicines for the Prevention of Metabolic Syndrome: A Pharmacological and Biopharmaceutical Review. *Front. Bioeng. Biotechnol.* **2020,** *8,* 425.

Oliver, M. J. Why We Need GMO Crops in a Agriculture. *Missouri Med.* **2014,** *111*(6), 492.

Ombódi, A.; Saigusa, M. Broadcast Application Versus Band Application of Polyolefin-Coated Fertilizer on Green Peppers Grown on Andisol. *J. Plant Nutr.* **2000,** *23*(10), 1485–1493.

Pandey, G. Challenges and Future Prospects of Agri-nanotechnology for Sustainable Agriculture in India. *Environ. Technol. Innov.* **2018,** *11,* 299–307.

Pérez-de-Luque, A.; Hermosín, M. C. Nanotechnology and its use in Agriculture. In *Bio-Nanotechnology: A Revolution in Food, Biomedical and Health Sciences*, 2013; pp 383–398.

Prasad, R.; Bhattacharyya, A.; Nguyen, Q. D. Nanotechnology in Sustainable Agriculture: Recent Developments, Challenges, and Perspectives. *Front. Microbiol.* **2017,** *8,* 1014.

Qian, Y.; Qin, C.; Chen, M.; Lin, S. Nanotechnology in Soil Remediation− Applications Vs. Implications. *Ecotoxicol. Environ. Aaf.* **2020,** *201,* 110815.

Rajput, V. D.; Minkina, T.; Sushkova, S.; Tsitsuashvili, V.; Mandzhieva, S.; Gorovtsov, A.; Nevidomskyaya, D.; Gromakova, N. Effect of Nanoparticles on Crops and Soil Microbial Communities. *J. Soils Sediment* **2018,** *18*(6), 2179–2187.

Rajput, V.; Singh, A.; Minkina, T.; Shende, S.; Kumar, P.; Verma, K.; Bauer, T.; Gorobtsova, O.; Deneva, S.; Sindireva, A. Potential Applications of Nanobiotechnology in Plant Nutrition and Protection for Sustainable Agriculture. In *Nanotechnology in Plant Growth Promotion and Protection*, 2021; pp 79–92.

Raman, R. The Impact of Genetically Modified (GM) Crops in Modern Agriculture: A Review. *GM Crops Food* **2017,** *8*(4), 195–208.

Rastogi, A.; Zivcak, M.; Sytar, O.; Kalaji, H. M.; He, X.; Mbarki, S.; Brestic, M. Impact of Metal and Metal Oxide Nanoparticles on Plant: A Critical Review. *Front. Chem.* **2017,** *5*, 78.

Schlich, K.; Hund-Rinke, K. Influence of Soil Properties on the Effect of Silver Nanomaterials on Microbial Activity in Five Soils. *Environ. Pollut.* **2015,** *196,* 321–330.

Sebastian, A.; Nangia, A.; Prasad, M. N. V. Advances in Agrochemical Remediation Using Nanoparticles. In *Agrochemicals Detection, Treatment and Remediation*; Butterworth-Heinemann: Oxford, 2020; pp. 465–485.

Sekhon, B. S. Nanotechnology in Agri-food Production: An Overview. *Nanotechnol. Sci. Appl.* **2014,** *7,* 31.

Shang, Y.; Hasan, M.; Ahammed, G. J.; Li, M.; Yin, H.; Zhou, J. Applications of Nanotechnology in Plant Growth and Crop Protection: A Review. *Molecules* **2019,** *24*(14), 2558.

Shennan, C.; Krupnik, T. J.; Baird, G.; Cohen, H.; Forbush, K.; Lovell, R. J.; Olimpi, E. M. Organic and Conventional Agriculture: A Useful Framing? *Ann. Rev. Environ. Resour.* **2017,** *42,* 317–346.

Shukla, P.; Chaurasia, P.; Younis, K.; Qadri, O. S.; Faridi, S. A.; Srivastava, G. Nanotechnology in Sustainable Agriculture: Studies From Seed Priming to Post-Harvest Management. *Nanotechnol. Environ. Eng.* **2019,** *4*(1), 11.

Siddiqui, M. H.; Al-Whaibi, M. H.; Firoz, M.; Al-Khaishany, M. Y. Role of Nanoparticles in Plants. In *Nanotechnology and Plant Sciences*, 2015; pp 19–35.

Silva, J. G. D. The Future of Food and Agriculture, Trends and challenges. Food and Agriculture Organization of the United Nations, 2017 [Online]. http://www.fao.org

Singh, A.; Rajput, V. D., Mehrotra, R.; Pal, N.; Singh, V. K.; Minkina, T.; Chokheli, V. A.; Singh, R. K. In *Sustainable Soil Fertility Management*; Nova. Sci. Publishers, Inc., 2020; vol 1, pp 73–100.

Singh, A.; Rajput, V. D., Rawat, S.; Sharma, R.; Singh, A. K.; Singh, A. K.; Tomar, R. S. In *Emerging Tools for SustainableAgriculture and Food Security*; Rajput, D.; Book Agency: New Delhi, Delhi, 2021a; vol 1, pp 1–15.

Singh, A.; Rajput, V.; Singh, A.; Sengar, R.; Singh, R.; Minkina, T. Transformation Techniques and Their Role in Crop Improvements: A Global Scenario of GM Crops. *Policy Issues Genetically Modified Crops* **2021**b, *1,* 515–542.

Sonkaria, S.; Ahn, S. H.; Khare, V. Nanotechnology and its Impact on Food and Nutrition: A Review. *Recent Pat. Food Nutr. Agric.* **2012,** *4*(1), 8–18.

Srivastava, A.; Rao, D. P. Enhancement of Seed Germination and Plant Growth of Wheat, Maize, Peanut and Garlic Using Multiwalled Carbon Nanotubes. *Eur. Chem. Bull.* **2014,** *3*(5), 502–504.

Tan, X.; Wang, X.; Fang, M.; Chen, C. Sorption and Desorption of Th (IV) on Nanoparticles of Anatase Studied by Batch and Spectroscopy Methods. *Colloids Surf. A Physicochem. Eng. Asp.* **2007,** *296*(1–3), 109–116.

Tilman, D.; Cassman, K. G.; Matson, P. A.; Naylor, R.; Polasky, S. Agricultural Sustainability and Intensive Production Practices. *Nature* **2002,** *418*(6898), 671–677.

Towell, M. G.; Browne, L. A.; Paton, G. I.; Semple, K. T. Impact of Carbon Nanomaterials on the Behaviour of 14C-Phenanthrene and 14C-Benzo-[a] Pyrene in Soil. *Environ. Pollut.* **2011,** *159*(3), 706–715.

Usman, M.; Farooq, M.; Wakeel, A.; Nawaz, A.; Cheema, S. A.; ur Rehman, H.; Ashraf, I.; Sanaullah, M. Nanotechnology in Agriculture: Current Status, Challenges and Future Opportunities. *Sci. Total Environ.* **2020,** *721*, 137778.

Waalewijn-Kool, P. L.; Ortiz, M. D.; van Straalen, N. M.; van Gestel, C. A. Ecotoxicological Assessment of ZnO Nanoparticles to 'Folsomia Candida'. Vrije Universiteit, 2013.

Wang, C.; Lu, J.; Zhou, L.; Li, J.; Xu, J.; Li, W.; Zhang, L.; Zhong, X.; Wang, T. Effects of long-term exposure to zinc oxide nanoparticles on development, zinc metabolism and biodistribution of minerals (Zn, Fe, Cu, Mn) in mice. *PloS One* **2016,** *11*(10), e0164434.

Wang, J.; Meng, G.; Tao, K.; Feng, M.; Zhao, X.; Li, Z.; Xu, H.; Xia, D.; Lu, J. R. Immobilization of Lipases on Alkyl Silane Modified Magnetic Nanoparticles: Effect of Alkyl Chain Length on Enzyme Activity. *PloS One* **2012,** *7*(8), 43478.

Wang, Y.; Pan, C.; Chu, W.; Vipin, A. K.; Sun, L. Environmental Remediation Applications of Carbon Nanotubes and Graphene Oxide: Adsorption and Catalysis. *Nanomaterials* **2019,** *9*(3), 439.

Wong, J. K.; Mohseni, R.; Hamidieh, A. A.; MacLaren, R. E.; Habib, N.; Seifalian, A. M. Will Nanotechnology Bring New Hope for Gene Delivery? *Trends Biotechnol.* **2017,** *35*(5), 434–451.

Xiong, T.; Dumat, C.; Dappe, V.; Vezin, H.; Schreck, E., Shahid, M.; Pierart, A.; Sobanska, S. Copper Oxide Nanoparticle Foliar Uptake, Phytotoxicity, and Consequences for Sustainable Urban Agriculture. *Environ. Sci. Technol.* **2017,** *51*(9), 5242–5251.

Xu, C.; Peng, C.; Sun, L.; Zhang, S.; Huang, H.; Chen, Y.; Shi, J. Distinctive Effects of TiO_2 and CuO Nanoparticles on Soil Microbes and Their Community Structures in Flooded Paddy Soil. *Soil Biol. Biochem.* **2015,** *86*, 24–33.

Yin, L.; Cheng, Y.; Espinasse, B.; Colman, B. P.; Auffan, M.; Wiesner, M.; Rose, J.; Liu, J.; Bernhardt, E. S. More Than the Ions: The Effects of Silver Nanoparticles on Lolium Multiflorum. *Environ. Sci. Technol.* **2011,** *45*(6), 2360–2367.

You, T.; Liu, D.; Chen, J.; Yang, Z.; Dou, R.; Gao, X.; Wang, L. Effects of Metal Oxide Nanoparticles on Soil Enzyme Activities and Bacterial Communities in Two Different Soil Types. *J. Soils Sediment.* **2018,** *18*(1), 211–221.

Yusefi-Tanha, E.; Fallah, S.; Rostamnejadi, A.; Pokhrel, L. R. Zinc Oxide Nanoparticles (ZnONPs) as a Novel Nanofertilizer: Influence on Seed Yield and Antioxidant Defense System in Soil Grown Soybean (Glycine max cv. Kowsar). *Sci. Total Environ.* **2020,** *738*, 140240.

PART IV

Emerging Nanotechnological Tools and Techniques for Crop Improvement

CHAPTER 9

Speed Breeding: Space-Inspired Crop Improvement in the Nano-Era

ANAMIKA KASHYAP[1], KUNAL TANWAR[1], POOJA GARG[1], SUJATA KUMARI[1], PHAM THI THU HA[2], SANJAY SINGH[1], and MAHESH RAO[1]

[1]ICAR-National Institute for Plant Biotechnology, Delhi, India

[2]Genomic Research Institute and Seed, Ton Duc Thang University, Vietnam

ABSTRACT

Crop improvement programs are enhanced regularly by introducing new technologies to produce enough for human consumption in the changing climatic conditions. One such recent approach as inspired by NASA's extra-terrestrial experiments is "speed breeding." The scientists have developed the speed breeding platform where crops can be grown at a faster rate in controlled conditions as compared with natural seasonal cycles. As speed breeding shortens the crop generation time, this strategy can be used for various research studies as well as breeding programs. The benefits harnessed from speed breeding are enormous over conventional breeding approaches. In this chapter, we describe in detail about speed breeding, how it was developed, methods used for speed breeding, how it functions, and where can it be applied for crop improvement.

9.1 INTRODUCTION

In order to satisfy the demand of the growing population, there is a need for increasing crop yield in the future. According to a prediction, by the

Nanotechnology for Sustainable Agriculture: An Innovative and Eco-Friendly Approach. Vishnu D. Rajput, Abhishek Singh, Tatiana M. Minkina, Krishan K. Verma, and Awani Kumar Singh (Eds.)

next 30 years, the human population is expected to grow by 25% and reach approximately 10 billion (Hickey et al., 2019). There are several methods for increasing the yield, but the present increasing scenario of genetic gain could not sufficiently meet the higher demand (Singh et al., 2020). It has been observed that breeding programs for crop improvement have somehow become stagnant and localized, and also, they have certain environmental influence (Singh et al., 2020). There is tremendous pressure on plant breeders to improve breeding programs in a particular duration of time and focus mainly on enhancing crop yields with better resistance to abiotic and biotic stress and resilience to climate change (Singh et al., 2021a). One of the major threats to world food security is climate change which is leading to an increase in temperature, CO_2 levels, pest and diseases, droughts, floods, decline of quality of the agricultural produce, and reduction in global crop yields (Peng et al., 2004; Nelson et al., 2009; Wassmann et al., 2009a; Newton et al., 2011; Lobell and Gourdji, 2012; IPCC, 2014; Sreenivasulu et al., 2015; Zhao et al., 2017; Van Oort and Zwart, 2018). Various improved cultivars or varieties via conventional breeding programs have been released worldwide in the last 5–6 decades. Traditional breeding programs are slow and time-consuming. It follows a fixed process of crossing between parents and then selfing in the next 4–6 generations to obtain homozygosity to evaluate agro-morphological traits further. To enhance the present level of crop productivity to meet the higher demand, an accelerated breeding process is required to shorten the generation time.

Climate change is unpredictable, and insufficient light during different crop development stages can hamper productivity, especially reduced yield and crop failure in the winter season in different parts of the world (Singh et al., 2021b). To overcome this yield loss and failure, the artificial light source was given to crops under controlled environment and in protected cultivation such as glasshouses. Plant tissue cultures were maintained under the artificial light sources in plant tissue culture laboratories (Mpelkas, 1980). Initial references of using artificial electric lamps, incandescent lamps, electric arc lamps, and light-emitting diode (LEDs) as artificial supplement to sunlight are also reported. Inspired by NASA's extraterrestrial experiments to grow wheat under constant light, a new concept of "speed breeding" was developed by the University of Queensland by utilizing controlled growth chambers that can shorten the life cycle of crops for commercial and research purposes as can be seen in Figure 9.1. Alongside, accelerated breeding additional growth parameters, which included immature seeds harvest was also investigated. The acceleration process is accomplished in fully enclosed, controlled environment growth chambers where mutant studies, phenotyping of adult plant traits, and transformation can be studied. At present, many generations of plants can be grown in a single year using speed breeding. For example, four generations per year of canola (*Brassica napus*), six generations in a year for barley (*Hordeum vulgare*), durum wheat (*Triticum durum*), spring wheat (*Triticum aestivum*), pea (*Pisum sativum*), and chickpea (*Cicer arietinum*) can be grown using speed breeding (Ghosh et al., 2018; Watson et al., 2018).

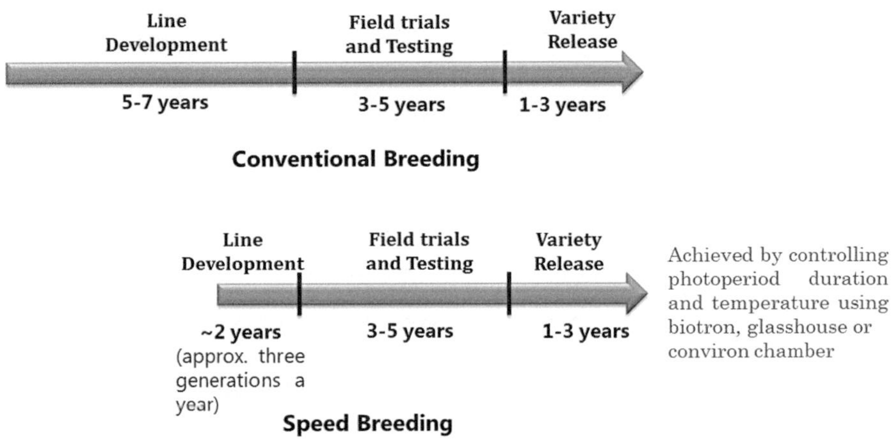

FIGURE 9.1 This figure represents the time duration of a variety development via conventional breeding with 1–2 generations per year and speed breeding with 3–6 generations per year. The time difference for line development is notable.

9.2 METHODS OF SPEED BREEDING

Speed breeding is a very old technique, but it has emerged significantly in the past few years. Thus, many different methods have been described that the breeders can employ depending on the crop, conditions, and cost. These methods are listed below. Tables 9.1 and 9.2 provide a brief overview of some of these methods.

- Speed breeding I (SB I)
- Speed breeding II (SB II)
- Speed breeding III (SB III)
- Greenhouse (Ochatt et al., 2002),
- Greenhouse RGT (O'Connor et al., 2013)
- In vitro only (Mobini et al., 2015)

- In vitro along with in vivo (Rizal et al., 2014; Ochatt et al., 2002)
- Biotron
- Modified controlled biotron
- Rapid generation advance (RGA) (Collard et al., 2017)

TABLE 9.1 Major Specifications and Differences in Speed Breeding I, II, and III Methods.

Parameters	SB I	SB II	SB III
Growth chamber	Conviron BDW chamber	Glasshouse	Homemade growth room design
Lights/lamps	The mixture of white LED bars, far-red LED lamps, and ceramic metal hydrargyrum quartz iodide lamps	High-pressure sodium vapor lamps	Sandwich paneling fitted with seven LB-8 LED light boxes
Light intensity	490–500 µmol m^{-2} s^{-1}(at adult plant height)	440–650 µmol m^{-2} s^{-1} (at adult plant height)	340–590 µmol m^{-2}(s^{-1} at adult plant height)
Photoperiod duration	22 h	22 h	12 h for 4 weeks, then increased to 18 h
Photoperiod temperature	22°C	22°C	21°C
Dark period duration	2 h	2 h	12 h for 4 weeks, then reduced to 6 h
Dark period temperature	17°C	17°C	18°C
Ramping	Yes. Natural dawn and dusk conditions were simulated by ramping up and down of light and temperature for 1 h and 30 min	No	No
Crops	Wheat, barley, *Brachypodium*, pea, and *Medicago truncatula* seeds	Wheat, barley, canola, and chickpea seeds	Wheat, barley, oat, and triticale
Reference	Watson et al., 2018	Watson et al., 2018	Watson et al., 2018

9.3 SPEED BREEDING IN CROPS

9.3.1 CEREALS

In temperate cereals, such as barley and wheat, speed breeding protocols are used to quicken growth and development by extending the photoperiods and controlling temperatures to produce six generations/year (Watson et

TABLE 9.2 This Table Highlights the Growth Conditions of the Two Breeding Systems. Biotron Breeding System was Developed as an Alternative Method to Grow Rice Under Artificial Climate Conditions, Further Modified by Rana et al. to Reduce Costs.

Parameters	Biotron	Modified controlled biotron
Photoperiod duration	11 h	14 h for 30 days 10 h afterwards
Photoperiod temperature	30°C	30°C
Dark period duration	13 h	10 h for 30 days 14 h afterwards
Dark period temperature	25°C	25°C
Relative humidity	70%	70%
CO2 level	20% (CO_2 gas cylinder)	No CO_2 supply to reduce cost
Crops	Rice cultivars	Rice cultivars (Kaijin, Yukinko-mai)
Other techniques used	Tiller removal, embryo rescue	Tiller removal, embryo rescue
Observations	Vigorous growth and sufficient seed production with accelerated flowering time. Nipponbare life cycle was shortened to about 2 months	"YNU31-2-4" salt-tolerant introgression line was developed in six generations within a short span of 17 months
Reference	Ohnishi et al., 2011	Rana et al., 2019

al., 2018). Protocols for speed breeding tropical short-day plants, such as cereal crops and orphan grass have yet to be developed. In *Amaranthus,* a short-day plant, flowering occurs between 30 and 50 days after sowing, so to synchronize the flowering, speed breeding protocols can be helpful in breeding strategies. It has been possible to promote early flowering in late-flowering lines with late flowering and promote production and quick development of new varieties by this technique. Speed breeding has been done in amaranth by providing the high temperature and long-day conditions to promote robust vegetative growth. Transfers to short-day conditions induce flowering almost immediately in short-day species. This technique is also helpful in synchronizing the flowering times and in promoting hybridization in amaranth. At the University of Queensland, Australia, this amaranth protocol has been trialed in sorghum (*Sorghum bicolor*), a short-day crop (Joshi et al., 2018; Stetter et al., 2016).

The phenotyping of a population for resistance in the field is reduced to once a year because it is affected by weather conditions. In wheat, identification of plant resistance for leaf rust at the adult stage under controlled growth conditions at both juvenile and adult growth stages was done. Genotypes of spring wheat having known APR genes, such as *Lr34 and Lr46* were used under field and controlled conditions (Riaz et al., 2016).

9.3.2 LEGUMES AND OILSEEDS

In orphan legumes, such as in lupin (*Lupinus sp.*), chickpea (*Cicer arietinum*), lentil (*Lens culinaris*), and faba bean (*Vicia faba*) generation time has been reduced by speed breeding protocols. Different light spectra, red to far-red were used to optimize early flowering, which speeds up generation turnover. It was also examined that the environment with R: FR ratio below 3.5 with the highest light intensity in the FR region was highly effective (Croser et al., 2016). In chickpea (*Cicer arietinum*), early flowering was induced by providing 24-h day-length incandescent light. Plants grown under 24 h day length showed slight increase in seed size, the number of pods, and plant height. This technique allows us to obtain more than one generation/year (Sethi et al., 1981). In the case of subterranean clover (*Trifolium subterraneum*), in vitro-assisted single seed descent (IVASSD) technique was used to obtain 2.7–6.1 generations per year. The IVASSD protocol was useful in accelerating the generation cycle by minimizing the time of floral initiation by growing under extended photoperiod and appropriate temperature, followed by reducing the seed filling period and germination of immature seed in vitro on B5. Normal and fertile F_7 pants were obtained (Pazos et al., 2017). Procedures have also been developed in broad beans (*Vicia faba*) and lentil (*Lens culinaris*) (Lulsdorf and Banniza, 2018; Mobini et al., 2015). These crops are of greater interest because they have become a part of the agro-based food supply chain, and cost-efficient breeding programs are required to combat disease or climate challenges (Pazos et al., 2017; Kole, 2007).

Speed breeding was performed in peanuts and was found to be successful in reducing generation time (O'Connor et al., 2013). In *Brassica napus*, speed breeding procedure was performed using two methods. The first method is the "all soil" method to achieve five generations/year, where plants were provided water stress, long lighting hours, and high temperature. Using the second method, "embryo culture plus soil," seven generations were achieved in a year. It is also used for RILs and NILs production in a research project, and the extensive use of this method could promote canola breeding along with biological studies (Yao et al., 2016).

In oilseed crops like soybean (*Glycine max*), in which the reproductive cycle contributes 50% of the total life cycle, immature seeds were used in speed breeding protocol to shorten the reproductive cycle by pretreatment of seeds and exposing pods to slight dehydration for 4 days at a temperature of 26 + 2°C, 14 h/10 h day/night, 70/100% relative humidity. Later on, excised embryos were provided germination conditions. This method is useful for the development of recombinant lines (Roumet and Morin, 1997). The same immature seed culturing was also performed in sunflowers for shortening the breeding cycle to produce four generations/annum. The immature embryos of 10–12 DAP were cut and cultured into MS medium. The developed young plantlets were then transferred in a viol containing peat, perlite, and soil mixture (v/v) in 1:1:2 ratio in the growth chamber, developed to maturity, self-pollinated, and set seed but plants developed reduced agronomic characters as compared with field experiment (Dagustu et al., 2012).

9.3.3 ROOTS, TUBERS, AND BANANAS

Being multiplied by clonal propagation, these crops lack diversity, and thus, only few clones are available for millions of people (Heslop and Schwarzacher, 2007). Therefore, these crops are highly at risk for disease and pest, and the livelihood of people depending on these crops is at risk because it has been reported that their incidence and impact increase due to a variety of socioeconomic and biophysical factors (Godfray et al., 2010; Ehrlich and Harte, 2015; Sundstrom et al., 2014). Speed breeding will be beneficial in introducing required traits in such crops to surmount the diseases like cassava brown streak disease and banana bacterial wilt. The seeds of these crops have no economic importance to farmers. These crops are mostly grown via vegetative propagation. These crops have either late flowering or sometimes they flower only under special conditions.

In root crop cassava, plants were provided no light at night along with extended photoperiod by using red light-emitting diodes with a range of 625–635 nm wavelength for three seasons. EP induced early flowering in erect-plant genotypes (Pineda et al., 2020). In cassava, speed breeding has also been performed by grafting to increase flowering. There was an increase in shoot production on grafted cassava plants. Grafting of genotypes with high flowering rates onto genotypes with low flowering rates can result in flowering in genotypes with low flowering rates (Souza et al., 2018). In cassava, if genomics-assisted breeding approaches estimate the breeding value of individuals plants, then progenies can be grown rapidly for the next generation using speed breeding (Wolfe et al., 2017). For clonally propagated crops and RTBs, the development of elite clones is the main aim of breeding programs, for example, in root crop Cocoyam, Wilson (1979) developed elite clones by foliar treatment of plants with an aqueous solution of the potassium salt of GA_3 at different concentration at 3–4 leaf stage. The large number of inflorescences was recorded after 120 days when dry season terminated the experiment. Hand pollination was performed, and after fruit development, seeds were collected and grown in MS medium, and then seedlings were transplanted into greenhouse, followed by field transfer.

In the potato breeding program, the *sli* gene was introduced into potato germplasm using speed breeding by crossing two donors D1 and D2, to generate F_1 progenies. The progenies were grown in the winter in a glass-house, and the temperature was set at 20/10°C for day/night. Self-compatible F_2 plants obtained were selfed to get F_3 progenies (Lindhout et al., 2011). The International Potato Centre developed new elite clones of potato showing heat tolerance, resistance against late blight, and potato virus Y by selecting clones where the night and day temperatures were 21°C and 27°C, respectively for heat tolerance. Further selection for late blight resistance by exposing them with the pathogen and selection for viral resistance were done by inoculation and grafting procedure (Gastelo et al., 2015). Suppose advanced technology, such as gene mapping, genetic analysis, genomic selection, and QTL are used in potato breeding programs. In that case, it will be possible to tag breeding traits to specific DNA markers for their use for selection to speed up hybrid seed development (Ortiz, 2020).

9.3.4 FRUIT TREES

In most fruit-bearing trees, flowering occurs only after a juvenile phase that can last from few to more than 20 years (Korbo et al., 2013). Therefore, to speed up fruit tree breeding, efforts are laid to reduce the juvenile period. Such techniques have been observed to increase vegetative development and flowering in a fragment of the period; for example, for apple (*Malus × Domestica*), in 10 months rather than 5 years (van Nocker and Gardiner, 2014) and chestnut (*Castanea sativa*) in 2 years rather than 7 years (Baier et al., 2012). Due to tree crops, extra resources and protocol modifications are needed so that the plants, at a certain height, can be managed in the controlled-environment facilities. In some species, the seed-to-seed interval can be removed by breaking seed dormancy (van Nocker and Gardiner, 2014). It has been reported that transgenic early flowering plants can be used for inducing early flowering in apple and combined with MAS process to speed up the breeding process. To achieve this, transgenic apple lines with early flowering were evaluated in glasshouse conditions showing overexpression of the BpMADS4 gene encoding tree morphology. The selected line was analyzed and used in fast-breeding (Flachowsky et al., 2011). Speed breeding has been applied in several crops for different traits. A summary of such crops has been given in Table 9.3.

9.4 APPLICATIONS

The most significant advantage of speed breeding technology is the "rapid advancement of generations." Due to shorter generation time, research and breeding programs are accelerated at great rates. This technology can be combined with other breeding and molecular approaches, such as genomic selection, marker-assisted selection, gene editing for faster outcomes (Shivakumar et al., 2018).

9.4.1 *SPEED BREEDING FOR VARIETY DEVELOPMENT*

The scientists at ICAR-Indian Institute of Soybean Research (ICAR-IISR), Indore are working on varietal development and are taking up off-season (January–April) generation at Bengaluru. The advancement of F_1 to F_2 generations is done at greenhouse and poly house facilities at ICAR-IISR, Indore, India. Further, the selected segregating material is advanced via speed breeding to pace the variety development (Shivakumar et al., 2018).

There was another such research conducted by O'Connor et al. in peanut. Variety release is usually a lengthy process of about 10–15 years, but speed breeding can be shortened. With this approach, the life cycle of maturity cultivars is reduced from 15 to 89 days, and the inbreeding of F_2, F_3, and F_4 generations can rapidly progress within a year. This can lead to the rapid commercial release of a new variety within 6 to 7 years (O'Connor et al., 2013).

TABLE 9.3 The Table Summarizes the Crops Grown Using Different Methods of Speed Breeding and Their Targeted Traits and Selection Methods.

Crop	Technique	Trait targeted	Gen./year	Selection method	References
Amaranth	Temperature and photoperiod	Early flowering	6	SSD	Stetter et al. (2016)
Apple	The apical portion of 10 months old plant was grafted and grown under optimal growth conditions.	Early flowering	1 generation in less than a year	–	Van nocker and Gardiner (2014)
	Culturing in medium and acclimatization of plantlets In glasshouse under long-day conditions (16 h light: 8 h darkness) at 22°C.	Genetic fixing of BpMADS4 gene encoding early flowering	1	Marker-assisted selection	Flachowsky et al. (2011)
Barley	Temperature, photoperiod, immature germination of seed, embryo rescue	Early flowering	9	SSD	Zheng et al. (2013)
	Temperature, photoperiod	Early flowering	6	SSD	Watson et al. (2018)
Canola	Temperature, light intensity, photoperiod	Early flowering	4	SSD	Watson et al. (2018)
Cassava	High temperature, long photoperiod, embryo culture, and water stress	Reduced generation time	5–7	All soil and SPD	Yao et al. (2016)
	Extended photoperiod, different light intensities, and temperature	Accelerated flowering	–	–	Marcela et al. (2020)
	Grafting followed by the development of plants in the nursery	Increase flowering	–	–	Souza et al. (2018), Ceballos et al. (2017)
Chestnut	Tissue cultured plantlets were grown under high-intensity light, with the temperature of 23°C and 60% relative humidity	Early flowering	1 generation in 2 years	–	Baier et al. (2012)

TABLE 9.3 *(Continued)*

Crop	Technique	Trait targeted	Gen./year	Selection method	References
Chickpea	Photoperiod and immature seed germination	Early flowering	7	SPD	Semineni et al. (2019)
	Temperature, photoperiod	Early flowering	2	–	Sethi et al. (1981)
	Photoperiod, plant hormones, light intensity, and immature seed	Early flowering	8	SPD	Mobini et al. (2015)
Cocoyam	Development of elite clones via GA3 application and hand pollination done and collected seeds cultured and plants transferred to the greenhouse	Flower induction and seed formation	–	–	Wilson (1979)
Faba bean	Photoperiod, light intensity, plant hormones, and immature seed	Early flowering	7	SPD	Mobini et al. (2015)
Lentil	Photoperiod, light intensity	Early flowering and Aphanomyces root rot selection	5	SSD	Lulsdorf et al. (2018)
Lupin	Temperature, photoperiod, immature seeds germination	Early flowering	5	–	Croser et al. (2016). Also used in chickpea, faba bean, lentil
Pea	Temperature, photoperiod, immature embryo culture	Rapid generation of progenies	5	–	Mobini and Warkent in (2016)
Peanut	Temperature, photoperiod	Reducing generation time	3	SSD	O'Connor et al. (2013)
Pigeon pea	Photoperiod, temperature, immature seed germination	Rapid flowering	4	SPD	Saxena et al. (2019)

TABLE 9.3 *(Continued)*

Crop	Technique	Trait targeted	Gen./year	Selection method	References
Potato	Crossing of parents, seeds germinated in the greenhouse under controlled temperature followed by selfing and marker selection	Sli gene fixing, which inhibits gametophytic self-incompatibility	2–3	Marker-assisted selection	Lindhout et al. (2011)
Rice	Photoperiod, high-density planting, and temperature	High yielding trait	4	SSD	Collard et al. (2017)
	Backcrossing, germination in controlled temperature and daylight and embryo rescue followed by selection	Hst1 gene introgression for salt tolerance	6	Marker-assisted selection	Rana et al. (2019)
Sorghum	Photoperiod, temperature, and immature seed germination	Early flowering	6	SSD	Forster et al. (2014)
Soybean	Temperature, photoperiod germination, pretreatment of pods at high temperature, immature seeds germinated	Early flowering	5	SSD	Roumet and Morin (1997)
	Temperature, long photoperiod, light intensity (low red to far-red)	Accelerated flowering	5	SSD	Jahne et al. (2020), also done in rice and amaranth
	Temperature, photoperiod, CO_2	Accelerated flowering	5	–	Nagatoshi and Fujita (2018)
Subterranean clover	Photoperiod, temperature, germination of immature seeds	Early flowering	2.7–6.1	SSD	Pazos-Navarro et al. (2017)
Sunflower	Temperature, photoperiod,	Accelerated flowering	4	–	Nazan et al. (2012)
	immature embryos were excised and cultured, and plantlets were transferred to pot at 24 ± 2°C in 16 h/8 h				

TABLE 9.3 *(Continued)*

Crop	Technique	Trait targeted	Gen./year	Selection method	References
Wheat	Temperature, photoperiod, immature germination of seed followed by embryo rescue	Early flowering	7.6	SSD	Zheng et al. (2013)
	Temperature, photoperiod	Leaf rust resistance	–	–	Riaz et al. (2016)
	Temperature, photoperiod, speed breeding system	Phenotyping for seminal root number, seminal root angle, plant height, resistance to leaf rust, and tolerance to crown rot	6	–	Alahmad et al. 2018
	F1 progeny produced by crossing two germplasms and growing under controlled temperature and humidity and water deficit conditions upto F7	Improvement of stay green and root traits	5–7	–	Christopher et al. (2015)

SPD, single pod descent; SSD, single seed descent.

9.4.2 SPEED BREEDING FOR YIELD

In 2015, a study was conducted by Jack Christopher et al. to investigate the combined effects of phenotyping, nested association mapping (NAM), and speed breeding on root adaptation amidst climate change and water limitation. This study was conducted to increase wheat yield. Using speed breeding technique, a NAM population of 1000 recombinant inbred lines were progressed to F_5 generations within 18 months. Multiple trials were done to identify the superior germplasm (Christopher et al., 2015).

9.4.3 SPEED BREEDING FOR MULTIPLE QUANTITATIVE TRAITS SELECTION

Speed breeding has been developed successfully in bread wheat (*T. aestivum* L.) to advance it to six generations per year. Such technique has also been deployed in durum wheat (*T. durum* Desf.). When combined with speed breeding, the novel multitrait phenotyping method leads to the selection of early filial generations and fixed lines out-of-season characterization. Moreover, it offers the efficient use of resources by assaying multiple traits on the same set of plants (Alahmad et al., 2018).

9.4.4 SPEED BREEDING FOR BIOTIC AND ABIOTIC STRESS

9.4.4.1 BIOTIC STRESS (MULTIPLE DISEASE RESISTANCE)

One of the most important parameters to enhance crop production is incorporating disease resistance to plants. Breeders are trying to achieve this via speed breeding to achieve faster and better outputs. One such research was performed by Hickey et al. in 2017 in which multiple disease resistance was transferred into two-rowed barley with scarlett genetic background via novel trait introgression methodology. The backcross-derived Scarlett introgression lines (ILs) with enhanced resistance to foliar diseases developed within just 2 years under controlled environment conditions. Likewise, other breeding technologies can also be combined with speed breeding technology for crop improvement programs (Hickey et al., 2017).

9.4.4.2 ABIOTIC STRESS (SALT TOLERANCE)

One of the major reasons that speed breeding technology is now being explored rapidly is climate change. Climate change is responsible for a number of plant abiotic stresses, such as drought, salinity, heat, heavy metals, submergence (Wassmannet al., 2009b). Rice is severely affected by salinity due to low salt tolerance. The research was performed by Rana et al., 2019 to improve salinity resistance in locally grown cultivars. In this study, the tolerant gene for salinity, *hst1*, was introgressed from "Kaijin" to "Yukinko-mai" (WT) variety of rice through SNP (single-nucleotide polymorphism)-based MAS (marker-assisted selection using biotron speed breeding system as given in Table 9.2. The population BC_3F_3 (YNU31-2-4) was developed in six generations within 17 months only. In the seedlings, survival rates were higher and increased shoot and root length. Avoiding accumulation of Na^+ in shoots under salinity stress, high tolerance was found in YNU31-2-4 against salt stress at both seedling and reproductive stages (Rana et al., 2019). This research is a great example of the potential that we can harness from the fusion of developed and upcoming technologies.

9.5 DRAWBACKS AND CHALLENGES OF SPEED BREEDING

For accelerating conventional breeding programs, the process of speed breeding is a valuable approach that requires appropriate infrastructure, plant phenomics facilities along with the expertise to handle with continuous financial support (Shimelis et al., 2019). Such resources can be developed easily only if approaches for speed breeding are given enough recognition for conventional plant breeding, MAS, and genetic engineering. Also, this process requires expertise and skills in biotechnology and plant breeding and support from government at policy and long-term financial requirements. However, challenges that most commonly hamper the use of speed breeding are: (1) direct access to proper lab facilities, (2) trained workers, (3) embracing major changes to breeding program design and operations, (4) species/crop-specific protocols, and (5) the requirement for long-term funding. Also, there are certain limitations of speed breeding. Firstly, the use of extended photoperiods may cause harmful results on plant growth. Extended photoperiod exposure can result in photooxidation and high starch production and can also lead to the production of various stress hormones. Secondly, this technique cannot be successfully applied in short-day crops. Thirdly, controlled environmental conditions can lead to the outbreak of diseases, such as leaf injury, chlorosis, and limited plant growth and productivity. Fourthly, losses of plants in single seed descent under greenhouse conditions, and lastly, the cost of speed breeding is much higher than conventional breeding (Wanga et al., 2021).

9.6 CONCLUSIONS

Speed breeding is a powerful technique that can be used to produce indirectly more food by directing the fast development of improved varieties for the growing population as it reduces the generation time of crops. When combined with other modern or conventional techniques like embryo rescue, marker-assisted selection, gene editing, quantitative analysis can accelerate the academic, and commercial research.

KEYWORDS

- **breeding program**
- **crop improvement**
- **speed breeding**

REFERENCES

Alahmad, S.; Dinglasan, E.; Leung, K. M.; Riaz, A.; Derbal, N.; Voss-Fels, K. P.; Able, J. A.; Bassi, F. M.; Christopher, J.; Hickey, L. T. Speed Breeding for Multiple Quantitative Traits in Durum Wheat. *Plant Methods* **2018,** *14*(1), 1–15.

Baier, K.; Maynard, C.; Powell, W. Chestnuts and Light. *J. Am. Chestnut Foundation* **2012,** 8–10.

Christopher, J.; Richard, C.; Chenu, K.; Christopher, M.; Borrell, A.; Hickey, L. Integrating Rapid Phenotyping and Speed Breeding to Improve Stay-Green and Root Adaptation of Wheat in Changing, Water-Limited, Australian Environments. *Procedia Environ. Sci.* **2015,** *29*(Agri), 175–176.

Collard, B. C. Y.; Beredo, J. C.; Lenaerts, B.; Mendoza, R.; Santelices, R.; Lopena, V.; Verdeprado, H.; Raghavan, C.; Gregorio, G. B.; Vial, L.; Demont, M.; Biswas, P. S.; Iftekharuddaula, K. M.; Rahman, M. A.; Cobb, J. N.; Islam, M. R. Revisiting Rice Breeding Methods–Evaluating the use of Rapid Generation Advance (RGA) for Routine Rice Breeding. *Plant Prod. Sci.* **2017,** *20*(4), 337–352.

Croser, J. S.; Pazos-Navarro, M.; Bennett, R. G.;Tschirren, S.; Edwards, K.; Erskine, W.; Creasy, R.; Ribalta, F. M. Time to Flowering of Temperate Pulses In Vivo and Generation Turnover In Vivo–In Vitro of Narrow-Leaf Lupin Accelerated by Low Red to Far-Red Ratio and High Intensity in the Far-Red Region. *Plant Cell Tissue Organ Culture (PCTOC)* **2016,** *127*(3), 591–599.

Dagustu, N.; Bayram, G.; Sincik, M.; Bayraktaroglu, M. The Short Breeding Cycle Protocol Effective on Diverse Genotypes of Sunflower (*Helianthus annuus*L.). *Turk. J. Field Crops* **2012,** *17*(2), 124–128.

Ehrlich, P. R.; Harte, J. Opinion: To feed the World in 2050 Will Require a Global Revolution. *Proc. Natl. Acad. Sci.* **2015,** *112*(48), 14743–14744.

Flachowsky, H.; Le Roux, P. M.; Peil, A.; Patocchi, A.; Richter, K.; Hanke, M. V. Application of a High-Speed Breeding Technology to Apple (Malus× domestica) Based on Transgenic Early Flowering Plants and Marker-Assisted Selection. *New Phytol.* **2011,** *192*(2), 364–377.

Forster, B. P.; Till, B. J.; Ghanim, A. M. A.; Huynh, H. O. A.; Burstmayr, H.; Caligari, P. D. S. Accelerated Plant Breeding. *Cab Rev*. **2014,** *9*(043), 1–16.

Gastelo, M.; Diaz, L.; Landeo, J. A.; Bonierbale, M. New Elite Potato Clones With Heat Tolerance, Late Blight and Virus Resistance to Address Climate Change. In *Potato and Sweet potato in Africa: Transforming the Value Chains for Food and Nutrition Security*; Low, J., Nyongesa, M., Quinn, S., Parker, M,. Eds.; CABI International: Oxfordshire, UK,2015; pp 143–152.

Ghosh, S.; Watson, A.; Hickey, L.T. Speed Breeding in Growth Chambers and Glasshouses for Crop Breeding and Model Plant Research. *Nat. Protoc*. **2018,** *13,* 944–2963.

Godfray, H. C. J.; Beddington, J. R.; Crute, I. R.; Haddad, L.; Lawrence, D.; Muir, J. F. Pretty, J.; Robinson, S.; Thomas, S.M.; Toulmin, C. Food Security: The Challenge of Feeding 9 Billion People. *Science* **2010,** *327*(5967), 812–818.

Heslop-Harrison, S.; Schwarzacher, T. Domestication, Genomics and the Future for Banana. *Ann. Botany* **2007,** *100*(5), 1073–1084.

Hickey, L. T.; Germán, S. E.; Pereyra, S. A.; Diaz, J. E.; Ziems, L. A.; Fowler, R. A.; Platz, G. J.; Franckowiak, J. D.;Dieters, M. J. Speed Breeding for Multiple Disease Resistance in Barley. *Euphytica* **2017,** *213*(3).

Hickey, L. T.; Hafeez, A. N.; Robinson, H.; Jackson, S. A.; Leal-Bertioli, S. C.; Tester, M.; Wulff, B. B. Breeding Crops to Feed 10 Billion. *Nat. Biotechnol.* **2019,** *37*(7), 744–754

Jähne, F.; Hahn, V.; Würschum, T.; Leiser, W. L. Speed Breeding Short-Day Crops by LED-Controlled Light Schemes. *Theor. Appl. Genet.* **2020,** *133*(8), 2335–2342.

Joshi, D. C.; Sood, S.; Hosahatti, R.; Kant, L.; Pattanayak, A.; Kumar, A.; Yadav, D.; Stetter, M. G. From Zero to Hero: The Past, Present and Future of Grain Amaranth Breeding. *Theor. Appl. Genet.* **2018,** *131*(9), 1807–1823.

Kole, C. (Ed.); *Pulses, Sugar and Tuber Crops*. Springer Science & Business Media, 2007; vol 3.

Korbo, A.; Kjær, E. D.; Sanou, H.; Ræbild, A.; Jensen, J. S.; Hansen, J. K. Breeding for High Production of Leaves of Baobab (*Adansonia digitata* L) in an Irrigated Hedge System. *Tree Genet. Genomes* **2013,** *9,* 779–793.

Lindhout, P.; Meijer, D.; Schotte, T.; Hutten, R. C.; Visser, R. G.; van Eck, H. J. Towards F1 Hybrid Seed Potato Breeding. *Potato Res*. **2011,** *54*(4), 301–312.

Lobell, D. B.; Gourdji, S. M. The Influence of Climate Change on Global Crop Productivity. *Plant Physiol.* **2012,** *160,* 1686–1697.

Lulsdorf, M. M.; Banniza, S. Rapid Generation Cycling of an F2 Population Derived From a Cross Between *Lens culinaris* Medik. And *Lens ervoides* (Brign.) Grande After Aphanomyces Root Rot Selection. *Plant Breed.* **2018,** *137*(4), 486–491.

Mobini, S. H.; Lulsdorf, M.; Warkentin, T. D.; Vandenberg, A. Plant Growth Regulators Improve In Vitro Flowering and Rapid Generation Advancement in Lentil and Faba Bean. *In Vitro Cell. Dev. Biol. Plant* **2015,** *51*(1), 71–79.

Mpelkas, C. C. Light Sources for Horticultural Lighting. *IEEE Trans. Ind. Appl.* **1980,** IA-16(4), 557–565.

Nagatoshi, Y.; Fujita, Y. Accelerating Soybean Breeding in a CO_2-Supplemented Growth Chamber. *Plant Cell Physiol.* **2019,** *60*(1), 77–84.

Nelson, G. C.; Rosegrant, M. W.; Koo, J.; Robertson, R.; Sulser, T.; Zhu, T.; Ringler, C.; Msangi, S.; Palazzo, A.; Batka, M. *Climate Change: Impact on Agriculture and Costs of Adaptation*. International Food Policy Research Institute, 2009; 21.

Newton, A. C.; Johnson, S. N.; Gregory, P. J. Implications of Climate change for Diseases, Crop Yields and Food Security. *Euphytica* **2011,***179,* 3–18.

Ochatt, S. J.; Sangwan, R. S.; Marget P.; Ndong, Y. A.; Rancillac, M.; Perney, P. Robbelen, G. New Approaches Towards the Shortening of Generation Cycles for Faster Breeding of Protein Legumes. *Plant Breed.* **2002,** *121*(5), 436–440. https://doi.org/10.1046/j.1439-0523.2002.746803.x

O'Connor, D. J.; Wright, G. C.; Dieters, M. J.; George, D. L.; Hunter, M. N.; Tatnell, J. R.; Fleischfresser, D. B. Development and Application of Speed Breeding Technologies in a Commercial Peanut Breeding Program. *Peanut Sci.* **2013,** *40*(2), 107–114.

Ohnishi, T.; Yoshino, M.; Yamakawa, H.; Kinoshita, T. The Biotron Breeding System: A Rapid and Reliable Procedure for Genetic Studies and Breeding in Rice. *Plant Cell Physiol.* **2011,** *52*(7), 1249–1257.

Ortiz, R. Genomic-Led Potato Breeding for Increasing Genetic Gains: Achievements and Outlook. *Genet. Genom. Breed. Crop Plants*. **2020,** *2*(e200010), 1–25.

Pazos-Navarro, M.; Castello, M.; Bennett, R. G.; Nichols, P.; Croser, J. In Vitro-Assisted Single-Seed Descent for Breeding-Cycle Compression in Subterranean Clover (*Trifolium subterraneum* L.). *Crop Pasture Sci.* **2017,** *68*(11), 958–966.

Peng, S.; Huang, J.; Sheehy, J. E.; Laza, R. C.; Visperas, R. M.; Zhong, X.; Centeno, G. S.; Khush, G. S.; Cassman, K. G. Rice Yields Decline With Higher Night Temperature From Global Warming. *Proc. Natl. Acad. Sci.* **2004,** *101,* 9971–9975.

Pineda, M.; Morante, N.; Salazar, S.; Cuásquer, J.; Hyde, P. T.; Setter, T. L. Ceballos, H. Induction of Earlier Flowering in Cassava Through Extended Photoperiod. *Agronomy* **2020,** *10*(9), 1273.

Rana, M. M.; Takamatsu, T.; Baslam, M.; Kaneko, K.; Itoh, K.; Harada, N.; Sugiyama, T.; Ohnishi, T.; Kinoshita, T.; Takagi, H.; Mitsui, T. Salt Tolerance Improvement in Rice Through Efficient SNP Marker-Assisted Selection Coupled With Speed-Breeding. *Int. J. Mol. Sci.* **2019,** *20*(10), 2585.

Riaz, A.; Periyannan, S.; Aitken, E.; Hickey, L. A Rapid Phenotyping Method for Adult Plant rResistance to Leaf Rust in Wheat. *Plant Methods* **2016,** *12*(1), 1–10.

Rizal, G.; Karki, S.; Alcasid, M.; Montecillo, F.; Acebron, K.; Larazo, N.; Garcia, R.; Slamet-Loedin, I. H.; Quick, W. P. Shortening the Breeding Cycle of Sorghum, A Model Crop for Research. *Crop Sci.* **2014,** *54,* 520–529.

Roumet, P.; Morin, F. Germination of Immature Soybean Seeds to Shorten Reproductive Cycle Duration. *Crop Sci.* **1997,** *37*(2), 521–525.

Samineni, S.; Sen, M.; Sajja, S. B.; Gaur, P. M. Rapid Generation Advance (RGA) in Chickpea to Produce up to Seven Generations per Year and Enable Speed Breeding. *Crop J.* **2020,** *8*(1), 164–169.

Saxena, K. B.; Saxena, R. K.; Hickey, L. T.; Varshney, R. K. Can a Speed Breeding Approach Accelerate Genetic Gain in Pigeonpea? *Euphytica* **2019,** *215*(12), 1–7.

Sethi, S. C.; Byth, D. E.; Gowda, C. L. L.; Green, J. M. Photoperiodic Response and Accelerated Generation Turnover in Chickpea. *Field Crops Res.* **1981,** *4,* 215–225

Shimelis, H.; Gwata, E. T.; Laing, M. D. Crop Improvement for Agricultural Transformation in Southern Africa. In *Transforming Agriculture in Southern Africa*, 2019; 97.

Shivakumar, M.; Nataraj, V.; Kumawat, G.; Rajesh, V.; Chandra, S.; Gupta, S.;Bhatia, V. S. Speed Breeding for Indian Agriculture: A Rapid Method for Development of New Crop Varieties. *Curr. Sci.* **2018,** *115*(7), 1241.

Singh, A.; Rajput, V. D., Mehrotra, R.; Pal, N.; Singh, V. K.; Minkina, T.; Chokheli, V. A.; Singh, R. K. In *Sustainable Soil Fertility Management*; Nova. Sci. Publishers, Inc., 2020; vol 1, pp. 73–100.

Singh, A.; Rajput, V. D., Rawat, S.; Sharma, R.; Singh, A. K.; Singh, A. K.; Tomar, R. S. In *Emerging Tools for Sustainable Agriculture and Food Security*; Rajput, D.; Book Agency: New Delhi, Delhi, 2021a; vol 1, pp. 1–15.

Singh, A., Rajput, V., Singh, A., Sengar, R., Singh, R., Minkina, T. Transformation Techniques and Their Role in Crop Improvements: A Global Scenario of GM Crops. *Policy Issues Genetically Modified Crops* **2021b**, *1,* 515–542.

Souza, L. S.; Diniz, R. P.; de Jesus Neves, R.; Alves, A. A. C.; de Oliveira, E. J. Grafting as a Strategy to Increase Flowering of Cassava. *Sci. Hortic.* **2018,** *240,* 544–551.

Sreenivasulu, N.; Butardo, V. M.; Misra, G.; Cuevas, R. P., Anacleto, R.; Kishor, P. B. K. Designing Climate-Resilient Rice With Ideal Grain Quality Suited for High-Temperature Stress. *J. Exp. Bot.* **2015,** *66,* 1737–1748.

Stetter, M. G.; Zeitler, L.; Steinhaus, A.; Kroener, K.; Biljecki, M.; Schmid, K. J. Crossing Methods and Cultivation Conditions for Rapid Production of Segregating Populations in Three Grain Amaranth Species. *Front. Plant Sci.* **2016,** *7,* 816.

Sundström, J. F.; Albihn, A.; Boqvist, S.; Ljungvall, K.; Marstorp, H.; Martiin, C.; Nyberg, K., Vågsholm, I.; Yuen, J.; Magnusson, U. Future Threats to Agricultural Food Production Posed by Environmental Degradation, Climate Change, and Animal and Plant Diseases–A Risk Analysis in Three Economic and Climate Settings. *Food Secur*. **2014,** *6*(2), 201–215.

Van Oort, P. A.; Zwart, S. J. Impacts of Climate Change on Rice Production in Africa and Causes of Simulated Yield Changes. *Glob. Change Biol. Bioenergy*. **2018,** *24,* 1029–1045.

vanNocker, S.; Gardiner, S. E. Breeding Better Cultivars, Faster: Applications of New Technologies for the Rapid Deployment of Superior Horticultural Tree Crops. *Hortic. Res*. **2014,** *1*(1), 1–8.

Wanga, M. A.; Shimelis, H.; Mashilo, J.; Laing, M. D. Opportunities and Challenges of Speed Breeding: A Review. *Plant Breed.* **2021,** *140*(2), 185–194.

Wassmann, R.; Jagadish, S. V. K.; Sumfleth, K.; Pathak, H.; Howell, G.; Ismail, A.; Serraj, R.; Redona, E.; Singh, R. K.; Heuer, S. Regional Vulnerability of Climate Change Impacts on Asian Rice Production and Scope for Adaptation. *Adv. Agron.* **2009a,** *102,* 91–133.

Wassmann, R.; Jagadish, S. V. K.; Heuer, S.; Ismail, A.; Redona, E.; Serraj, R.; Singh, R. K.; Howell, G.; Pathak, H.; Sumfleth, K. Climate Change Affecting Rice Production: The Physiological and Agronomic Basis for Possible Adaptation Strategies. *Adv. Agron.* **2009b,** *101,* 59–122.

Watson, A., Ghosh, S., Matthew, J., et al. Speed Breeding is a Powerful Tool to Accelerate Crop Research and Breeding. *Nature Plants* **2018,** *4,* 23–291. https://doi.org/10.1038/s41477-017-0083-8.

Wilson, J. E. Promotion of Flowering and Production of Seed in Cocoyam (Xanthosoma and Colocasia). In *International Symposium on Tropical Root and Tuber Crops*, IITA, Ibadan, Nigeria, Sept, 1979.

Wolfe, M. D.; Del Carpio, D. P.; Alabi, O.; Ezenwaka, L. C.; Ikeogu, U. N.; Kayondo, I. S.; Lozano, R.; Okeke, G. U.; Ozimati, A. A.; Williams, E.; Egesi, C.; Kawuki, S. R.; Kulakow, P.; Rabbi, Y. I.; Jannink, J. L. Prospects for Genomic Selection in Cassava Breeding. *Plant Genome*. **2017,** *10*(3).

Yao, Y.; Zhang, P.; Wang, H. B.; Lu, Z. Y.; Liu, C. J.; Liu, H.; Yan, G. J. How to Advance up to Seven Generations of Canola (*Brassica napus* L.) per Annum for the Production of Pure Line Populations? *Euphytica* **2016,** *209*(1), 113–119.

Zhao, C.; Liu, B.; Piao, S.; Wang, X.; Lobell, D. B.; Huang, Y.; Huang, M. Temperature Increase Reduces Global Yields of Major Crops in Four Independent Estimates. *Proc. Natl. Acad. Sci.* **2017,** *114*(35), 9326–93312.

Zheng, Z.; Wang, H. B.; Chen, G. D.; Yan, G. J.; Liu, C. J. A Procedure Allowing up to Eight Generations of Wheat and Nine Generations of Barley Per Annum. *Euphytica* **2013,** *191*(2), 311–316.

CHAPTER 10

Nanotechnological Approaches in the Second Green Revolution

SUKH VEER SINGH[1], RAKHI SINGH[1], AND SADHAN JYOTI DUTTA[2]

[1]*Department of Food Science and Technology, National Institute of Food Technology Entrepreneurship and Management, Kundli, Sonipat, Haryana, India*

[2]*National Institute of Food Technology Entrepreneurship and Management, Kundli, Sonipat, Haryana, India*

ABSTRACT

Nanotechnology is considered as "one out of the six key permitting technologies that can add to sustainable competitiveness and growth in numerous fields of industrial application." The agricultural sector is one of those fields, where nanotechnology has proven to have a profound future. The advent of nanofertilizers, nanopesticides, nanosensors, and other nanoagrochemicals has brought the nanotechnology applications in agriculture to a more abutting affair. In this study, various applications of nanotechnology have been explained, specifically in the agriculture sector. The pros and cons of the first Green Revolution have been examined as the crucial issues, that must be taken care of during the second Green Revolution. The study concludes that despite enormous potentials, the agricultural and the food sector are hesitant about the applications of nanotechnology due to its possible adverse effects. Hence, establishment of regulatory bodies will be an important aspect to formulate stringent rules and regulations for an effective impact of nanotechnology in the second Green Revolution.

Nanotechnology for Sustainable Agriculture: An Innovative and Eco-Friendly Approach. Vishnu D. Rajput, Abhishek Singh, Tatiana M. Minkina, Krishan K. Verma, and Awani Kumar Singh (Eds.)

10.1 INTRODUCTION

Agriculture shows a major and decisive role in the economic prosperity of developed and developing countries (Singh et al., 2020a; Singh et al., 2021a). In the past few decades, the agricultural sector has noticed an astounding growth due to the revolutions, such as use of hybrid varieties, use of pest-resistant cultivars, synthetic fertilizers, and genetically modified crops (Sunding and Zilberman, 2001; Singh et al., 2020b). Apart from these, Green Revolution was another persuasive factor for the momentous development in the agricultural sector. Green Revolution in the early 1960s brought in the amalgamation of mechanization, agricultural expansion, use of genetically modified crops, high-yielding varieties (HYV) of crops, synthetic fertilizers, pesticides, and high crop research investment rates (Pingali, 2012). The success of this revolution resulted in the usage of plentiful food supplies and lower prices of agricultural products. However, the climate changes due to the greenhouse effect and global warming, environmental issues, urbanization, and non-judicious use of nonrenewable resources have brought about many challenges in the world of agriculture. The alarming rate of increase in the world population further intensified these problems (Ditta and Arshad, 2016). The United Nations Department of Economic and Social Affairs (UNDESA) published a report in 2017, which showed that the current world population (7.7 billion) is expected to reach about 9.8 billion in 2050 and 11.2 billion in 2100 (United Nations, 2017). The report further mentioned that the increase in the population will be concentrated in just nine such as countries. India, Nigeria, the Democratic Republic of the Congo, Pakistan, Ethiopia, the United Republic of Tanzania, the United States of America, Uganda, and Indonesia. This will bring a huge burden of feeding to the rapidly growing population of the world. This keeps us wondering with a question. How to feed these rapidly and exponentially growing populations while keeping intact the environment, climate, and the world? The answer lies in the way modern technologies and methods, which must be utilized in a more interactive and sustainable manner. Nanotechnology and agriculture can be combined in one such profound interactive mode that will unfold this question to feed the world, sustainably.

The European Commission (EC) has recognized "nanotechnology" as one out of its six basic enabling technology that pay off to sustainable competitiveness and progress in several fields of industrial application (Parisi et al., 2015). Back in 1974, the term "Nanotechnology" was first defined by Norio Taniguchi of the Tokyo Science University as "the study of manipulating matter on an atomic and molecular scale." It deals with the study of 1–100 nm sized structures and involves developing materials or devices within that size. The applications of nanotechnology have helped to develop various new materials and devices in a wide range of application. From medicine and electronics to food production and energy creation, nanotechnology has found applications in almost every field (Agrawal and Rathore, 2014). The unique properties of particles at a nanoscale facilitate their applications in many areas such as electronics, medicine, material science, biotechnology, and energy sectors. Agriculture is one such sector wherein nanotechnology has been recently introduced. This has brought promising results, especially in providing solutions to challenges faced to solve global food security, climate change, and environmental degradation. Out of these three challenges, the first Green Revolution was able to resolve the issue of global food security. But, the issues of climate change and environmental degradation were not addressed. Therefore, nanoscale science and nanotechnology which addressed to have potential to revolutionize agriculture and food systems, emerges as a promising tool to address all three challenges altogether in the form of second Green Revolution (Hakeem and Pirzadah, 2020; Agrawal and Rathore, 2014). In the present study, previous studies have been reviewed to understand the several applications of nanotechnology in agriculture. The main objective of this study is to review the possible role of nanotechnology in the second Green Revolution. Accordingly, we have evaluated the various effects of the first Green Revolution, the need for a second Green Revolution, and numerous applications of nanotechnology in agriculture and food sector.

10.2 FIRST GREEN REVOLUTION AND POST-FIRST GREEN REVOLUTION

The Green Revolution began with the foundation of an institute by the Rockefeller Foundation in 1944. The primary objective of the institute was to improve the agricultural output of Mexican farms. The results were astonishing as Mexico was importing half of its wheat prior to 1956, when it went on to become self-sufficient in 1956, and to exporting wheat by 1964 (Kendall and Pimentel, 1994). During the same period in the 1950s and 1960s, countries in Asia and Africa were facing food shortages. This led the Rockefeller Foundation to develop the International Rice Research Institute (IRRI) in the Philippines in 1960. The main objective of the IRRI was to develop high-yielding rice seeds for Asian countries. This led to booming growth in countries, such as India and Pakistan and this marked the beginning of the Green Revolution in Asia. This period of Green Revolution from 1960 to 2000 pronounced an astounding era of global food security (Zeigler and Mohanty, 2010).

Notwithstanding this enormous achievement, the Green Revolution also brought many side effects. One of the most perceptible effects in the post-Green Revolution era was the negative effect on the environment due to extreme use of fertilizers, pesticides, and irrigation water. It led to the pollution of ground water and other water ways, and land degradation. This also resulted in the killing of beneficial insects and other wild life, thereby, weakening the ecosystem. These chemicals were also toxic and hazardous, which adversely affected the health of the farmers. Genetic erosion was another ill-effect of green revolution (International Food Policy Research Institute, 2002). This occurred because only a few high-yielding varieties of crops were subjected to wide-spread cultivation. Another front where the green revolution is criticized the socioeconomic aspect. It is argued that the Green Revolution served only those regions, that had abundant resources, whereas, the resource-scarce regions were left behind. This is easily perceivable from the regions of Saharan Africa and eastern India. These are the regions where poverty density is the highest in the world and hugely needed the favor of the Green Revolution, but were less impacted by it (Zeigler and Mohanty, 2010).

10.3 NEED OF THE SECOND GREEN REVOLUTION

It has been estimated that the agricultural production must be increased globally up to 60–110 % by 2050 (FAO, 2009; OECD et al., 2012; Tilman et al., 2011). The increase in agricultural production must meet the demands of future populations. Additionally, it must also fulfill the requirements by providing food security to the 870 million people, who are undernourished (FAO WFP and IFAD, 2012). Nevertheless, many studies have shown that the crop yields are not increasing at the expected rate (Cassman, 1999; Finger, 2010; Peltonen-Sainio et al., 2009; Ray et al., 2012). A study was conducted using approximately 2.5 million agricultural statistics, collected from nearly 13,500 regions across the world to investigate yields per year of four global crops. maize, rice, wheat, and soybean (Ray et al., 2013). It was found that these crops were increasing at 1.6%, 1.0%, 0.9%, and 1.3% per year, respectively with non-compounding rates. It is less than the 2.4%

per year rate which is required to double the global production by 2050. These four crops would increase only by ~67%, ~42%, ~38%, and ~55%, respectively with the current increasing trend of the crop (Ray et al., 2013). This expected increase is unfit for what is forecasted to meet the demands by 2050 and beyond. Hence, there is a need of innovative strategies and tools to meet the future needs sustainably for ensuring unbroken supply of the food. This gives the fundamental requirement for a second Green Revolution which must be implemented based on modern technologies. In addition, it has to be kept in mind about the nonrecurrence of the various ill-effects that the first Green Revolution bestowed.

10.4 AGRICULTURE, FOOD, AND NANOTECHNOLOGY

The combined works of agriculture and nanotechnology are elucidated by many terms, such as "Nano Agriculture" and "Agricultural Nanotechnologies" which illustrate the applications of nanomaterials in agriculture. Shivendu et al., 2016 and Shivendu et al., 2017 also used the term "Agri-nanotechniques" to exemplify the terms agriculture, nanotechnology, and techniques. The applications of these agri-nanotechniques have huge potential in precision agriculture (Duhan et al., 2017), but the utilization of agri-nano techniques has not yet been fully developed in the global markets. Bradley et al., 2011; Chaudhry and Castle, 2011; Kuan et al., 2012; Meetoo, 2011; Momin et al., 2013; Rashidi and Khosravi-Darani, 2011; Ravichandran, 2010; Sekhon, 2010; Sekhon, 2014; Sonkaria et al., 2012 have reported the massive opportunities of nanotechnology in agriculture and in all facets of the food industry, including preservation, processing, packaging, and monitoring (Fig. 10.1). This can be blamed on the absence of directives on the uses of nanomaterials in agriculture and food (Parisi et al., 2015). One primary reason for this lack of regulations is due to the deficit of knowledge we possess about the various effects on health and environment. There have been enough shreds of evidences to support the claim that the agri-nanotechniques result in a significant increase in crop yields and a decrease in other losses from external conditions which include biotic stresses, such as insects and diseases, and abiotic stresses, such as drought and heat. The applications of functional nanomaterials in plants resulted in a conclusive remark that the interactions between nanoparticles and agriculture will increase food production (Swift et al., 2019). Nanofertilizers, nanopesticides, and nanosensors also improve the efficiency of food production in a more sustainable way (McClements, 2020). Some of the applications of nanotechnology like nanosensors and nano-based smart delivery systems which are now exploited in the agricultural industry to help with combating viruses and other pathogens (crop infection) (Singh et al., 2021b). These techniques increase agrochemicals efficiency at lower dosage rates. Besides this, engineered nanomaterials are being developed for a variety of agri-food applications for example food additives, feed additives, flavorings, food packaging, novel foods, and pesticides (Handford et al., 2014). Moreover, applications may be classified as nano-inside and nano-outside. Nano-inside and nano-outside refer to application in the food product as in primary production or food processing and food packaging, respectively (Henchion et al., 2013).

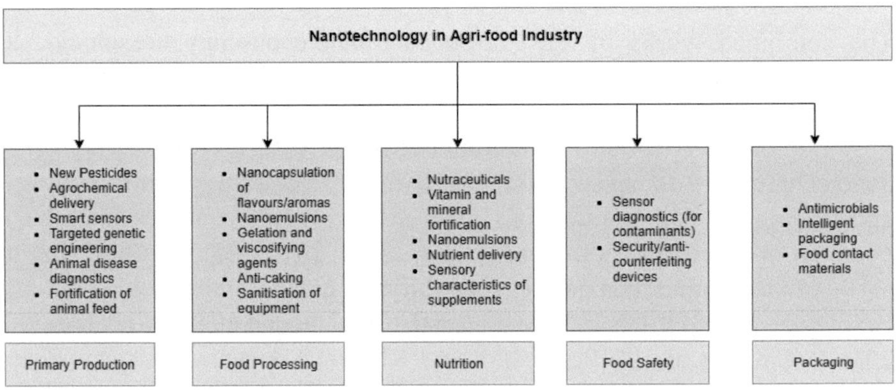

FIGURE 10.1 This schematic shows summarized illustration of nanotechnological application in agri-food sector. (Modified from Handford et al. (2014)).

10.5 APPLICATION OF NANOTECHNOLOGY IN AGRICULTURE

The present study broadly classified the applications of nanotechnology in agriculture into five categories: sensors, pest control, water and nutrient control, genetic engineering, and nanomaterials (Fig. 10.2). The five categories have been created based on the paramount scope of nanotechnology in agricultural field. These applications give an idea about the role of nanotechnology in the near future to ensure the food security for rapidly growing population (Hakeem and Pirzadah, 2020).

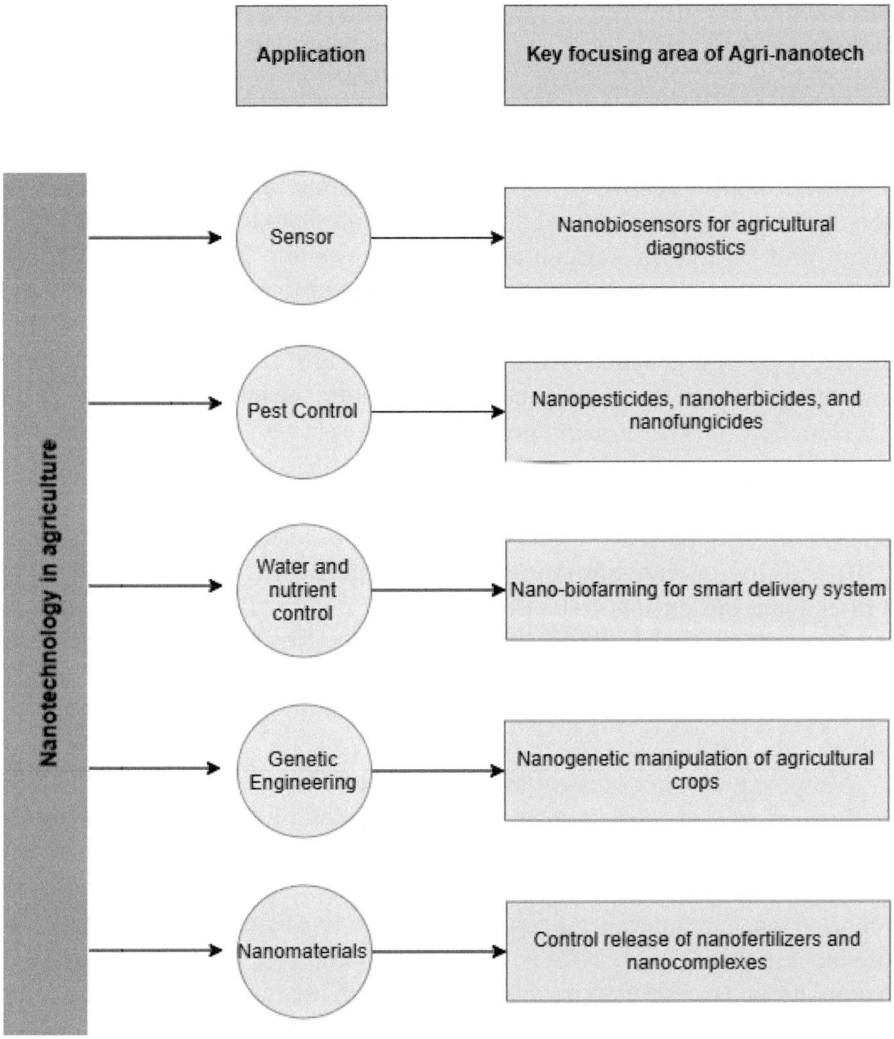

FIGURE 10.2 This schematic shows illustrated applications of nanotechnology in agriculture. (Modified from Hakeem and Pirzadah 2020 and Agrawal and Rathore 2014).

10.5.1 SENSORS

Nanosensors are advantageous due to their characteristics of being "small, transferable, sensitive with real-time monitoring, precise, quantitative, reliable, accurate, reproducible and robust, and stable which can overcome the deficits of present sensors" (Agrawal and Rathore, 2014). The nanotechnology-enabled sensors or nanosensors facilitate to track the data and monitor the growth of crops and conditions of agriculture fields which include (1) soil conditions, such as pH, moisture level, and fertility, (2) temperature, (3) crop nutrients, and (4). detection of phytopathogens and weeds (Mousavi and Rezaei, 2011; Prasad et al., 2017). Collection of these data will provide important information that will be helpful in decreasing the losses and increasing the yield. One possible application of nanosensors is using it as an identification tool to spot pathogens, such as bacteria, fungi, and viruses in agriculture (Boonham et al., 2008). Etefagh et al. (2013) developed nanosensors of CuO nanoparticles and nanostructural layer biosensors, which were used to detect *Aspergillus niger* by detecting the toxic gases it produced. Another application of nanosensors is the real-time detection of the pathogens in agriculture. Neethirajan et al. (2010) developed a CO_2 sensor using polyaniline boronic acid conducting polymer, which simplified the real-time detection of spoilage in stored grain. The use of nanoparticles along with enzyme-based biosensors resulted in the development of nanoparticles-based enzymatic biosensors. Organophosphorus (OP) and non-organophosphorus (non-OP) pesticides were detected using a bi-enzyme biosensing system consisting of multiwalled carbon nanotubes (CNTs) (Zhang et al., 2015). Atomic force microscopy (AFM) was used to develop a nanobiosensor equipped with the acetolactate synthase (ALS) enzyme, which aided in the detection of herbicide metsulfuron-methyl (da Silva et al., 2013).

Moreover, another class of biosensors are known as aptasensors. These are biosensors that consists aptamers and nanomaterials. The aptamers are the target-recognition element and the nanomaterials are the signal transducers and/or signal enhancers. Omanović-Mikličanina & Maksimović (2016) reported that "aptamers are single-stranded nucleic acid or peptide molecules of size less than 25 kDa with natural or synthetic origin." A wide variety of nanomaterials can be used in aptasensors which includes metal nanoparticles and nanoclusters, semiconductor nanoparticles, carbon nanoparticles, and magnetic nanoparticles (Sharma et al., 2015). Table 10.1 depicts a broad range of transducing systems that have been employed in aptasensors for food quality assessment and safety. The principle involved in the working of these aptasensors are based on the property of the nanoparticles that have been used. These aptameters are classified into optical and electrochemical systems on the basis of the detection mechanism (Omanović-Mikličanina and Maksimović, 2016). Another study was performed where copper oxide nanoparticles were utilized to develop an antibacterial agent by loading them on the surfaces of graphene oxide sheets (GO–Cu NPs). The result showed that GO–Cu NPs were very effective in reducing the bacterial speck disease caused by *Pseudomonas syringae* pv. tomato (Pst) in tomatoes without phytotoxicity (Li et al., 2017).

TABLE 10.1 Particular of Sensors, Designing Nanomaterial, and its Application.

Sensors particulars	Nanomaterial compositions	Applications	References
Nanosensors	Copper-doped montmorillonite	Propineb fungicide detection in aquatic environment	Abbaci et al., 2014
Nanosensors	Graphene	Pathogen detection in wastewater	Wibowo et al., 2018
Smart sensor technology	–	Monitoring the quality of grain, dairy products, fruits, and vegetables in a storage environment in order to detect the source and the type of spoilage	Omanović-Mikličanina and Maksimović, 2016
Aptasensor	Aptamer and nanomaterial	Determination of pesticides and insecticides (phorate, acetamiprid, isocarbophos)	Omanović-Mikličanina and Maksimović, 2016
Aptasensor	Aptamer and nanomaterial	Determination of antibiotics, drugs, and their residues (cocaine, oxytetracycline, tetracycline, kanamycin)	Omanović-Mikličanina, and Maksimović, 2016
Aptasensor	Aptamer and nanomaterial	Determination of heavy metals (Hg^{2+}, As^{3+}, Cu^{2+}), and microbial toxins (OTA, Fumonisin B_1)	Omanović-Mikličanina, and Maksimović, 2016
Nanoparticles and nanochips	–	Selectively bind and remove chemicals or pathogens from food. For identity preservation and tracking	Omanović-Mikličanina, and Maksimović, 2016

10.5.2 PEST CONTROL

Protection of crops against insect pests is one of the most challenging problems in every environmental condition. Nanotechnology can be used as a functional tool to solve this problem through its pest detection and management abilities. The nanopesticides and nano-encapsulated pesticide formulations are eco-friendly and also reduce harmful exposure to humans. These have been taken as very constructive techniques to avoid the spread of dangerous chemicals to the environment (Alfadul, 2017; Nuruzzaman et al., 2016). Amphiphilic zein-based nanomaterials have been reported to be efficacious when applied as aqueous nanocarriers for hydrophobic pesticides (Hao et al., 2020). An eco-friendly polyurethane delivery system with enhanced pesticide retention on plant surfaces was developed by encapsulating biopesticide (azadirachtin) in their hydrophobic cores. This study showed the development of an efficient pesticide delivery system (Zhang et al., 2020). The nanocarriers can have other beneficial effects such as an increase in the efficiency of the activity and stability of the nanopesticides under environmental pressures, such as UV and rain. This decreases the toxicity and also reduces their costs due to significant reduction in the number of applications (Worrall et al., 2018) as shown in Figure 10.3

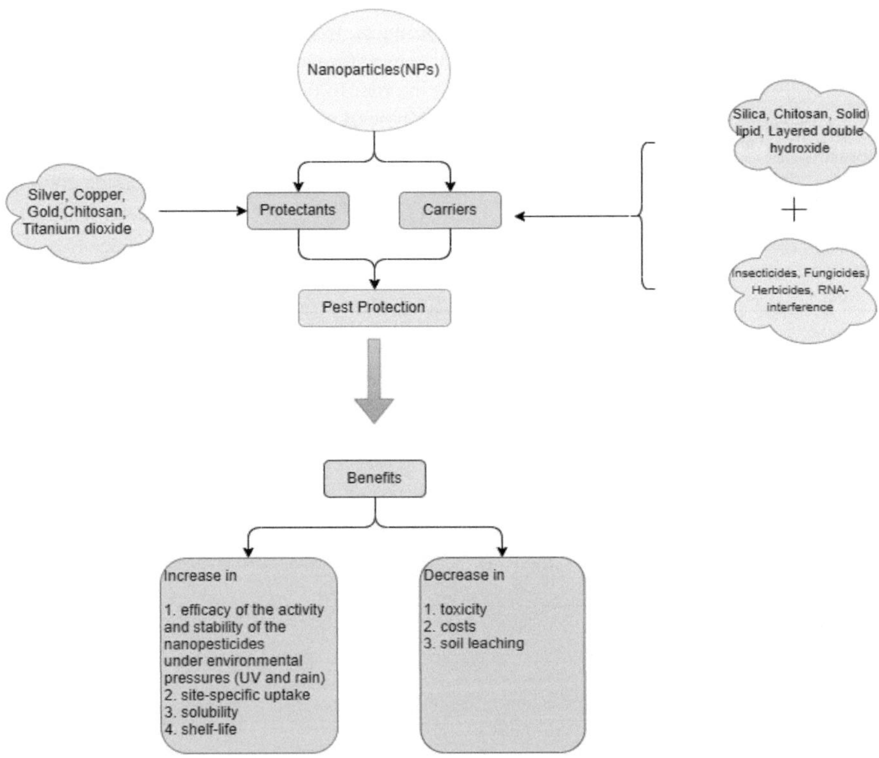

FIGURE 10.3 This schematic shows nanomaterials (protectants or carriers). These are for actives, such as insecticides, fungicides, herbicides, or RNA-interference molecules. Nanomaterials target a wide spectrum (pests to pathogens) that deals with crop protection. (Modified from Worrall et al. (2018)).

10.5.3 WATER AND NUTRIENT CONTROL

In precision farming, it is necessary to ensure the efficient use of water and nutrients. It can be achieved by using nanosensors and nano-based smart delivery systems. The employment of smart delivery systems improves the yield and quality in an eco-friendly manner (Mousavi and Rezaei, 2011). Kottegoda et al. (2011) developed a N-carrying nanomaterials, that is, hydroxyapatite NPs ($Ca_{10}(PO_4)_6(OH)_2$), HA) using a hybrid nanostructure based on urea molecules, which had a slower release of N. Similar studies were also carried out to study the intake of macronutrient fertilizers, such as phosphorus, calcium, magnesium, etc., and micronutrient fertilizers, such as copper, iron, manganese, etc. Hydroxyapatite ($Ca_{10}(PO_4)_6(OH)_2$) nanoparticles (*n*HA) were found to have clear features for an effective nanofertilizer (Marchiol et al., 2019). Nanoparticle-based nanofertilizer regulates the different aspect related to nutrients in agriculture as shown in Figure 10.4. On the other hand, the increase in concentrations of *n*HA did not hold the percentage germination of *S. lycopersicum,* while it strongly influenced root elongation. Copper (Cu) nanoparticles facilitated the quality of *Moringa oleifera* Lam. Significant improvements were observed in the leaves of *M. oleifera* due to an increase in the bioactive compounds and increase in antioxidant capacity on the application of Cu nanoparticles (Juárez-Maldonado et al., 2018).

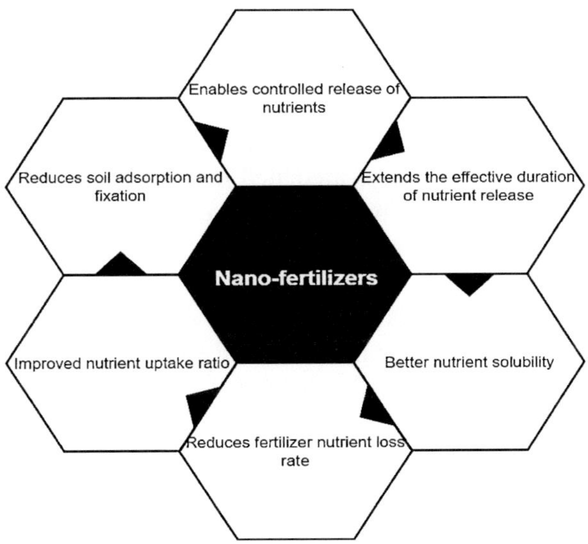

FIGURE 10.4 This schematic shows the better nutrient regulation and utilization through nanofertilizers in agricultural farming system.

Sources: Modified from Pudake et al. (2019).

10.5.4 GENETIC ENGINEERING

Genetic engineering can be coupled with nanotechnology to boost the process of crop improvement and plant transformation (Singh et al., 2020b). The application of genetic engineering in agriculture has resulted in momentous outcomes, which include the development of crops with high yields and good nutritional value, and crops with resistance against herbicides. This also led crops to resist against biotic stresses, such as insects and diseases, and against abiotic stresses, such as drought and heat (Shivendu et al., 2017). Despite surprising results, there were limitations in genetic engineering such as the lack of versatile and high-throughput tools for the delivery of biomolecules into plant cells. This hurdle can be overcome by using nanobiotechnology. It is described to offer "a new set of tools to manipulate the genes using nanoparticles, nanofibers, and nanocapsules." One example of nanobiotechnology is nanofiber arrays. For the quick and efficient delivery of genetic materials to cells, nanofiber arrays find their applications in drug delivery, crop engineering, and environmental monitoring. These nanofibers are used for the controlled biochemical manipulations in cells (Agrawal and Rathore, 2014). Another example of nanobiotechnology in agriculture is the transfer of plasmid DNA by gene gun method using gold-capped nanoparticles. This method is used for the direct delivery of DNA into intact plant cells. This proves to be beneficial as there is simultaneous delivery of both DNA and effector molecules to the specific sites (Nair et al., 2010).

On the other hand, DNA and chemicals were transported into isolated plant cells and intact leaves using a honeycomb mesoporous silica nanoparticle (MSN) system with 3-nm pores (Torney et al., 2007). This was one of the initial studies that showed nanoparticle-mediated delivery in plants. A recent study on DNA nanostructures describes the effective role of DNA nanostructures in the delivery of exogenous biomolecules such as siRNA to plants (Zhang et al., 2019). Similar studies were carried out for force-independent internalization through cell walls using MSNs for plasmid DNA delivery (Chang et al., 2013), layered double hydroxide clay nanosheets for the RNAi molecules delivery (Mitter et al., 2017), CNTs for the unassisted delivery of plasmid DNA and siRNA (Demirer et al., 2019a; Demirer et al., 2019b; Kwak et al., 2019).

10.5.5 NANOMATERIALS

Nanomaterials are defined as "an ingredient containing particles with at least one dimension that measures about 1–100 nm" (United States Environmental

Protection Agency) (Ditta and Arshad, 2016). There is an efficient way for synthesizing nanomaterials using agricultural-based plant material. It has been a challenge to control the nanomaterials in response to immediate environmental changes at such high-resolution along with precision. In addition to this, there is limitation in the study of bioinspired nanomaterials in food and agriculture sectors. Still, there has been success in studying about some artificial bioinspired devices for pesticide delivery, chemical and pathogen detection, and environmental sensing (He et al., 2019), as listed in Table10.2. The utilization of plants for synthesizing nanomaterials is found to be eco-friendly, cost-effective, and less hazardous (Shivendu et al., 2017). Since, this method involves plant phytochemicals, such as terpenoids, quinines, carboxylic acids, flavonoids, ketones, aldehydes, and amides for synthesizing nanomaterials, it is generally known as a green synthesis method. Numerous studies have been performed, which have revealed magnificent results. Silver nanoparticles (AgNPs) were produced from silver nitrate aqueous using aqueous extract of *Alternanthera sessilis* Linn. (Amaranthaceae) (Niraimathi et al., 2013). These synthesized AgNPs possessed good antimicrobial and antioxidant activity with a coating of proteins that indicated a dual role of biomolecules for capping and efficient stabilization. A similar study was performed where AgNPs were synthesized using aqueous garlic, green tea, and turmeric extracts. but AgNPs from turmeric extract demonstrate excellent antioxidant and cytotoxicity activity than AgNPs synthesized from other extracts (Arumai Selvan et al., 2018). The leaves extract of *Mirabilis jalapa* was used to synthesize monometallic ZnO/Ag nanoparticles (NPs) and bimetallic ZnO/Ag NPs. The biological characterization of these nanoparticles displayed higher total antioxidant, free radical scavenging, and reducing power potentials (Sumbal et al., 2019).

TABLE 10.2 Specific Bioinspired Nanomaterials in Agriculture and Allied Sectors.

Bioinspired template	Nanomaterials	Features	Applications
Silk	Graphene	Battery-free sensors for remote monitoring of pathogenic bacteria at single-cell level	Shed light on wireless nanosensors for food pathogen detection
Biological cilia	Polyvinylidene fluoride piezoelectric nanofiber	Flow velocity and flow direction	May assist in taste sensors or real-time sensing in food safety, such as food pathogen, allergens, and food quality monitoring

TABLE 10.2 *(Continued)*

Bioinspired template	Nanomaterials	Features	Applications
Insect tentacles	Nanoporous Prussian blue (PB) nanocube heads/ TiO_2 nanowire (NW) arms	Sensitive detection of H_2O_2 at a low detection limit (~20 nM), broad detection range (10^{-8} to 10^{-5} M), short response time (~5 s), and long-term biocatalytic activity (up to 6 months)	Show potential for biomolecule detection in food safety
Zwitterion	Fluorescent biomimetic carbon quantum dots	Detection limit for vitamin B12 at 81 nM; highly biocompatible	Show potential in bioanalaysis in nutritional and dietary supplement
Mussel	Polydopamine (PDA)-coated molecularly imprinted SiO_2 NPs	Specific recognition of the trace quantities of papain with low detection limit of 0.63 nM	Show potential in bioanalysis in nutritional and dietary supplement
Candida lusitaniae	Silver/silver chloride NPs	Antimicrobial activity	Yeast isolated from termite gut
Aloe vera plant	Nanoscale zero-valent iron	Removal of arsenic (As) and Selenium (Se) from water	Plant extract as reducing agent

Data modified from He et al. (2019).

10.6 DRAWBACK OF NANOTECHNOLOGY IN AGRICULTURE AND FOOD SECTOR

Although nanotechnology has enormous opportunities and potential applications in the agri-food sector, there are increasing concerns related to safety and health (Handford et al., 2014). Consumer safety and adverse environmental effects are the two major concerns of nanotechnology in agriculture and are shown in Figure 10.5. Nanotechnology is still in the primitive stage because only a few nanofertilizers are commercially available to be used in agriculture sector. Hence, there is a lack of risk assessment of these nanomaterials being used in agriculture and the food sectors. The lack of risk assessment is also primarily due to the complexity of environmental conditions. It further restricts the determinateness of the nanomaterials, thereby increasing the complexity to monitor the dissemination of nanomaterials (He et al., 2014a; He et al., 2015a; He et al., 2018). Assessment of nanotoxicology on human health has recently attracted public concerns. The nanomaterials which are used as food additives and other functional and nutritional ingredients pose a threat to human health due to intentional or unintentional addition, and moreover due to migration of nanomaterials from food or agricultural products (He et al., 2015a; He et al., 2015b; He et al., 2014b). Many studies have also shown the low toxic effects of nanomaterials in food and agricultural products from short-term exposure. Titanium dioxide (TiO_2) particles were evaluated for their adverse effects on health (Warheit et al., 2015). It was found that there was no TiO_2-related hazards. TiO_2 is used as food-grade additive E171 which is included in candies, sweets, and chewing for whitening or brightening. Another study was performed to investigate the genotoxicological effects of Fe_2O_3-30 nm and Fe_2O_3-bulk in female Wistar rats. The results showed that Fe_2O_3-30 nm and Fe_2O_3-bulk was not genotoxic with the single doses of 500, 1000, and 2000 mg/kg (Singh et al., 2013). On the contrary, increasing scientific evidence suggest that exposure to nanoparticles can lead to oxidative damage and inflammatory reactions of the gastrointestinal tract (i.e., carbon black, silicates, titanium dioxide, and iron oxide) (Handford et al., 2014) as shown in Figure 10.5. In contrast, a study concludes that there were accumulations of Fe nanomaterials in body organs, such as liver, spleen, kidney, heart, and bone marrow. But, this accumulation did not cause any significant genotoxicity. Even though these results proved the less toxic effects of nanomaterials, however, more studies must be conducted to evaluate the safety limits of certain nanomaterials for their commercial applications. This is a necessity because public concerns must be on priority for the successful applications of nanomaterials in the food and agricultural sector (Handford et al., 2014).

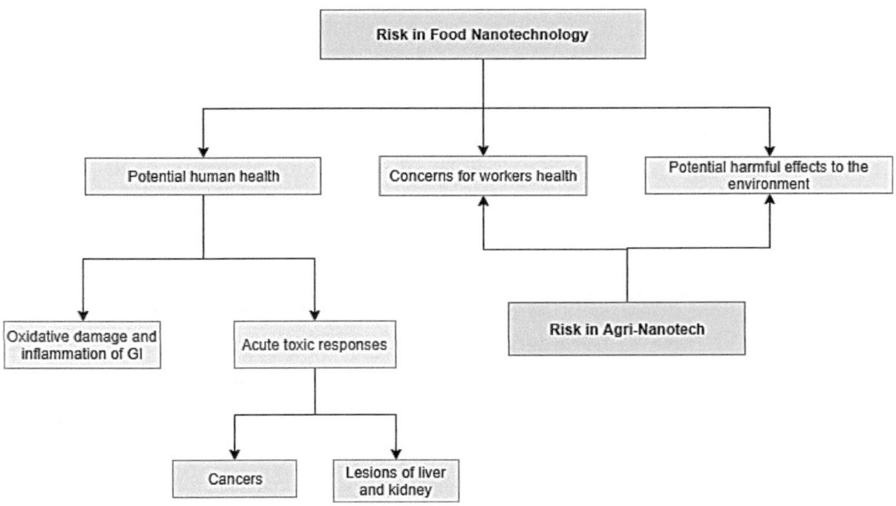

FIGURE 10.5 This schematic shows risk in agri-food nanotechnology.

Source: Modified from Handford et al. (2014).

On the other hand, nanotechnology faces drawback in the dimensions of environmental effects, health and safety issues, economic and societal issues. It is a sure implication that the acquisition of nanotechnology in agriculture will result in environmental regulations due to the compulsion of sustainable agriculture in the future. The subject of the environmental safety is one area, where research and development must be focused. Another drawback is related to the health and safety of the public. There are safety concerns over the use of nanofertilizers, nanopesticides, and nanoherbicides in agricultural products. Issues have also been raised on the use of nanomaterials in food packaging over the fear of migrations of nanomaterials from food package to the foods. In terms of socioeconomic aspects, the adoption of new agricultural technology is a challenge as it will depend on two factors like employment opportunities and an increase in wage rate (Zeigler and Mohanty, 2010). These factors will exert higher impact in developing countries, such as India, where agriculture is the prime occupation of majority of the population. Hence, these challenges must be addressed for full-fledged utilization of nanotechnology in the agriculture sector.

10.7 URGENT NEED FOR LEGISLATION

Nanomaterials behave differently from conventional materials because they follow a different set of laws that defines their properties, functions, and interactions. Hence, they require new rules and legislation to address food safety, risks, and applications. The present legislations to maintain the applications of nanotechnology/nanomaterials are limited to only for the general aspect of application. The current trends in the application of nanotechnology in the food industry promote to make us understand that the food industry should invest hugely for the utilization of nanotechnology in food products and processes. In food industry, nanotechnology already has food applications as nanomaterials-based food additives, food packages, detectors, and sensors. They promote the application of nanotechnology in food products which raise questions about the safety of the foods for the human health and wellness. Thus, there is the need for new legislations for specific nanomaterials apart from the general legislation for the use of nanotechnology in the food industry so that novel food products can have a safe standpoint in the market (He et al., 2019). One suggestion which can be implemented by the governments across the world is that they can form common and strict norms and monitor the use of these technologies before commercialization and bulk use of these nanomaterials (Agrawal and Rathore, 2014).

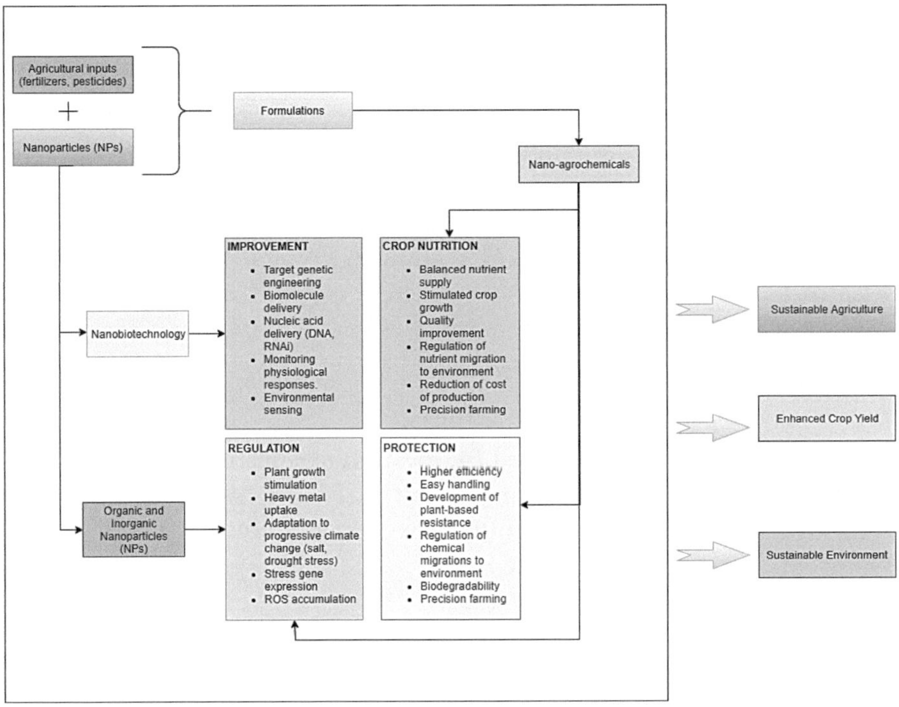

FIGURE 10.6 Simplified overview of nanomaterials applications in agricultural production for sustainability. Improvement in crop production is through nanoparticles-based pesticides and fertilizers. Regulation of metal uptake and plant growth is by using organic and inorganic nanoparticles. Protection of crops using nano-agrochemicals. Providing crop nutrition through formulations of nanoparticles. Modified from Shang et al. (2019).

10.8 CONCLUSIONS

The application of nanotechnology in agriculture has shown significant results by boosting the production of crops and enriching their quality. This is done by bringing improvements in the farming systems as schematically presented in Figure 10.6. Nanotechnology holds promising possibilities for the second Green Revolution. We can be optimistic about the role that nanotechnology will play to solve crucial global issues, such as food security, climate changes, and environmental degradation. The fusion of nanotechnology in agriculture generated various applications, such as sensors, pest control, water and nutrient control, genetic engineering, and nanomaterials. Despite such huge potentials, the maximum utilization about the applications of nanotechnology in the agricultural and food sector has not been tapped to its fullest. The primary reason which holds this fact is due to the uncertainties that may hold in terms of environmental effects, health and safety issues, and economic and societal issues. To solve these issues, a significant role can be played upon by regulatory bodies. The proper establishments of these regulatory bodies will view under its guidance the adherence of all the rules and regulations. Hence, they can play an important role during the second Green Revolution for its effective implementation. Nanotechnology thus has provoked its enormity in agriculture as a promising interdisciplinary research field.

ACKNOWLEDGMENT

This book chapter was supported by the National Institute of Food Technology Entrepreneurship and Management (NIFTEM), set up by Ministry of Food Processing Industries (MOFPI), Government of India, Kundli, Sonipat district, Haryana, under Delhi NCR.

AUTHOR DISCLOSURES

The authors declare no conflict of interest, financial or otherwise.

KEYWORDS

- **agriculture**
- **green revolution and application**
- **nanotechnology**

REFERENCES

Abbaci, A.; Azzouz, N.; Bouznit, Y. A New Copper Doped Montmorillonite Modified Carbon Paste Electrode for Propineb Detection. *Appl. Clay Sci.* **2014,** *90*, 130–134.

Abbaci, A.; Azzouz, N.; Bouznit, Y. A New Copper Doped Montmorillonite Modified Carbon Paste Electrode for Propineb Detection. *Appl. Clay Sci. 90*, 130–134. https://doi.org/https://doi.org/10.1016/j.clay.2013.12.036

Abrol, D. P.; Shankar, U. Integrated Pest Management. In *Breeding Oilseed Crops for Sustainable Production: Opportunities and Constraints*, 2015. https://doi.org/10.1016/B978-0-12-801309-0.00020-3

Agrawal, D. S.; Rathore, D. P. Review Article Nanotechnology Pros and Cons to Agriculture: A Review. *Int. J. Curr. Microbiol. Appl. Sci. (ISSN: 2319-7706)* **2014,** *3*, 43–55. https://doi.org/10.13140/2.1.3352.1283

Aguila, S. A.; Shimomoto, D.; Ipinza, F.; Bedolla-Valdez, Z. I.; Romo-Herrera, J.; Contreras, O. E.; Farías, M. H.; Alonso-Núñez, G. A Biosensor Based on Coriolopsis Gallica Laccase Immobilized on Nitrogen-Doped Multiwalled Carbon Nanotubes and Graphene Oxide for Polyphenol Detection; 2015. *Http://Www.Tandfonline.Com/Action/JournalInformation?Show=aimsScope&journalCode=tsta20#.VmBmuzZFCUk*, *16*(5). https://doi.org/10.1088/1468-6996/16/5/055004

Alahmad, S.; Dinglasan, E.; Leung, K. M.; Riaz, A.; Derbal, N.; Voss-Fels, K. P.; Able, J. A.; Bassi, F. M.; Christopher, J.; Hickey, L. T. Speed Breeding for Multiple Quantitative Traits in Durum Wheat. *Plant Methods* **2018,** *14*(1), 1–15. https://doi.org/10.1186/s13007-018-0302-y

Alessio, P.; Martin, C. S.; De Saja, J. A.; Rodriguez-Mendez, M. L. Mimetic Biosensors Composed by Layer-by-Layer Films of Phospholipid, Phthalocyanine and Silver Nanoparticles to Polyphenol Detection. *Sens. Actuators B Chem.* **2016,** *233*, 654–666. https://doi.org/10.1016/J.SNB.2016.04.139

Alfadul, S. Application of Nanotechnology in the Field of Food Production. *Acad. J. Sci. Res.* **2017,** *5*, 143–154. https://doi.org/10.15413/ajsr.2017.0220

Allison, S. D.; Martiny, J. B. H. In Resistance, Resilience, and Redundancy in Microbial Communities. Proceedings of the National Academy of Sciences of the United States of America, 2008. https://doi.org/10.1073/pnas.0801925105

Arumai Selvan, D.; Mahendiran, D.; Senthil Kumar, R.; Kalilur Rahiman, A. Garlic, Green Tea and Turmeric Extracts-Mediated Green Synthesis of Silver Nanoparticles: Phytochemical, Antioxidant and In Vitro Cytotoxicity Studies. *J. Photochem. Photobiol. B Biol.* **2018,** *180*, 243–252. https://doi.org/10.1016/j.jphotobiol.2018.02.014

Augustin, M. A.; Sanguansri, P. Chapter 5 Nanostructured materials in the food industry. *Adv. Food Nutr. Res.* **2009,** *58*, 183–213. https://doi.org/10.1016/S1043-4526(09)58005-9

Bass, C.; Denholm, I.; Williamson, M. S.; Nauen, R. The Global Status of Insect Resistance to Neonicotinoid Insecticides. *Pestic. Biochem. Physiol.* **2015,** *121,*78–87. https://doi.org/10.1016/j.pestbp.2015.04.004

Bergman, J. M.; Tingey, W. M. Aspects of Interaction Between Plant Genotypes and Biological Control. *Bulletin of the Entomological Society of America,*1979. https://doi.org/10.1093/besa/25.4.275

Bilir, K.; Weil, M.-T.; Lochead, J.; Kök, F. N.; Werner, T. Construction of an Oxygen Detection-Based Optic Laccase Biosensor for Polyphenolic Compound Detection. *Turk. J. Biol.* **2016,** *40*(6), 1303–1310.

Boonham, N.; Glover, R.; Tomlinson, J.; Mumford, R. Exploiting Generic Platform Technologies for the Detection and Identification of Plant Pathogens. *Euro. J. Plant Pathol.* **2008,** *121*(3), 355. https://doi.org/10.1007/s10658-008-9284-3

Burrell, A. M.; Taylor, K. G.; Williams, R. J.; Cantrell, R. T.; Menz, M. A.; Pepper, A. E. A Comparative Genomic Map for Caulanthus Amplexicaulis and Related Species (Brassicaceae). *Mol. Ecol.* **2011**. https://doi.org/10.1111/j.1365-294X.2010.04981.x

Butnariu, M.; Butu, A. Plant Nanobionics: Application of Nanobiosensors in Plant Biology. In *Nanotechnology in the Life Sciences*, 2019; pp 337–376. https://doi.org/10.1007/978-3-030-16379-2_12

Cai, H.; Xu, C.; He, P., chemistry, Y. F.-J. of electroanalytical, & 2001, undefined. (n.d.). Colloid Au-enhanced DNA immobilization for the electrochemical detection of sequence-specific DNA. *Elsevier*.

Cassman, K. G. Ecological Intensification of Cereal Production Systems: Yield Potential, Soil Quality, and Precision Agriculture. *Proc. Natl. Acad. Sci.* **1999,** *96*(11), 5952–5959. https://doi.org/10.1073/pnas.96.11.5952

Chamarthi, S. K.; Sharma, H. C.; Sahrawat, K. L.; Narasu, L. M.; Dhillon, M. K. Physico-chemical Mechanisms of Resistance to Shoot Fly, Atherigona soccata in sorghum, Sorghum bicolor. *J. Appl. Entomol.* **2011**. https://doi.org/10.1111/j.1439-0418.2010.01564.x

Cho, U. H.; Park, J. O. Mercury-Induced Oxidative Stress in Tomato Seedlings. *Plant Sci.* **2000**. https://doi.org/10.1016/S0168-9452(00)00227-2

Christopher, J.; Richard, C.; Chenu, K.; Christopher, M.; Borrell, A.; Hickey, L. Integrating Rapid Phenotyping and Speed Breeding to Improve Stay-Green and Root Adaptation of Wheat in Changing, Water-Limited, Australian Environments. *Procedia Environ. Sci.* 2015, *29*(Agri), 175–176. https://doi.org/10.1016/j.proenv.2015.07.246

Cloonan, K. R. The Navel Orangeworm, Amyelois transitella: An Examination of its Biology, Pest Ecology in Almonds, and Development of Screening Bioassays to Identify Compounds for Reducing Oviposition. ProQuest Dissertations and Theses, 2013.

Cockbain, A. J. Low Temperature Thresholds for Flight in Aphis Fabae Scop. *Entomologia Experimentalis et Applicata,* 1961. https://doi.org/10.1111/j.1570-7458.1961.tb02136.x

Collard, B. C. Y.; Beredo, J. C.; Lenaerts, B.; Mendoza, R.; Santelices, R.; Lopena, V.; Verdeprado, H.; Raghavan, C.; Gregorio, G. B.; Vial, L.; Demont, M.; Biswas, P. S.; Iftekharuddaula, K. M.; Rahman, M. A.; Cobb, J. N.; Islam, M. R. Revisiting Rice Breeding Methods–Evaluating the Use of Rapid Generation Advance (RGA) for Routine Rice Breeding. *Plant Prod. Sci.* **2017,** *20*(4), 337–352. https://doi.org/10.1080/1343943X.2017.1391705

Cook, S. M.; Khan, Z. R.; Pickett, J. A. The use of push-pull strategies in integrated pest management. *Ann. Rev. Entomol.* (2007). https://doi.org/10.1146/annurev.ento.52.110405.091407

Cranston, P. J. G.; P. S. In *The Insects: An Outline of Entomology*, 4th ed.. Journal of Insect Conservation, 2010.

Crowder, D. W.; Northfield, T. D.; Strand, M. R.; Snyder, W. E. Organic Agriculture Promotes Evenness and Natural Pest Control. *Nature* **2010**. https://doi.org/10.1038/nature09183

da Silva, A. C. N.; Deda, D. K.; da Róz, A. L.; Prado, R. A.; Carvalho, C. C.; Viviani, V.; Leite, F. L. Nanobiosensors Based on Chemically Modified AFM Probes: A Useful Tool for Metsulfuron-Methyl Detection. *Sensors (Basel, Switzerland)*, **2013,** *13*(2), 1477–1489. https://doi.org/10.3390/s130201477

Davies, T. G. E.; Field, L. M.; Williamson, M. S. The Re-emergence of the Bed Bug as a Nuisance Pest: Implications of Resistance to the Pyrethroid Insecticides. *Med. Vet. Entomol*. **2012**. https://doi.org/10.1111/j.1365-2915.2011.01006.x

Davis, J. J.; Coleman, K. S.; Azamian, B. R.; Bagshaw, C. B.; Green, M. L. H. Chemical and Biochemical Sensing With Modified Single Walled Carbon Nanotubes. *Chem. Euro. J.* 2003, *9*(16), 3732–3739. https://doi.org/10.1002/CHEM.200304872

Delphia, C. M.; Mescher, M. C.; De Moraes, C. M. Induction of Plant Volatiles by Herbivores With Different Feeding Habits and the Effects of Induced Defenses on Host-Plant Selection by Thrips. *J. Chem. Ecol.* **2007**. https://doi.org/10.1007/s10886-007-9273-6

Dermody, O.; O'Neill, B. F.; Zangerl, A. R.; Berenbaum, M. R.; DeLucia, E. H. Effects of Elevated CO_2 and O_3 on Leaf Damage and Insect Abundance in a Soybean Agroecosystem. *Arthropod-Plant Interact.* **2008**. https://doi.org/10.1007/s11829-008-9045-4

Després, L.; David, J. P.; Gallet, C. The Evolutionary Ecology of Insect Resistance to Plant Chemicals. *Trends Ecol. Evol.* **2007**. https://doi.org/10.1016/j.tree.2007.02.010

Ditta, A.; Arshad, M. Applications and Perspectives of Using Nanomaterials for Sustainable Plant Nutrition. *Nanotechnol. Rev.* **2016,** *5*(2), 209–229. https://doi.org/doi:10.1515/ntrev-2015-0060

Dong, Y.; Phillips, K. S; Cheng, Q. Immunosensing of Staphylococcus Enterotoxin B (SEB) in Milk With PDMS Microfluidic Systems Using Reinforced Supported Bilayer Membranes (r-SBMs). *Lab Chip* **2006,** *6*(5), 675–681.

Duggan, A.; Ma, C.; Chalfie, M. Regulation of Touch Receptor Differentiation by the Caenorhabditis Elegans mec-3 and unc-86 Genes. *Development* **1998.**

Duhan, J. S.; Kumar, R.; Kumar, N.; Kaur, P.; Nehra, K.; Duhan, S. Nanotechnology: The New Perspective in Precision Agriculture. *Biotechnol. Rep.* **2017,** *15*, 11–23. https://doi.org/https://doi.org/10.1016/j.btre.2017.03.002

Dunse, K. M.; Kaas, Q.; Guarino, R. F.; Barton, P. A.; Craik, D. J.; Anderson, M. A. Molecular Basis for the Resistance of an Insect Chymotrypsin to a Potato Type II Proteinase Inhibitor. *Proc. Natl. Acad. Sci. U S A*. https://doi.org/10.1073/pnas.1009327107

Durán, N.; Marcato, P. D. Nanobiotechnology Perspectives. Role of Nanotechnology in the Food Industry: A Review. *Int. J. Food Sci. Technol.* **2013,** *48*(6), 1127–1134. https://doi.org/10.1111/IJFS.12027

Eigenbrode, S. D. The Effects of Plant Epicuticular Waxy Blooms on Attachment and Effectiveness of Predatory Insects. *Arthrop. Struct. Dev.* **2004**. https://doi.org/10.1016/j.asd.2003.11.004

Etefagh, R.; Azhir, E.; Shahtahmasebi, N. Synthesis of CuO Nanoparticles and Fabrication of Nanostructural Layer Biosensors for Detecting Aspergillus Niger Fungi. *Scientia Iranica* **2013,** *20*(3), 1055–1058. https://doi.org/https://doi.org/10.1016/j.scient.2013.05.015

FAO. *Global Agriculture Towards 2050.* Rome; 2009.

FAO WFP, & IFAD. The State of Food Insecurity in the World 2012. *Economic Growth Is Necessary but Not Sufficient to Accelerate Reduction of Hunger and Malnutrition.* Rome, FAO; 2012

Fathipour, Y.; Maleknia, B.; Bagheri, A.; Soufbaf, M.; Reddy, G. V. P. Functional and Numerical Responses, Mutual Interference, and Resource Switching of Amblyseius Swirskii on Two-Spotted Spider Mite. *Biol. Control.* **2020**. https://doi.org/10.1016/j.biocontrol.2020.104266

Fatouros, N. E.; Dicke, M.; Mumm, R.; Meiners, T.; Hilker, M. Foraging Behavior of Egg Parasitoids Exploiting Chemical Information. *Behav. Ecol.* **2008**. https://doi.org/10.1093/beheco/arn011

Ferreira, N. S..; Cruz, M. G. N..; Gomes, M. T. S. R.; Rudnitskaya, A. Potentiometric Chemical Sensors for the Detection of Paralytic Shellfish Toxins. *Talanta* **2018,** *181,* 380–384.

Finger, R. Evidence of Slowing Yield Growth – The Example of Swiss Cereal yields. *Food Pol.* **2010,** *35*(2), 175–182. https://doi.org/https://doi.org/10.1016/j.foodpol.2009.11.004

Fuertes, G.; Soto, I.; Carrasco, R.; Vargas, M.; Sabattin, J.; Lagos, C. Intelligent Packaging Systems: Sensors and Nanosensors to Monitor Food Quality and Safety. *J. Sens.* **2016**. https://doi.org/10.1155/2016/4046061

Gadanakis, Y.; Bennett, R.; Park, J.; Areal, F. J. Evaluating the Sustainable Intensification of Arable Farms. *J. Environ. Manag.* **2015**. https://doi.org/10.1016/j.jenvman.2014.10.005

Gale, M. D.; Devos, K. M. Comparative Genetics in the Grasses. In *Proceedings of the National Academy of Sciences of the United States of America*, 1998. https://doi.org/10.1073/pnas.95.5.1971

Ghorbanpour, M.; Fahimirad, S. Plant Nanobionics a Novel Approach to Overcome the Environmental Challenges. *Med. Plants Environ. Chall.* **2017,** 247–257. https://doi.org/10.1007/978-3-319-68717-9_14

Gould, F.; Kennedy, G. G.; Johnson, M. T. Effects of Natural Enemies on the Rate of Herbivore Adaptation to Resistant Host Plants. *Entomol. Exp. Appl.* **1991**. https://doi.org/10.1111/j.1570-7458.1991.tb01445.x

Hahlbrock, K.; Grisebach, H. Enzymic Controls in the Biosynthesis of Lignin and Flavonoids. *Ann. Rev. Plant Physiol.* **1979**. https://doi.org/10.1146/annurev.pp.30.060179.000541.

Hakeem, K. R.; Pirzadah, T. B. Nanobiotechnology in Agriculture.Springer International Publishing. **2020**.

Hall, J.; Matos, S. Incorporating Impoverished Communities in Sustainable Supply Chains. *Int. J. Phys. Distrib. Logist. Manag.* **2010**. https://doi.org/10.1108/09600031011020368.

Handford, C. E.; Dean, M.; Henchion, M.; Spence, M.; Elliott, C. T.; Campbell, K. Implications of nanotechnology for the agri-food industry: opportunities, benefits and risks. *Trends in Food Science & Technology*. **2014**, *40*(2), 226–241.

Hao, L.; Lin, G.; Wang, H.; Wei, C.; Chen, L.; Zhou, H; Chen, H.; Xu, H.; Zhou, X. Preparation and Characterization of Zein-Based Nanoparticles via Ring-Opening Reaction and Self-Assembly as Aqueous Nanocarriers for Pesticides. *J. Agric. Food Chem.* **2020,** *68*(36), 9624–9635. https://doi.org/10.1021/acs.jafc.0c01592

Hare, J. D.; Andreadis, T. G. Variation in the Susceptibility of Leptinotarsa decemlineata (Coleoptera: Chrysomelidae) When Reared on Different Host Plants to the Fungal Pathogen, Beauveria Bassiana in the Field and Laboratory. *Environ. Entomol.* **1983**. https://doi.org/10.1093/ee/12.6.1892

He, X.; Aker, W. G.; Fu, P. P.; Hwang, H.-M. Toxicity of Engineered Metal Oxide Nanomaterials Mediated by Nano–Bio–Eco–Interactions: A Review and Perspective. *Environ. Sci. Nano* **2015,** *2*(6), 564–582. https://doi.org/10.1039/C5EN00094G

He, X.; Aker, W. G.; Huang, M. -J; Watts, J. D.; Hwang, H. M. Metal Oxide Nanomaterials in Nanomedicine: Applications in Photodynamic Therapy and Potential Toxicity. *Curr. Top. Med. Chem.* **2015,** *15*(18), 1887–1900). https://doi.org/http://dx.doi.org/10.2174/15680266 15666150506145251

He, X.; Aker, W. G.; Hwang, H.-M. An In Vivo Study on the Photo-enhanced Toxicities of S-doped TiO2 Nanoparticles to Zebrafish Embryos (Danio rerio) in Terms of Malformation, Mortality, Rheotaxis Dysfunction, and DNA Damage. *Nanotoxicology* **2014,** *8*(supl), 185–195. https://doi.org/10.3109/17435390.2013.874050

He, X.; Aker, W. G.; Leszczynski, J.; Hwang, H.-M. Using a Holistic Approach to Assess the Impact of Engineered Nanomaterials Inducing Toxicity in Aquatic Systems. *J. Food Drug Anal.* **2014,** *22*(1), 128–146. https://doi.org/https://doi.org/10.1016/j.jfda.2014.01.011

He, X.; Deng, H.; Hwang, H. The Current Application of Nanotechnology in Food and Agriculture. *J. Food Drug Anal.* **2019,** *27*(1), 1–21. https://doi.org/https://doi.org/10.1016/j.jfda.2018.12.002

He, X.; Fu, P.; Aker, W. G.; Hwang, H.-M. Toxicity of Engineered Nanomaterials Mediated by Nano–Bio–Eco Interactions. *J. Environ. Sci. Health Part C* **2018,** *36*(1), 21–42. https://doi.org/10.1080/10590501.2017.1418793

Hickey, L. T.; Germán, S. E.; Pereyra, S. A.; Diaz, J. E.; Ziems, L. A.; Fowler, R. A.; Platz, G. J.; Franckowiak, J. D.; Dieters, M. J. Speed Breeding for Multiple Disease Resistance in Barley. *Euphytica* **2018,** *213*(3). https://doi.org/10.1007/s10681-016-1803-2

Iglesias, A.; Rosenzweig, C.; Pereira, D. Agricultural Impacts of Climate Change in Spain: Developing Tools for a Spatial Analysis. *Global Environ. Change*. **2000**. https://doi.org/10.1016/S0959-3780(00)00010-8

International Food Policy Research Institute. Green Revolution: Curse or Blessing? *Int. Food Policy Res. Inst*. 2002.

Johnson, M. T.; Gould, F.; Kennedy, G. G. Effects of Natural Enemies on Relative Fitness of Heliothis Virescens Genotypes Adapted and not Adapted to Resistant Host Plants. *Entomol. Exp. Appl*. https://doi.org/10.1046/j.1570-7458.1997.00133.x

Jönsson, A. M.; Appelberg, G.; Harding, S.; Bärring, L. Spatio-temporal Impact of Climate Change on the Activity and Voltinism of the Spruce Bark Beetle, Ips Typographus. *Global Change Biol.* **2009**. https://doi.org/10.1111/j.1365-2486.2008.01742.x

Joyner, J.; Nanosensors and their applications in food analysis: a review. *Search.Proquest. Com*.

Juárez-Maldonado, A.; Ortega-Ortíz, H.; Cadenas-Pliego, G.; Valdés-Reyna, J; Pinedo-Espinoza, J.; López-Palestina, C.; Hernández-Fuentes, A. Foliar Application of Cu Nanoparticles Modified the Content of Bioactive Compounds in Moringa oleifera Lam. *Agronomy* **2018,** *8*(9), 167. https://doi.org/10.3390/agronomy8090167

Kalita, D.; Baruah, S. The Impact of Nanotechnology on Food. In *Nanomaterials Applications for Environmental Matrices: Water, Soil and Air*. Elsevier Inc.; 2019 https://doi.org/10.1016/B978-0-12-814829-7.00011-2

Karp, A.; Kresovich, S.; Bhat, K. V.; Ayad, W. G.; Hodgkin, T. Molecular Tools in Plant Genetic Resources Conservation: A Guide to the Technologies. In *IPGRI Technical Bulletin,* 1997.

Kasili, P. M.; Cullum, B. M.; Griffin, G. D.; Vo-Dinh, T. Nanosensor for In Vivo Measurement of the Carcinogen Benzo[a]pyrene in a Single Cell. *J. Nanosci. Nanotechnol.* **2002,** *2*(6), 653–658. https://doi.org/10.1166/JNN.2002.155

Kasili, P. M.; Song, J. M.; Vo-Dinh, T. Optical Sensor for the Detection of Caspase-9 Activity in a Single Cell. *J. Am. Chem. Soc.* **2004,** *126*(9), 2799–2806. https://doi.org/10.1021/JA037388T

Kendall, H. W.; Pimentel, D. Constraints on the Expansion of the Global Food Supply. *Ambio* **1994,** *23*(3), 198–205.

Kennedy, G. G.; Gould, F.; Deponti, O. M. B.; Stinner, R. E. Ecological, Agricultural, Genetic, and Commercial Considerations in the Deployment of Insect-resistant Germplasm. *Environ. Entomol*. **1987**. https://doi.org/10.1093/ee/16.2.327

Kissinger, P. T. Biosensors—A Perspective. *Biosens. Bioelectron.* **2005,** *20*(12), 2512–2516. https://doi.org/10.1016/J.BIOS.2004.10.004

Knipling, E. F. No TitleThe basic Principles of Insect Population Suppression and Management. US Department of Agriculture, No. 512. 1979.

Kottegoda, N.; Munaweera, I.; Adassooriya, N.; Karunaratne, V. A Green Slow-Release Fertilizer Composition Based on Urea-Modified Hydroxyapatite Nanoparticles Encapsulated Wood. *Curr. Sci.* **2011,** *101*, 73–78.

Kumar, A.; Sinha, R. P.; Häder, D. P. Effect of UV-B on Enzymes of Nitrogen Metabolism in the Cyanobacterium Nostoc Calcicola. *J. Plant Physiol.* **1996**. https://doi.org/10.1016/S0176-1617(96)80298-7

Kwak, S. Y.; Lew, T. T. S.; Sweeney, C. J.; Koman, V. B.; Wong, M. H.; Bohmert-Tatarev, K.; Snell, K. D.; Seo, J. S.; Chua, N. H.; Strano, M. S. Chloroplast-Selective Gene Delivery and Expression in Planta Using Chitosan-Complexed Single-Walled Carbon Nanotube Carriers. *Nature Nanotechnol.* **2019,** *14*(5), 447–455. https://doi.org/10.1038/S41565-019-0375-4

Lee, C.; Itoh, T.; Sasaki, G.; Suga, T. Sol-gel Derived PZT Force Sensor for Scanning Force Microscopy. *Mater. Chem. Phys.* **1996,** *44*(1), 25–29. https://doi.org/10.1016/0254-0584(95)01647-D

Lew, T. T. S.; Koman, V. B.; Gordiichuk, P.; Park, M.; Strano, M. S. The Emergence of Plant Nanobionics and Living Plants as Technology. *Adv. Mater. Technol.* **2020,** *5*(3), 1900657. https://doi.org/10.1002/ADMT.201900657

Li, X.; Huang, Q.; Yuan, J.; Tang, Z. Fipronil Resistance Mechanisms in the Rice Stem Borer, Chilo Suppressalis Walker. *Pestic. Biochem. Physiol*. **2007**. https://doi.org/10.1016/j.pestbp.2007.06.002

Li, Y.; Yang, D.; Cui, J. Graphene Oxide Loaded With Copper Oxide Nanoparticles as an Antibacterial Agent Against Pseudomonas Syringae pv. Tomato. *RSC Adv.* **2017,** *7*(62), 38853–38860. https://doi.org/10.1039/C7RA05520J

Liebhold, A. M.; Tobin, P. C. Population Ecology of Insect Invasions and Their Management. *Ann. Rev. Entomol.***2008**. https://doi.org/10.1146/annurev.ento.52.110405.091401

Lim, T. C.; Ramakrishna, S. A Conceptual Review of Nanosensors. *Zeitschrift Fur Naturforschung A* **2006,** *61*(7–8), 402–412. https://doi.org/10.1515/zna-2006-7-815

Liu, T.; Tang, J.; Jiang, L. The Enhancement Effect of Gold Nanoparticles as a Surface Modifier on DNA Sensor Sensitivity. *Biochem. Biophys. Res. Commun.* **2004,** *313*(1), 3–7.

Lu, J.; Buwles, M. How Will Nanotechnology Affect Agricultural Supply Chains? Int. Food Agribus. Manag. Ass. **2013,** *16*(2):21–42

Luechinger, N. A.; Loher, S.; Athanassiou, E. K.; N. Grass, R.N; Stark, W. J. Highly Sensitive Optical Detection of Humidity on Polymer/Metal Nanoparticle Hybrid Films. *Langmuir* **2007,** *23*(6), 3473–3477. https://doi.org/10.1021/LA062424Y

Malik, P.; Katyal, V.; Malik, V.; Asatkar, A.; Inwati, G.; Mukherjee, T. K. Nanobiosensors: Concepts and Variations. *ISRN Nanomater.* **2013,** *2013*, 1–9. https://doi.org/10.1155/2013/327435

Marchiol, L.; Filippi, A.; Adamiano, A.; Degli Esposti, L.; Iafisco, M.; Mattiello, A.; Petrussa, E.; Braidot, E. Influence of Hydroxyapatite Nanoparticles on Germination and Plant Metabolism of Tomato (Solanum lycopersicum L.): Preliminary Evidence. *Agronomy* **2019,** *9*(4), 161. https://doi.org/10.3390/agronomy9040161

McClements, D. J. Nanotechnology Approaches for Improving the Healthiness and Sustainability of the Modern Food Supply. *ACS Omega* **2020,** *5*(46), 29623–29630. https://doi.org/10.1021/acsomega.0c04050

Mello, L. D.; Kubota, L. T. Review of the Use of Biosensors as Analytical Tools in the Food and Drink Industries. *Food Chem.* **2002,** *77*(2), 237–256.

Mobini, S. H.; Lulsdorf, M.; Warkentin, T. D.; Vandenberg, A. Plant Growth Regulators Improve In Vitro Flowering and Rapid Generation Advancement in Lentil and Faba Bean. *In Vitro Cell. Develop. Biol. Plant* **2015,** *51*(1), 71–79. https://doi.org/10.1007/s11627-014-9647-8

Mousavi, S. R.; Rezaei, M. Nanotechnology in Agriculture and Food Production. *J. Appl. Environ. Biol. Sci.* **2011,** *1*, 414–419.

Narang, J.; Chauhan, N.; Rani, P.; Pundir, C. S. Construction of an Amperometric TG Biosensor Based on AuPPy Nanocomposite and Poly (indole-5-carboxylic acid) Modified Au Electrode. *Bioproc. Biosyst. Eng.* **2012,** *36*(4), 425–432. https://doi.org/10.1007/S00449-012-0799-9

Nauen, R.; Denholm, I. Resistance of Insect Pests to Neonicotinoid Insecticides: Current Status and Future Prospects. *Arch. Insect Biochem. Physiol.* **2005**. https://doi.org/10.1002/arch.20043

Neethirajan, S.; Freund, M. S.; Jayas, D. S.; Shafai, C.; Thomson, D. J.; White, N. D. G. Development of Carbon Dioxide (CO2) Sensor for Grain Quality Monitoring. *Biosyst. Eng.* **2010,** *106*(4), 395–404. https://doi.org/https://doi.org/10.1016/j.biosystemseng.2010.05.002

Niraimathi, K. L.; Sudha, V.; Lavanya, R.; Brindha, P. Biosynthesis of Silver Nanoparticles Using Alternanthera Sessilis (Linn.) Extract and Their Antimicrobial, Antioxidant Activities. *Coll. Surf. B Biointerf.* **2013,** *102*, 288–291. https://doi.org/10.1016/j.colsurfb.2012.08.041

Nuruzzaman, M.; Rahman, M. M.; Liu, Y.; Naidu, R. Nanoencapsulation, Nano-guard for Pesticides: A New Window for Safe Application. *J. Agric. Food Chem.* **2016,** *64*(7), 1447–1483. https://doi.org/10.1021/acs.jafc.5b05214

O'Connor, D. J.; Wright, G. C.; Dieters, M. J.; George, D. L.; Hunter, M. N.; Tatnell, J. R.; Fleischfresser, D. B. Development and Application of Speed Breeding Technologies in a Commercial Peanut Breeding Program. *Peanut Sci.* **2013,** *40*(2), 107–114. https://doi.org/10.3146/ps12-12.1

Ochatt, S. J.; Sangwan, R. S.; Marget, P.; Assoumou Ndong, Y.; Rancillac, M.; Perney, P. New Approaches Towards the Shortening of Generation Cycles for Faster Breeding of Protein Legumes. *Plant Breed.* **2002,** *121*(5), 436–440. https://doi.org/10.1046/j.1439-0523.2002.746803.x

OECD, Food, & of the United Nations, A. O.. *OECD-FAO Agricultural Outlook 2012,* 2002.. https://doi.org/https://doi.org/https://doi.org/10.1787/agr_outlook-2012-en

Olaitan, A. F.; Abiodun, T. A.; Foluke, A. O.; Oluwaseyi, O. E. Comparative Assessment of Insect Pests Population Densities of Three Selected Cucurbit Crops. *Acta Fytotechnica et Zootechnica* **2017**. https://doi.org/10.15414/afz.2017.20.04.78-83

Omanović-Mikličanina, E.; Maksimović, M. *Nanosensors Applications in Agriculture and Food Industry.*

Omanović, E.; Maksimović, M. Nanosensors Applications in Agriculture and Food Industry. *Bull. Chem. Technol. Bosnia Herzegovina* **2016,***47*, 59–70.

Ong, B.; Lin-Heng, L.; Chun, J. Biological Diversity Conservation Laws in South East Asia and Singapore: A Regional Approach in Pursuit of the United Nations' Sstainable Development Goals? *Asia Pacific J. Environ. Law,* **2000**. https://doi.org/10.4337/apjel.2016.01.05

Palevitz, B. A. Genetic Parasites and a Whole Lot More. *Scientist* **2000**.

Parisi, C.; Vigani, M.; Rodríguez-Cerezo, E. Agricultural Nanotechnologies: What are the Current Possibilities? *Nano Today* 2016, *10*(2), 124–127. https://doi.org/https://doi.org/10.1016/j.nantod.2014.09.009

Peltonen-Sainio, P.; Jauhiainen, L.; Laurila, I. P. Cereal Yield Trends in Northern European Conditions: Changes in Yield Potential and its Realisation. *Field Crops Res.* 2009, *110*(1), 85–90. https://doi.org/https://doi.org/10.1016/j.fcr.2008.07.007

Peshin, R.; Dhawan, A. K. Integrated pest management. In *Integrated Pest Management,* 2009. https://doi.org/10.1007/978-1-4020-8992-3

Pingali, P. L. Green Revolution: Impacts, limits, and the Path Ahead. *Proc. Natl. Acad. Sci.* 2012, *109*(31) 12302 LP – 12308. https://doi.org/10.1073/pnas.0912953109

Piškur, J.; Langkjær, R. B. Yeast Genome Sequencing: The Power of Comparative Genomics. *Molecul. Microbiol.* **2004**. https://doi.org/10.1111/j.1365-2958.2004.04182.x

Prasad, S. Nanobiosensors: The Future for Diagnosis of Disease? *Nanobiosens. Dis. Diagn.* **2014,** *3*, 1–10. https://doi.org/10.2147/NDD.S39421

Prasad, R.; Bhattacharyya, A.; Nguyen, Q. D. Nanotechnology in Sustainable Agriculture: Recent Developments, Challenges, and Perspectives. *Front. Microbiol.* 2017, *8*, 1014. https://doi.org/10.3389/fmicb.2017.01014

Prasanna de Silva, A.; Nimal Gunaratne, H. Q.; Gunnlaugsson, Thorfinnur; Huxley, Allen J. M.; McCoy, Colin P.; Rademacher, Jude T.; Rice, T. E. Signaling Recognition Events With Fluorescent Sensors and Switches. *Chem. Rev.* **1997,** *97*(5), 1515–1566. https://doi.org/10.1021/CR960386P.

Pudake, R. N.; Chauhan, N.; Kole, C. (Eds.). Nanoscience for Sustainable Agriculture. Springer International Publishing. **2019**.

Rana, M. M.; Takamatsu, T.; Baslam, M.; Kaneko, K.; Itoh, K.; Harada, N.; Sugiyama, T.; Ohnishi, T.; Kinoshita, T.; Takagi, H.; Mitsui, T. Salt Tolerance Improvement in Rice Through Efficient SNP Marker-Assisted Selection Coupled With Speed-Breeding. *Int. J. Mol. Sci.* **2019,** *20*(10). https://doi.org/10.3390/ijms20102585

Rashidi, L.; Khosravi-Darani, K. The Applications of Nanotechnology in Food Industry. **2011,** *51*(8), 723–730. https://doi.org/10.1080/10408391003785417

rathbun 2013 nanosensors - Google विद्वान. (n.d.).

Ray, D. K.; Mueller, N. D.; West, P. C.; Foley, J. A. Yield Trends Are Insufficient to Double Global Crop Production by 2050. *PLoS One* **2013,** *8*(6), e66428.

Ray, D. K.; Ramankutty, N.; Mueller, N. D.; West, P. C.; Foley, J. A. Recent Patterns of Crop Yield Growth and Stagnation. *Nature Commun.* **2012,** *3*(1), 1293. https://doi.org/10.1038/ncomms2296

Richardson, J.; Hawkins, P.; Luxton, R. The Use of Coated Paramagnetic Particles as a Physical Label in a Magneto-Immunoassay. Biosens. Bioelectron. **2001,** *16*(9–12) 989–993.

Rivelli, A. R.; Trotta, V.; Toma, I.; Fanti, P.; Battaglia, D. Relation Between Plant Water Status and Macrosiphum Euphorbiae (Hemiptera: Aphididae) Population Dynamics on Three Cultivars of Tomato. *Euro. J. Entomol.***2013.** https://doi.org/10.14411/eje.2013.084

Rizal, G., Karki, S., Alcasid, M., Montecillo, F., Acebron, K., Larazo, N., Garcia, R., Slamet-Loedin, I. H.; Quick, W. P. Shortening the Breeding Cycle of Sorghum, A Model Crop for Research. *Crop Sci.* 2014, *54*(2), 520–529. https://doi.org/10.2135/cropsci2013.07.0471

Rodriguez-Saona, C., R., B.; Isaacs, R. Manipulation of Natural Enemies in Agroecosystems: Habitat and Semiochemicals for Sustainable Insect Pest Control. In *Integrated Pest Management and Pest Control - Current and Future Tactics*; 2012. https://doi.org/10.5772/30375

Russell, E. P. Enemies Hypothesis: A Review of the Effect of Vegetational Diversity on Predatory Insects and Parasitoids. *Environ. Entomol.* **1989**. https://doi.org/10.1093/ee/18.4.590

Samways, M. J. Insect Conservation: A Synthetic Management Approach. *Annu. Rev. Entomol*. **2007**. https://doi.org/10.1146/annurev.ento.52.110405.091317

Shang, Y.; Hasan, M. K.; Ahammed, G. J.; Li, M.; Yin, H.; Zhou, J. Applications of Nanotechnology in Plant Growth and Crop Protection: A Review. *Molecules (Basel)* **2019,** *24*(14). https://doi.org/10.3390/molecules24142558

Sharma, K.; Lavanya, M. Recent Developments in Transgenics for Abiotic Stress in Legumes of the Semi-arid Tropics. JIRCAS Working Report No. 23; 2003.

Sharma, H. C.; Ortiz, R. Host Plant Resistance to Insects: An Eco-friendly Approach for Pest Management and Environment Conservation. *J. Environ. Biol*. **2002**.

Sharma, S.; Ruud, A. On the Path to Sustainability: Integrating Social Dimensions Into the Research and Practice of Environmental Management. Business Strategy and the Environment, 2003. https://doi.org/10.1002/bse.366

Sharma, H. C.; Sujana, G.; Manohar Rao, D. Morphological and Chemical Components of Resistance to Pod Borer, Helicoverpa Armigera in Wild Relatives of Pigeonpea. *Arthropod Plant Interact*. **2009,** *3,* 151–161. https://doi.org/10.1007/s11829-009-9068-5

Sharma, H. C.; Venkateswarulu, G.; Sharma, A. Environmental Factors Influence the Expression of Resistance to Sorghummidge, Stenodiplosis sorghicola. *Euphytica* **2003**. https://doi.org/10.1023/A:1023041713713

Shivakumar, M.; Nataraj, V.; Kumawat, G.; Rajesh, V.; Chandra, S.; Gupta, S.; & Bhatia, V. S. (2018). Speed breeding for Indian agriculture: A rapid method for development of new crop varieties. *Current Science*, *115*(7), 1241. https://doi.org/10.18520/cs/v115/i7/1241-1241

Shivendu, R.; Dasgupta, N.; Lichtfouse, E. *Nanoscience in Food and Agriculture 2*; Springer International Publishing, 2016.

Shivendu, R.; Dasgupta, N.; Lichtfouse, E. *Nanoscience in Food and Agriculture 5*; Springer International Publishing, 2017.

Singh, S. P.; Rahman, M. F.; Murty, U. S. N.; Mahboob, M.; Grover, P. Comparative Study of Genotoxicity and Tissue Distribution of Nano and Micron Sized Iron Oxide in Rats After Acute Oral Treatment. *Toxicol. Appl.Pharmacol.* **2013,** *266*(1), 56–66. https://doi.org/ https://doi.org/10.1016/j.taap.2012.10.016

Singh, A.; Rajput, V. D.; Mehrotra, R.; Pal, N.; Singh, V. K.; Minkina, T.; Chokheli, V. A.; Singh, R. K. In *Sustainable Soil Fertility Management*; Nova. Sci. Publishers, Inc.; 2020a, vol 1, pp 73–100.

Singh, A.; Rajput, V.; Rawat, S.; Kumar Singh, A.; Bind, A.; Kumar Singh, A.; Chernikova, N.; Voloshina, M.; Lobzenko, I. Monitoring Soil Salinity and Recent Advances in Mechanism of Salinity Tolerance in Plants. *Biogeosyst. Techniq.* **2020b**, *7*(2). https://doi.org/10.13187/bgt.2020.2.66

Singh, A.; Rajput, V. D.; Rawat, S.; Sharma, R.; Singh, A. K.; Singh, A. K.; Tomar, R. S. In *Emerging Tools for Sustainable Agriculture and Food Security*; Rajput, D.; Book Agency: New Delhi, Delhi, **2021a**, vol 1, pp 1–15.

Singh, A.; Rajput, V.; Singh, A.; Sengar, R.; Singh, R.; Minkina, T. Transformation Techniques and Their Role in Crop Improvements: A Global Scenario of GM Crops. *Policy Issues Genetically Modified Crops* **2021b**, *1,* 515–542.

Slosser, J. E.; Price, J. R.; Puterka, G. J. Evaluation of Furrow Diking and Early-Season Insecticide Applications on Boll Weevils (Coleoptera: Curculionidae), Bollworms (Lepidoptera: Noctuidae), and Cotton Yield in the Texas Rolling Plains. *J. Eco. Entomol.* **1989**. https://doi.org/10.1093/jee/82.2.599

Smith, S. L.; Slywka, G. W.; Krueger, R. J. Anthocyanins of Strobilanthes Dyeriana and Their Production in Callus Culture. *J. Natl. Prod.* **1981**. https://doi.org/10.1021/np50017a020

Song, J. M.; Kasili, P. M.; Griffin, G. D.; Vo-Dinh, T. Detection of Cytochrome c in a Single Cell Using an Optical Nanobiosensor. *Anal. Chem.* **2004,** *76*(9), 2591–2594. https://doi.org/10.1021/AC0352878

Sotiropoulou, S., Gavalas, Vamvakaki, V.; Chaniotakis, N. A. Novel Carbon Materials in Biosensor Systems. *Biosens. Bioelectron.* **2003,** *18*(2–3), 211–215.

Southwood, T. R. E. Habitat, the Templet for Ecological Strategies? *J. Anim. Ecol.* **1977**. https://doi.org/10.2307/3817

Sparks, T. C.; Nauen, R. IRAC: Mode of Action Classification and Insecticide Resistance Management. *Pestic. Biochem. Physiol.* **2015**. https://doi.org/10.1016/j.pestbp.2014.11.014

Stoner, K. A. Plant Resistance to Insects: A Resource Available for Sustainable Agriculture. *Biol. Agric. Hortic.* **1996**. https://doi.org/10.1080/01448765.1996.9754764

Su, X.; Chew, F. T.; Sam, F.Y. Li.Design and Application of Piezoelectric Quartz Crystal-Based Immunoassay. *Anal. Sci.* **2000**.

Sumbal; Nadeem, A.; Naz, S.; Ali, J. S.; Mannan, A.; Zia, M. . *Biotechnol. Rep. (Amst)* **2019,** *22*, e00338. https://doi.org/10.1016/j.btre.2019.e00338

Sunding, D.; Zilberman, D. Chapter 4 The agricultural innovation process: Research and technology adoption in a changing agricultural sector. In Agricultural Production; Elsevier, 2001; vol 1, pp 207–261https://doi.org/https://doi.org/10.1016/S1574-0072(01)10007-1

Swift, T. A.; Oliver, T. A. A.; Galan, M. C.; Whitney, H. M. Functional Nanomaterials to Augment Photosynthesis: Evidence and Considerations for Their Responsible Use in Agricultural Applications. *Interface Focus* **2019,** *9*(1), 20180048. https://doi.org/10.1098/rsfs.2018.0048

Tabashnik, B. E.; Carrière, Y. Successes and Failures of Transgenic bt crops: Global Patterns of Field-Evolved Resistance. In *Bt Resistance: Characterization and Strategies for GM Crops Producing Bacillus thuringiensis Toxins*; 2015. https://doi.org/10.1079/9781780644370.0001

Tamhane, V. A.; Giri, A. P.; Sainani, M. N.; Gupta, V. S. Diverse Forms of Pin-II Family Proteinase Inhibitors From Capsicum Annuum Adversely Affect the Growth and Development of Helicoverpa Armigera. *Gene*. **2007.** https://doi.org/10.1016/j.gene.2007.07.024

Tefera, T.; Mugo, S.; Beyene, Y. Developing and Deploying Insect Resistant Maize Varieties to Reduce Pre-and Post-harvest Food Losses in Africa. *Food Secur*. **2016**. https://doi.org/10.1007/s12571-015-0537-7

Tilman, D.; Balzer, C.; Hill, J.; Befort, B. L. Global Food Demand and the Sustainable Intensification of Agriculture. *Proc. Natl. Acad. Sci.* **2011,** *108*(50), 20260 LP–20264. https://doi.org/10.1073/pnas.1116437108

United Nations, D. of E. and S. A. (UN-D. *Population Division, 2017. World Population Prospects: The 2017 Revision, Key Findings and Advance Tables. Working Paper No. ESA/P/WP/248.*; 2017.

van Emden, H. F. Host Plant-Aphidophaga Interactions. *Agric. Ecosyst. Environ*. **1995**. https://doi.org/10.1016/0167-8809(94)09001-N

van Lenteren, J. C.; Hua, L. Z.; Kamerman, J. W.; Rumei, X. The Parasite-Host Relationship Between Encarsia Formosa (Hym., Aphelinidae) and Trialeurodes Vaporariorum (Hom., Aleyrodidae) XXVI. Leaf Hairs Reduce the Capacity of Encarsia to Control Greenhouse

Whitefly on Cucumber. *J. Appl. Entomol.* **1995**. https://doi.org/10.1111/j.1439-0418.1995.tb01335.x

Vo-Dinh, T.; Cullum, B. Biosensors and Biochips: Advances in Biological and Medical Diagnostics. *Fresenius' J. Anal. Chem.* **2000,** *366*(6), 540–551. https://doi.org/10.1007/S002160051549

Warheit, D. B.; Brown, S. C.; Donner, E. M. Acute and Subchronic Oral Toxicity Studies in Rats With Nanoscale and Pigment Grade Titanium Dioxide Particles. *Food Chem. Toxicol.* **2015,** *84*, 208–224. https://doi.org/https://doi.org/10.1016/j.fct.2015.08.026

Wassmann, R.; Jagadish, S. V. K.; Heuer, S.; Ismail, A.; Redona, E.; Serraj, R.; Singh, R. K.; Howell, G.; Pathak, H.; Sumfleth, K. Chapter 2 Climate Change Affecting Rice Production. The Physiological and Agronomic Basis for Possible Adaptation Strategies. In *Advances in Agronomy*,1st ed.; Elsevier Inc., 2009; vol 101, Issue January 2009). https://doi.org/10.1016/S0065-2113(08)00802-X

Worrall, E. A.; Hamid, A.; Mody, K. T.; Mitter, N.; Pappu, H. R. Nanotechnology for plant disease management. *Agronomy*. **2018**, *8*(12), 285.

Zeigler, R. S.; Mohanty, S. Support for International Agricultural Research: Current Status and Future Challenges. *New Biotechnol.* 2010, *27*(5), 565–572. https://doi.org/10.1016/j.nbt.2010.08.003

Zhang, Y.; Arugula, M. A.; Wales, M.; Wild, J.; Simonian, A. L. A Novel Layer-by-Layer Assembled Multi-Enzyme/CNT Biosensor for Discriminative Detection Between Organophosphorus and Non-Organophosphorus Pesticides. *Biosens. Bioelectron.* **2015,** *67*, 287–295. https://doi.org/10.1016/j.bios.2014.08.036

Zhang, Yi; Liu, B.,;Huang, K.; Wang, S.; Quirino, R. L.; Zhang, Z.; Zhang, C. Eco-Friendly Castor Oil Based Delivery System With Sustained Pesticide Release and Enhanced Retention. *ACS Appl. Mater. Interfaces*, 2020, *12*(33), 37607–37618. https://doi.org/10.1021/acsami.0c10620

CHAPTER 11

Nanopore DNA Sequencing: A New Era for Crop Improvement

MAINAK BARMAN[1], SURACHITA DAS[2], SUBHRA MUKHERJEE[3], and SATISH KUMAR SINGH[4]

[1]*Department of Genetics and Plant Breeding, Bidhan Chandra Krishi Viswavidyalaya, West Bengal, India*

[2]*Department of Human Physiology, Vidyasagar University, West Bengal, India*

[3]*Department of Genetics and Plant Breeding, Bidhan Chandra Krishi Viswavidyalaya, West Bengal, India*

[4]*Department of Plant Breeding and Genetics, Dr. Rajendra Prasad Central Agricultural University, Bihar, India*

ABSTRACT

DNA sequencing technologies have shed the light of success to all the current-era discoveries of biological science to achieve a tremendous height. Nanopore sequencing is the most recent in the queue of DNA sequencing techniques. This approach reliably senses the nucleotides with no active DNA production. The nanopore DNA sequencing principles rely on nucleotide-detecting mechanisms. The standard protocol of DNA extraction for nanopore sequencing is always one of the critical steps to achieve success from the procedure. Biological nanopore and solid-state nanopore are the two primary groups of nanopore technologies along with the most recently developed hybrid nanopores. Oxford Nanopore Technologies has

Nanotechnology for Sustainable Agriculture: An Innovative and Eco-Friendly Approach. Vishnu D. Rajput, Abhishek Singh, Tatiana M. Minkina, Krishan K. Verma, and Awani Kumar Singh (Eds.)

been constructing nanopore-based systems to sequence DNA for commercial utilization. Several plant species have been used for the Oxford Nanopore Sequencing. Several sensing methods have been suggested and discovered with a great success rate in the last few years, which are strong determining factors for a successful DNA nanopore sequencing method. When a DNA strand translocates through the nanopore, speed control is a major challenge for this DNA sequencing approach. It is being handled with great innovative researches nowadays. The major constraint of the nanopore sequencer is the relatively low reading precision. Many bioinformatics tools have already been released, and several further are in progress, focused, or optimized for nanopore sequencing.

11.1 INTRODUCTION

Since the last decade, innovative technologies envisaging the unseen or unravelling the imperceptible have always extended helping hands to the advances in scientific inventions. The quick arrival of high-throughput and reasonable machinery of sequencing DNA has unquestionably served as the significant driving strength in advancing life science studies (Goodwin et al., 2016). Genomic information has always played the role of one of the hearts of molecular biology, supplying as well as aiding means for genome probing, its configuration, expression of genes, epigenetics, and a large number of other relevant fields (Singh et al., 2020; Singh et al., 2021). Nanopore sequencers are the most recent in the queue of DNA sequencers (Jain et al., 2016). Although a range of sequencing methods are accessible like single-molecule real-time sequencing (SMRT), sequencing by ligation, sequencing by synthesis, and pyrosequencing, approaches of DNA sequencing since Sanger sequencing largely depended upon the procedure of DNA synthesis (Schadt et al., 2010). DNA nanopore sequencing differentiates itself from these preceding techniques. This approach unswervingly senses the nucleotides with no active DNA production. The stretch of DNA moves through a protein nanopore as an elongated single-stranded DNA stretch and is then stabilized in an electrically resistant membrane of polymer (Feng et al., 2015). A voltage is fixed across this membrane, and detection of ionic current alterations is done by these sensors caused by shifting by nucleotides dwelling in the pore in real time as the DNA molecule passes through. The exceptional sequencing mode of nanopore offers manifold benefits over other conventional approaches. Firstly, the sequencing does not entail imaging tools for detecting the nucleotides, permitting the system to rationalize in size to a convenient degree. Secondly, as the sequencing method unswervingly senses the input molecule exclusive of DNA synthesis or amplification, there is no prominent extent edge of the DNA we can sequence. In addition, ionic current alterations recognition reallocated by the nucleotides passing across the nanopore is not restricted to the four canonical nucleo-bases of A,T,G, C. Captivating pros of the character of unamplified unswerving sequencing, nanopore sequencing is potent of straight observing the changes of bases like methylations (Simpson et al., 2017), and even can directly sequence the molecules of RNA that contain uracil bases (Garalde et al., 2018). In this chapter, we will discuss how the method of nanopore sequencing actually functions and the present scenario of this mesmerizing scientific innovation.

11.2 PRINCIPLE OF NANOPORE SEQUENCING

The protocols of nanopore DNA sequencing are dependent upon the nucleotide-detecting mechanisms. Optical readout (i.e., fluorescent detection) (Huang et al., 2015) and electrical signature (Branton et al., 2008) are the two categories of principles of detection presently relied on. To be exact, the earliest discovered nanopore sequencing principle used is the ionic current blockade (Howorka et al., 2001), which is basically acquired from the usually known "Coulter counter" technique. Figure 11.1 represents this principle, and currently, this technology is explored in the most extensive manner as its ease permits lessening the cost as well as the size of instruments (Iqbal et al., 2011; Edel and Albrecht, 2013). In this method, current patterns are caused by the diverse nucleotide bases along a strand of DNA across the nanopore for the duration of its translocation in a linear manner under a field of electricity. These prototypes are specific for different nucleotides and are ultimately translated into the original string of DNA (Pennisi, 2012). When two tiny detached compartments get in touch with a fragile membrane with a nanopore, the field of electricity across the membrane generates a flow of ionic current. The ion flow is disturbed by the single-stranded DNA when it traverses through the nanopore. Thus, the pattern of blockade current is traced as a function of time which contains the information regarding the nucleotides. The patterns in the ionic currents show a sturdy relationship with the sequence of a number of nucleotide bases positioned at the nanopore all through the translocation- this is the hypothesis on which the central idea of this principle stands. This hypothesis has already been established to be applicable with protein nanopores (Manrao et al., 2012; Laszlo et al., 2014). Although, it has not so far been authenticated utilizing solid-state nanopores. Subsequently, tough confronts must be addressed concurrently for demonstrating concrete sequencing of DNA. In exacting, the following major necessities must be resolved essentially: an ultra-small nanopore as sequencing needs nanopore diameter less than 2 nm (Lindsay, 2016), an ultrathin membrane as detection of a distinctive signal at a level of single-nucleotide would turn out to be complicated if the membrane is much wider than several nucleotides, control of DNA speed as well as recognition of four nucleotides as the quick speed of traversing across the nanopore of ssDNA (usually faster than 1 μs/nt (Meller, 2003; Akahori et al., 2014) is excessively hasty to detain the signal at the level of single-nucleotides by the amplifiers that are existing at present.

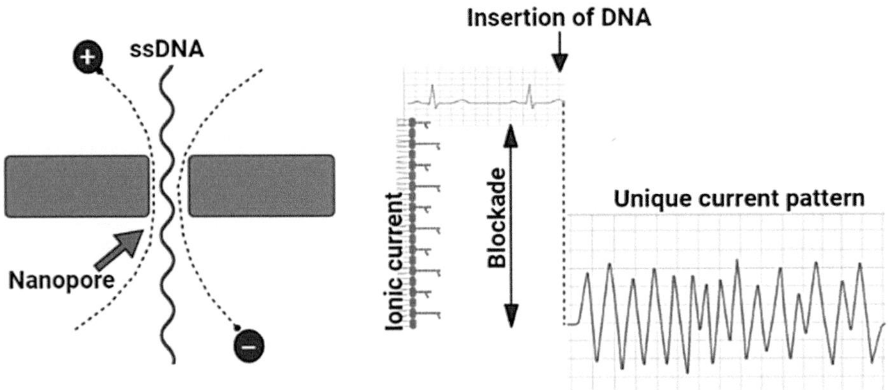

FIGURE 11.1 Schematic representation of DNA nanopore sequencing (adapted and modified from Goto et al., 2019).

11.3 TYPE OF NANOPORES

We can broadly classify nanopore technologies into two groups: (1) biological nanopores and (2) solid-state nanopores. The first category of nanopores, that is, biological ones, has been used extensively in detecting single molecules, diagnosing a range of diseases, and sequencing of DNA. Recent progress in the field of nanotechnology has smoothened the advancement of solid-state nanopore sensors. These synthetic nanopores can be incorporated additionally on a circuit chip together with other tools, for instance, a field-effect transistor offering the prospective for portable, miniature devices

TABLE 11.1 Key Events of Nanopore Sequencing (Deamer, 1998; Howorka et al., 2001; Quick et al., 2015; Quick et al., 2016; Quick et al., 2017).

Year	Key events	Scientists	Institutes
1992	First-time experiment that shows the capacity of alpha-hemolysin pore in nanopore sequencing	Deamer and Kasianowicz	National Institute of Standards and Technology
1998	For nanopore sequencing, the first patent was granted	Branton, Balderelli, Kasianowicz, Church, and Deamer	National Institute of Standards and Technology, University of California, Harvard University
2001	By using an engineered nanopore, up to 30 nucleotides length individual DNA strands were recognized	Howorka, Cheley, and Bayley	Texas A&M University
2006	Engineered alpha-hemoplysin pore with a molecular adaptor was used for easy recognition of the four bases of DNA for the very first time	Bayley, Braha, and Astler	Oxford University
2011	Hand-held DNA sequencer (MinION) effectively sequenced the first piece of DNA	Clive Brown	Oxford Nanopore Technology
2014	Successful use of MinION for sequencing the bacterium *Pseudomonas aeruginosa* genome	Nick Loman	University of Birmingham
2015	Successful use of MinION to sequence 142 Ebola virus samples in Guinea to combat outbreak of the disease	Quick and Nick Loman	University of Birmingham
2016	For sequencing and preventing the spread of Zika virus in Brazil, the mobile laboratory equipped MinION was used	Faria, Loman, Quick, de Jesus, Goodfellow, and Ramabut	FIOCRUZ Bahia, InstitutoEvandro Chagas, ARTIC Network, Oxford Nanopore Technology
2018	MinION appeared to be potential means to sequence human genome	Koren, Miga, Rand, Olsen, O'Grady, Nieto, Marriot, Malla, Fiddes, Dilthey, Simpson and Loose Beggs, Philippy, Paten, Tee, Snutch, Quinlan, Richardson, Rhie, Pedersen, Tyson, Sasani, Jain, Quick, and Loman	University of Nottingham, University of Birmingham, University of Utah, University of British Columbia, Ontario Institute for Cancer Research, Santa Cruz, University of East Anglia, University of California, National Human Genome Research Institute

TABLE 11.1 *(Continued)*

Year	Key events	Scientists	Institutes
2019	For the first time, sequencing tools are opted for program of population genome genomics (Abu Dhabi Genome Program)		Oxford Nanopore Technology
2020	Nanopore sequencers started using ARCTIC protocol for decoding the SARS-Cov2	Loman	Oxford Nanopore Technology, ARTIC Network
2020	Oxford instigated a highly precise COVID-19 test termed as LamPORE which was its first IVD controlled diagnostic		Oxford Nanopore Technology

for sequencing DNA. To capture the benefits of the attributes of both the solid-state as well as biological nanopores, hybrid nanopores have been put forward (Derrington et al., 2010). The technology of nanopore DNA sequencing is being rapidly developed.

11.3.1 BIOLOGICAL NANOPORES

These kinds of nanopores are termed transmembrane protein channels. These are typically introduced into a substrate, for example, liposomes, planar lipid bilayers, or other polymer films. Well-defined as well as highly reproducible nanopore structure and size are the major leads of biological nanopores. More significantly, we can modify these types of nanopores easily with up to date methods of molecular biology like mutating the sequence of nucleotides for altering the amino acid residue at an exact position. In this segment, we will discuss three biological nanopores that are well-studied.

11.3.1.1 A-HEMOLYSIN

α-Hemolysin also termed as α-toxin. Among the biological nanopores, it is the first and most frequently utilized type that holds remarkable importance in the DNA sequencing field. It is an exotoxin produced by a human pathogenic bacterium *Staphylococcus aureus*. This mushroom-structured heptamer is a transmembrane channel of 232.4-kDa which consists of a cap of 3.6-nm diameter and a transmembrane β-barrel of diameter 2.6-nm (Song et al.,1996). The outside measurements of the apertures are 10 nm × 10 nm. The thickness of the α-Hemolysin channel from inside and the extent of a molecule of ssDNA are extremely close in size. Hence, within the nanopore, this α-Hemolysin is capable of discriminating single nucleotides employing ionic current (Cherf et al., 2012). This provides this biological nanopore a skillful means to analyze biomolecular communications as well as configurations at the level of single molecule. Moreover, the structure of the nanopore can dwell steady at high temperature close to 100ºC within an ample range of pH (2–12) (Kang et al., 2005). Even though these pores are broadly exploited in biological researches, the restricted size of pore (1.4 nm) confines its utility in analyzing RNA, ssDNA, or small molecules. In addition, the b-barrel is excessively lengthy for directly differentiating each nucleotide from single long-chain molecules of DNA.

11.3.1.2 MSPA

Branton et al. (2008) reported *Mycobacterium smegmatis* porin A or MspA to be a hopeful and potent biological nanopore to read information from four nucleotides concurrently. The diameter of the MspA octamer channel is 1 nm at the minimum end, which is comparatively petite as well as slender than α-Hemolysin. As a result, it has the potential of improving the spatial resolution of sequencing single-stranded DNA. Additionally, this biological nanopore is vigorous and maintains the channel dynamic under tremendous experimental circumstances, for instance, varying pH range from 0 to 14 and maintaining the temperature at 100°C for 30 min (Abiola et al., 2003).

11.3.1.3 BACTERIOPHAGE PHI29

Phi29 is one other well-studied biological nanopore that has created immense attention. For the first time, Wendell et al. (2009) revealed that through the phi29 pore, dsDNA could pass. The bacteriophage phi29 DNA packaging motor has a 12-subunit gp10 connector (Guasch et al., 2002), six replicas of ATP-binding DNA packaging RNA (Guo et al., 1998), and an ATPase protein, gp16 (Lee et al., 2006), which offers the chemical force entailed for translocation of DNA. The connector protein is capable of effortlessly self-assembling to shape an unwavering and recurring dodecameric organization in solution. In contrast to the previously discussed two nanopores, this phi29 pore holds a greater diameter, which permits measuring molecules, such as proteins, DNA complexes, and double-stranded DNA that are larger. In addition, a larger phi29 pore offers more litheness for biochemical alterations.

11.3.2 SOILD-STATE NANOPORE

Even though biological nanopores are reported to show exclusively stirring test results for sequencing single-stranded DNA, these protein pores include an invariable size of the pore, profile, in addition to be deficient in steadiness. Additionally, they survive from conventionally supported lipid membrane frailty. To fiddle with these paucities, a range of synthetic nanopores have been constructed employing diverse techniques and are put in an application to the analysis of DNA as well as RNA. Li et al. (2001) are credited for confirming that we can use these solid-state nanopores for studying the procedure on how different molecules translocate (Cheley et al., 2002). The advancement of microfabrication technologies has drawn the rising interest of solid-state nanopores. This category of nanopores holds several advanced benefits over biological equivalents, such as mechanical, chemical and thermal stability, adjustability of size, and incorporation. Solid-state nanopores can accurately function under a broad range of experimental states in addition to being able to be produced in mass through conventional semiconductor practices. In recent times, solid-state nanopores have been applied as an innovative technique in diverse areas, counting disease diagnosis, molecule translocation process, protein detection, and DNA sequencing. A number of primary skills are frequently exercised for constructing nanopores in polymer membranes (Menestrina et al., 2014), graphene (Garaj et al., 2010), boron nitride (BN) (Liu, 2013), aluminum oxide (Al_2O_3) (Venkatesan and Bashir, 2011), silicon dioxide (SiO_2) (Storm et al., 2003), silicon nitride (Si_3N_4) (Heng et al., 2004), and hybrid materials (Bai et al., 2014). The elite geometric along with electrical characteristics of these nanopores offer them a distinctive lead over the biological ones. Still, it is necessary to improve their thermal and chemical steadiness to craft trustworthy mechanisms (Kwok et al., 2014).

11.3.2.1 SI_3N_4 AND SIO_2 NANOPORES

High chemical stability and low mechanical stress have put Si_3N_4 and SiO_2 films into light to be extensively used as substrates. We can manufacture them using balancing metal oxide semiconductor-attuned industrial integrated circuit procedures (Gao and Xie, 2012; Dai et al., 2012; Gao et al., 2013). Ion or electron beam, carving in a free-positioning membrane window, frequently drill nanopores, and we can regulate them employing wet-etching methods with micrometer accuracy as well as standard photolithography (Dai et al., 2012; Gao et al., 2013). Both these substrates demonstrate to be excellent also in high concentrations solution of an electrolyte.

11.3.2.2 AL_2O_3 MEMBRANES

In contrast to the previous two, the electrical performance of Al_2O_3 films is superior in having an advanced ratio of signal-to-noise and lesser noise for the period of DNA translocation (Venkatesan et al., 2009). We can use atomic layer deposition for fabricating these at a single atomic-level thickness. We can take transmission electron microscopy and focused ion beam into a grant for constructing nanopores in metal oxide membranes (Venkatesan et al., 2009; Haque et al., 2013). The speed of DNA translocation is sluggish through Al_2O_3 nanopores compared with the speed through Si3N4 nanopores, which is endorsed to the well-built electrostatic dealings among the positively charged ones Al_2O_3 surface and the negatively charged molecules of double-stranded DNA.

11.3.3 SINGLE-LAYER MEMBRANES

Even though the solid-state nanopores engineered in insulating films have been functional extensively in protein and DNA translocation procedures, they are not seen as having an adequate spatial and temporal resolution for obtaining structural data of molecules at the single-base level. Graphene membrane has been used as a substitute in recent times to conventional solid-state membranes. It is a single atomic layer of carbons with astonishing electrical as well as mechanical characteristics (Traversi et al., 2013). Molybdenum disulfide (MoS_2) has also created huge attention. We can fabricate nanopores in suspended single-layer membranes by organized electron beam exposure via transmission electron microscopy (Liu et al., 2012). One meticulous benefit of utilizing nanopores in ultrathin membranes is the negligible membrane breadth (0.335 nm) comparable to the two DNA chain-bases distance (Traversi et al., 2013). Single-layer membranes possibly will grip the prospect of achieving astonishingly towering spatial resolution for the sequencing of DNA.

11.3.4 HYBRID BIOLOGICAL/SOILD-STATE NANOPORES

The lacuna of chemical discrimination from the target molecules of roughly identical size is presently the critical disadvantage of solid-state nanopores. By functionalizing surfaces, we can improve this chemical specificity (Bai et al., 2014); affixing precise sequences of recognition and receptors to the nanopores can also assist (Iqbal et al., 2007; Venkatesan and Bashir, 2011). Nanopores programmed with hairpin DNAs or other receptors are capable in distinctively recognizing nucleotides in sequencing applications (Iqbal et al., 2007; Branton et al., 2008). A fluid lipid bilayer can be used to cover the synthetic nanopores for controlling the translocations of protein (Yusko et al., 2011). A range of lipids can be employed to precisely manage the breadth and surface chemistry of the coating surface. These nanopore sensors demonstrate tremendous electrical attributes and improved mechanical strength, and thus, may uncover wider uses in nanobiotechnology.

TABLE 11.2 Comparative Structural Features of the Major Nanopore Categories (Yanxiao et al., 2015).

Types of nanopore	Shape of the channel	Fabrication membrane	Nanopore connecter	Channel diameter (nm)	Channel length (nm)	Special feature
Biological nanopore						
α-hemolysin	Mushroom-shaped, heptamer (β-barrel)	Lipid bilayer	Self-assemble	1.4–2.6	5 2	Discriminate single nucleotide using ionic current within nanopore
MspA	Octamer		Self-assemble	1.2	3.7	Potent to advance spatial resolution of ssDNA sequencing and maintains the channel dynamic under extreme conditions
Phi29	Dodecamer		Self-assemble	3.6–6	7	Allows to measure large molecules, that is, ssDNA, DNA complexes, and provide more flexibility for chemical modification
Solid-state nanopore						
Silicon-based nanopore		Si_3N_4/ SiO_2membrane	Electron beam-based decomposition sputtering, laser ablation, ion milling track-etch method, helium ion microscopy, FIB techniques, dielectric breakdown, electron-beam lithography	Measured by sub-nm scale	Membrane thickness	Perform in high concentration of electrolyte solution to detect DNA molecule
Al_2O_3 nanopore		Al_2O_3	FIB, TEM	Measured by sub-nm scale	45–60	Slow translocation speed is characteristic to well-built electrostatic interaction between (+ve) charged Al_2O_3 surface and (-ve) charged ds DNA

TABLE 11.2 *(Continued)*

Types of nanopore	Shape of the channel	Fabrication membrane	Nanopore connecter	Channel diameter (nm)	Channel length (nm)	Special feature
Single-layer membranes		BN, Graphene,MoS_2	TEM	Measured by sub-nm scale	0.335	Minimal thickness of the membrane is comparable to the space between the bases in a DNA chain and may grasp prospective to accomplish high spatial resolution to sequence DNA
Hybrid nanopore		Si_3N_4 Al_2O_3	TEM	Measured by sub-nm scale	Membrane thickness	Can function with hairpin DNA or other receptor have the capacity to distinctively recognize nucleotides in sequencing application

TABLE 11.3 Plant Species Used for the Oxford Nanopore Sequencing (Kathryn et al., 2020).

Year	Name of the plant species	Technology used for sequencing	Size of genome/N50	Assembler
2017	*Solanum pennellii* (wild tomato)	Illumina, ONT	1.0 Gbp/N50 2.45 Mbp (contig)	Pilon, SMARTdenovo, Canu
2018	*Sorghum bicolor*(sorghum)	Illumina, ONT, bionano	732 Mbp/N50 33.28 Mbp (scaffold), 3.05 Mbp (contigs)	Bionano, nanopolish, Pilon, SMARTdenovo, Canu
2018	*Oryza coarctata* (wild rice)	Illumina, ONT, illumina mate-pair	665 Mbp/N50 1.86 Mbp (scaffold), 15.13 kbp (contig)	PLATANUS, SSPACE, GapCloser
2018	*Brassica oleracea*	Illumina, ONT, bionano	630 Mbp N50 29.5 Mbp (scaffold), 7.3 Mbp (contig)	Bionano Solve and Access, Pilon, Racon, Ra (SMART denovo, wtdbg)
2018	*Arabidopsis thaliana*	Illumina, ONT	119.5 Mbp/N50 12.3 Mbp (contig)	Canu, Miniasm, Pilon

TABLE 11.3 *(Continued)*

Year	Name of the plant species	Technology used for sequencing	Size of genome/N50	Assembler
2019	*Juglans regia* (walnut)	ONT, illumina short read, Hi-C	547 Mbp/N50 31.49 Mbp (scaffold), 1.36 Mbp (contig)	MaSuRCA, HiRise
2019	*Lupinus albus* (white lupin)	ONT, PacBio, illumina, bionano optical mapping	451 Mbp/N50 9.88 Mbp (scaffold), 7.11 Mbp (contig)	Canu, Falcon (for PacBio data only), Pilon, Bionano Solve
2019	*Corylus avellana* L. (European hazel)	Illumina, ONT, Hi-C	370 Mbp/N50 36.65 Mbp (scaffold)	MaSuRCA, HiRise
2020	*Lonicera japonica* (Japanese honeysuckle)	ONT, illumina, Hi-C	843.2 Mbp N50 84.4 Mbp (scaffold)	LACHESIS, SMARTdenovo, Pilon; SLR, SALSA (for Hi-C data), Canu
2020	*Euryale ferox* (prickly water lily)	ONT, illumina, Hi-C	725.2 Mbp/N50 4.75 Mbp (contig)	LACHESIS (for Hi-C data); Pilon, Canu
2020	*Asparagus setaceus* (asparagus fern)	ONT, illumina, 10× genomics, Hi-C	710.15 Mbp/N50 2.19 Mbp (scaffold)	Canu, Pilon; LACHESIS (for Hi-C data)
2020	*Juglans sigillata* (iron walnut)	ONT, illumina, bionano, Hi-C	536.5 Mbp/N50 16.43 Mbp (scaffold), N50 4.34 Mbp (contig)	Canu, wtdbg, Pilon
2020	*Oryza sativa* (rice)	ONT, illumina	386.5 Mbp N50 6.32 Mbp (contig) (Basmati 334); 383.6 Mbp/N50 10.53 Mbp (contig) (Dom Sufid)	Canu, fly, Medaka, Pilon
2020	*Oryza sativa* (rice) Carolina Gold Select	ONT, illumina	377 Mbp/N50 1.72 Mbp (scaffold), N50 1.63 Mbp (contig)	Flye, MaSuRCA
2020	*Spirodela polyrhiza* (common duckweed)	ONT, Hi-C	138.49 Mbp/N50 3.34 Mbp (contig), 7.68 (scaffold)	Miniasm; Proximo (for Hi-C data)

11.3.5 COMMERCIAL BIOLOGICAL NANOPORE SEQUENCERS

Gordon Sangheraand Hagan Bayley founded Oxford Nanopore Technologies (ONT). This has been constructing nanopore-based systems of DNA sequencing for utilization in commercial purpose. In 2012, ONT revealed the preliminary investigational outcomes from its larger GridION system (Eisenstein, 2012). This GridION system can be expanded with supplementary cartridges, each of which includes assortments of nanopores. Numerous gigabytes of raw data can be produced per day by each GridION node and cartridge. The method is planned for stretchy run times having a wide range from a few minutes up to some days based upon the data prerequisites of the tests. The MinION is a one-time usable DNA sequencing USB memory stick-sized apparatus intended for wide-ranging DNA sequencing functions. Many research teams have taken the MinION for sequencing amplicons from a snake venom gland transcriptome, ʎ-phage genome, and *Escherichia coli* K-12 substrate (Quick et al., 2014). Nevertheless, the inaccuracy rates of these investigations were beyond 90% (Mikheyev and Tin, 2014). Even though it is still a lengthy approach from exploitation in a broad array of functions, these outcomes are incredibly cheering for nanopore-based DNA sequencing skills.

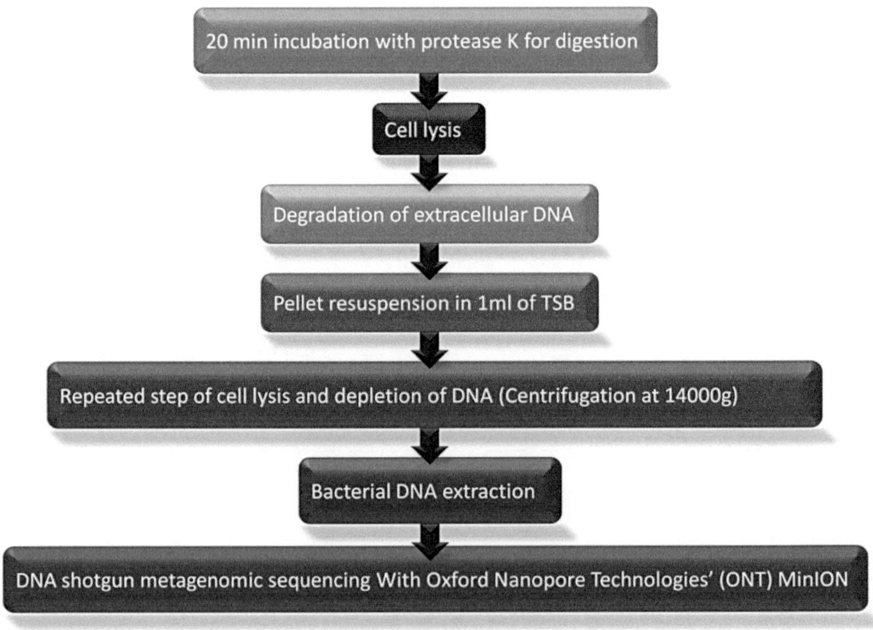

FIGURE 11.2 Standard protocol of DNA extraction for nanopore sequencing (Karin and Hege, 2020).

11.4 MECHANISMS OF DNA TRANSLOCATION SENSING

The means of sensing, that is, on which way we can generate the signal and by which means we can exclusively connect it with the DNA molecules' structure that is under translocation, is a strong determining factor for the accomplishment of DNA nanopore sequencing methods. A number of sensing methods have been suggested as well as discovered in last few years, though with diverse scales of triumph. In this section, we will talk about and evaluate fundamental techniques to read off the DNA sequences in the translocation via nanopores.

11.4.1 IONIC CURRENT BLOCKADES

Quantification of ionic currents for the period of translocation is the most extensively practiced method in detecting sequence (Dekker, 2007; Venkatesan and Bashir, 2011). The assumption is that the threading of biopolymer through the channel chunks the small ions' fluxes by geometrical omission, and recording is kept as a function time of this alteration in the current. Modulations in the ionic currents that sturdily associate with the chemical characteristics of the nucleotide at the constricted part of the nanopore are the major thought of this sequencing technique is the hypothesis. Massive studies have been executed on DNA nanopore sequencing exercising ion–current blockages (Yang et al., 2013; Liang and Zhang, 2015). However, there are a lot of challenges in the field that are to be resolved yet.

11.4.2 TRANSVERSE CURRENT MEASUREMENTS OF DNA TRANSLOCATION

In the current-blockade technique, the ionic fluxes corresponding to the translocating DNA chains are assessed and interpreted. In recent times, a method has been suggested as a substitute for DNA nanopore sequencing, which is rooted in calculating the electrical current at a 90-degree angle to the direction of DNA transportation all through the translocation. Transverse electrical conductivity is anticipated to be exceptionally dependent upon the character of each of the nucleotides in the nanopore. Scanning tunneling microscopy has helped to hypothetically inspect and investigationally envisage the physical characteristics of these currents (Nelson et al., 2010). Suggestions have been made regarding these currents that tunneling effects cause these, or they can be linked with the characteristics of semiconductor materials attached to the nanopores.

11.4.2.1 TRANSVERSE TUNNELING CURRENT

Electrodes are positioned (two in number) in the nanopore, and the transverse voltage is provided across them in transverse tunneling current measurements. Inside the nanopore, the electrons jump between electrodes along the nanometer distance. Passing DNA base sequences amend the impending obstruction among the two electrodes and result in a tunneling current that quickly crumbles with distance. This quantum effect of current alteration directs to the increased spatial resolution, and in addition, it offers an advanced molecular explicit. Measurement of electron tunneling current flow among vertical electrodes for the period of translocation of DNA is believed a capable move for the nanopore sequencing procedures. Still, a number of challenges exist. Firstly, nanoelectrode fabrication within a nanopore is a tremendously complicated technical stage, which is the foremost obstacle to apply these DNA sequencing devices. Additionally, the technique has to make sure that the DNA molecule constantly traverses the pore in an explicit direction because of the high sensitivity of tunneling current to atomic-scale alterations of directions along with distances (Agah et al., 2016). In alternative expression, the relative orientation of the nucleotide concerning the electrodes will be the depending factor for the tunneling current. The resolution of this DNA nanopore sequencing technique is considerably lessened by all these effects.

11.4.2.2 TRANSVERSE SEMICONDUCTOR CURRENT

An associated method exploits the alterations in the semiconductor material's conductance that is linked with the nanopore when translocation of DNA molecules takes through the path. Due to high intrinsic speeds and elevated sensitivity, carbon nanotube field-effect transistors (FETs) and nanowires are often exploited as biological and chemical sensors. Lieber's team proposed to link FET in a straight line with the nanopores to analyze how DNA translocates. Synchronized notes of the nanowire FET conductance and the corresponding ionic flux when *cis* and *trans* chambers are packed with solutions of diverse ionic potency validate that this technique accurately assesses the translocation incidents. Nevertheless, there is a considerable noise intensity in FET transistors which shrinks the worth of this move. It has been proposed to incorporate graphene nanogaps and/or graphene nanoribbons (GNR) with SiNx membranes into the nanopore to investigate the phenomenon of translocation to get better performance from this approach (Agah et al., 2016).

11.4.3 OPTICAL RECOGNITION

It is an exceedingly well-designed, recently developed technique that depends not on the measurements all through the translocation. As an alternative, optical sensing machinery is used here, as Figure 11 depicts. The immense benefit of this technique is the capability of recognizing DNA sequences by means of extremely analogous delivery. This is possibly the most capable arrangement to sequence DNA since it coalesces a high-speed parallel readout of the signals, elevated sensitivity, and the aptitudes to advance by exploiting multicolor schemes. Still, this technique has numerous confronts prior to being professionally applied in the DNA nanopore sequencing. The inaccuracies in nucleotide detection are of 10% order, and the target DNA switch is somewhat intricate (Venkatesan and Bashir, 2011). The capability of converting and reading hefty genomic pieces with high reliability applying this method has not been exhibited so far (Venkatesan and Bashir, 2011).

11.5 SPEED CONTROL OF DNA TRANSLOCATION

A major confront for this method is to control the speed while a DNA strand translocates through the nanopore. The classic reside period of DNA channel is less than 1 μs/nt across the nanopore. This time length is excessively minute for recording the ionic current signature, including nucleotide data utilizing existing amplifiers (Wanunu et al., 2008; Venkatesan and Bashir, 2011). Preferably, the settling time of DNA in a nanopore ought to be lengthened to greater than 100–1000 μs/nt in any case for satisfactory documentation of the electric signal at single-nucleotide level. Many research groups have stepped forward on developing competent approaches to trim down the speed of DNA (Wanunu, 2012; Carson and Wanunu, 2015; Wang et al., 2014; Pungetmongkol, 2018). Establishing a molecular motor close to the nanopore doorway is the most triumphant technique for protein nanopores (Fig. 11.3B). The functions as a molecular ratchet of DNA strands can be achieved by molecular motors like DNA helicase or DNA polymerase. This ratchet is capable of considerably lessening the DNA speed, in addition, to accurately controlling DNA relocation at the single-nucleotide level (Laszlo et al., 2014). Regrettably, a thriving effort on a molecular motor having a solid-state nanopore has not been noted until now as per our acquaintance. The compatibility of biological proteins with a nanoporous substance might be the reason of this insufficiency. Concerning solid-state nanopores, altering the gradient concentration (Wanunu et al., 2009), temperature (Verschueren et al., 2015), electrolyte species (Kowalczyk et al., 2012), and viscosity (Fologea et al., 2005) has been demonstrated to be trouble-free means. Many researchers have tried an approach in which different obstructions like nanocylinder (Yoshida et al., 2016), polymer gel (Tang et al., 2015), nano-bead array (Goto et al., 2015), or nanofiber mesh (Squires et al., 2013) are positioned near a nanopore opening for averting DNA movement. Relying on these incessant attempts, the single-stranded DNA translocation dwelling period might be expanded to ~10–100 μs/nt with no outside implements. DNA-immobilized probe or bead is one other approach to controlling speed. Manoeuvring of DNA movement is accomplished by regulating the bead or probe with an optical potential or piezoactuator (Akahori et al., 2017). In this direct regulating technique, we can slow down the dwelling time of DNA to 100 μs/nt. Even though the speed of DNA has effectively been slowed down, each documented regulatory process for solid-state nanopores still is not capable of stifling the vast disparity in dwell times.

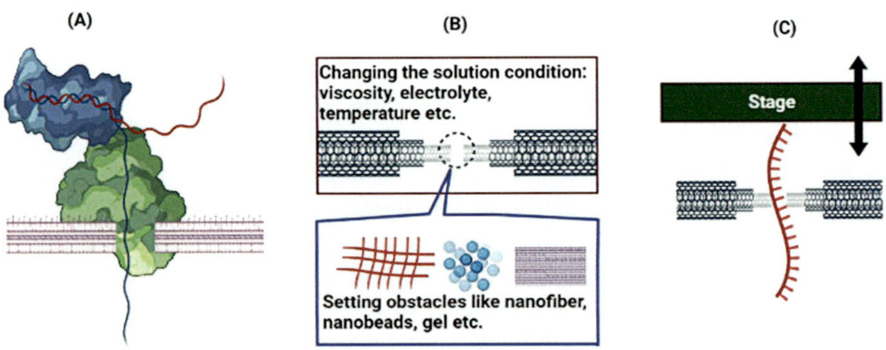

FIGURE 11.3 Schematic representation of DNA speed control mechanisms. (A) Molecular motor for DNA speed control, (B) alterations in a solution, and (C) DNA motion control directly via a piezo actuator or AFM (Adapted and modified from Goto et al., 2019).

TABLE 11.4 Instrument and Data Comparison Between Three Generations of Nanopore Sequencing (Kchouk et al., 2017).

Year	Generation	Instrument	Reading per run	Reading type	Avg length of reading (pb)	Rate of error	Category of error	Generated data per run (Gb)
First generation								
2002	ABI Sanger	3730xl	96	SE	400–900[a]	0.3	NA	0.00069–0.0021
Second generation								
2005	454	GS20	200	SE, PE	100	1	Indel	0.02
2007		GS FLX	400		250	1	Indel	0.01
2009		GS FLX Titanium	1 M		450	1	Indel	0.45
2010		GS junior	100		400	1	Indel	0.04
2011		GS FLX	1 M		700	1	Indel	–
		Titanium+						0.7
2014		GS Junior+	100		700	1	Indel	0.07
2011	Illumina	Miseq	25 M (max)	SE,PE	300	0.1	Mismatch	15 (max)
2012		Hiseq	5 B (max)		150	0.1	Mismatch	1.5 Tb (max)
2013		Miniseq	25 M (max)		150	1	Mismatch	7.5 (max)
2014		Nextseq	400 M (max)		150	1	Mismatch	120 (max)
2014		Hiseq	6 B (max)		150	0.1	Mismatch	1.8 Tb (max)
2011	SOLiD	5500 W	3 B	SE	75	0.1	Mismatch	160
2013		5500xl W	6 B		75	0.1	Mismatch	320
2011	Ion Torrent	PGM 316 chip v2	2–3 M		200	1	Indel	0.6–1
2011		PGM 314 chip v2	400–500		400	1	Indel	0.06–0.01
2012		Ion proton	60–80 M		200	1	Indel	10
2013		PGM 318 chip v2	4–5.5 M		400	1	Indel	1.2–2
2015		Ion S5/S5XL 540	60–80 M		400	1	Indel	–
2015		Ion S5/S5XL 520	3–5 M		400	1	Indel	1.2–5
2015		Ion S5/S5XL 530	15–20 M		400	1	Indel	3–5

TABLE 11.4 *(Continued)*

Year	Generation	Instrument	Reading per run	Reading type	Avg length of reading (pb)	Rate of error	Category of error	Generated data per run (Gb)
Third generation								
2011	PacBio	RS C1	932	SE	1300	15	Indel	0.54
2012		RS C2	932		2500	15	Indel	0.5–1
2012		RS C2 XL	932		4300	15	Indel	0.5–1
2013		RSII C2 XL	564		4600	15	Indel	0.5–1
2014		RS II P5 C3	528		8500	13	Indel	0.5-–1
2014		RS II P6 C4	660		13,500	12	Indel	0.5–1
2016		Sequel	350		10,000	_	_	7
2015	Oxford nanopore	MinION Mk	100	1 D, 2 D	9545	12	indel/mismatch	1.5
2016		PromethION	_		9846	_	_	2–4 Tb

[a]As per the run module. B, billion; Gb, gigabytes; M, million; PE, paired end; SE, single end; Tb, terabytes.

11.6 LIMITATIONS

The major constraint of the nanopore sequencer is the relatively inferior reading precision when evaluated against small read sequencers. The broad range of existing errors is from 5 to 20%, depending upon the molecule type as well as library preparation techniques (Rang et al., 2018). As insertions and deletions are embraced in errors, per se, nanopore reads are not the finest to detect single-nucleotide variation (SNV). Even though SNV genotyping with nanopore-sequenced reads are exhibited, high-exposure reads were entailed (Ebler et al., 2018), and HLA genotyping has hitches as well; it is not able to differentiate precise alleles because of a deficiency of read precision (Jain et al., 2018). Both the base-calling algorithm and the pore chemistry are the depending factors to improve the precision of nanopore sequencers (Patel et al., 2018). The most recent R9.X nanopore is based on the *E. coli* Curlin sigma S-dependent growth gene (Goyal et al., 2014) and is attaining a considerably decreased rate of error (Rang et al., 2018). A novel protein R10 nanopore was released in 2019, as said by the Nanopore Community Meeting 2018 to further advance the precision for homopolymers. Many tools of bioinformatics are already released that are focused or optimized for nanopore sequencing, and several further are in vigorous progress. Figure 11.4 lists representative software helpful to analyze nanopore data.

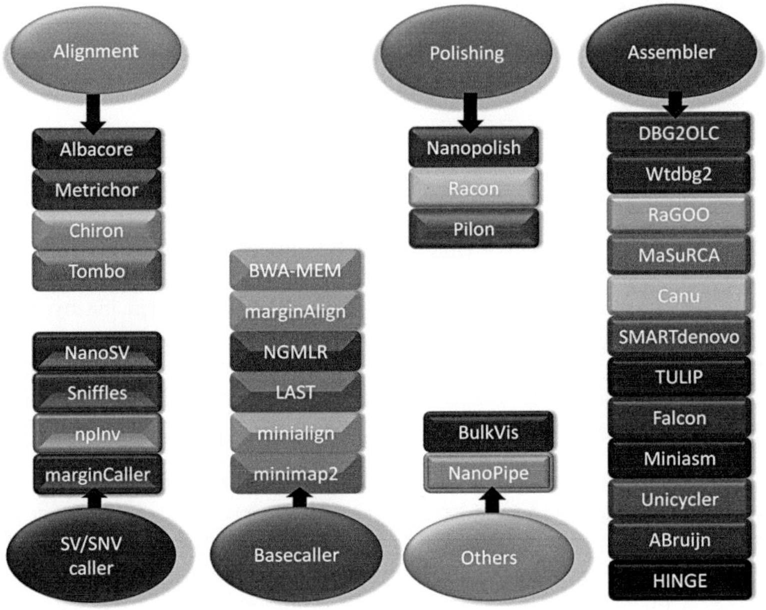

FIGURE 11.4 Softwares used for nanopore data analysis.

11.7 CONCLUSIONS

To conclude, we can state that this unique technology of DNA sequencing has opened up a new vista in the era of modern science and technology. The method wraps up vast potential supremacy and scope starting from its speed, the minimal operating cost, to the capacity of directly recognizing bases. Comparative low-grade reading precision is a backdrop of this glamorous scientific innovation that is in the way of recovery. Interventions of potential bioinformatics tools, including the use of numerous software to work together with this novel technology, is definitely going to show a new path of success to the researchers. The actual long-term success of this unique discovery relies on the combined efforts of geneticists, biotechnologists, physicists, and chemists.

KEYWORDS

- **bioinformatics tools**
- **DNA extraction**
- **DNA nanopore sequencing principle**
- **sensing methods**
- **speed control**

REFERENCES

Abiola, O.; Angel, J. M.; Avner, P.; Bachmanov, A. A.; Belknap, J. K.; Bennett, B. The Nature and Identification of Quantitative Trait Loci: A Community's View. *Nat. Rev. Gene*. **2003,** *4*, 911–916.

Agah, S.; Zheng, M.; Pasquali, M.; Kolomeisky, A. B. DNA Sequencing by Nanopores: Advances and Challenges. *J. Physics D App. Phy*. **2016,** *49*, 413001.

Akahori, R.; Haga, T.; Hatano, T.; Yanagi, I.; Ohura, T.; Hamamura, H. Slowing Single-Stranded DNA Translocation Through a Solid state nanopore by Decreasing the Nanopore Diameter. *Nanotechnol*. **2014,** *25*, 275501.

Akahori, R.; Yanagi, I.; Goto, Y.; Harada, K.; Yokoi, T.; Takeda, K. Discrimination of Three Types of Homopolymers in Single-stranded DNA With Solid-State Nanopores Through External Control of the DNA Motion. *Sci. Rep*. **2017,** *7*, 9073.

Bai, J. W.; Wang, D. Q.; Nam, S. W.; Peng, H. B.; Bruce, R.; Gignac, L. Fabrication of Sub-20 nm Nanopore Arrays in Membranes With Embedded Metal Electrodes at Wafer Scales. *Nanoscale* **2014,** *6*, 8900–8906.

Branton, D.; Deamer, D. W.; Marziali, A.; Bayley, H.; Benner, S. A.; Butler, T. The Potential and Challenges of Nanopore Sequencing. *Nat. Biotechnol.* **2008,** *26*, 1146–1153.

Carson, S.; Wanunu, M. Challenges in DNA Motion Control and Sequence Readout Using Nanopore Devices. *Nanotechnology* **2015,** *26*, 074004.

Cheley, S.; Gu, L. Q.; Bayley, H. Stochastic Sensing of Nanomolar Inositol 1,4,5-trisphosphate With an Engineered Pore. *Chem. Biol.* **2002,** *9*, 829–838.

Cherf, G.; Lieberman, K.; Rashid, H.; Lam, C.; Karplus, K.; Akeson, M. Automated Forward and Reverse Ratcheting of DNA in a Nanopore at 5-a Precision. *Nat. Biotechnol.* **2012,** *30*, 344–348.

Dai, L.; Gao, X.; Guo, Y.; Xiao, J.; Zhang, Z. Bioinformatics Clouds for Big Data Manipulation. *Biol. Direct.* **2012,** *7*, 43.

Deamer, D. Daniel Branton and Freeze-Fracture Analysis of Membranes. *Trend. Cell. Biol.* **1998,** *8*, 460–462.

Dekker, C. Solid-state Nanopores. *Nat. Nanotechnol.* **2002,** *2*, 209–915.

Derrington, I. M.; Butler, T. Z.; Collins, M. D.; Manrao, E.; Pavlenok, M.; Niederweis, M. Nanopore DNA Sequencing With MspA. *Proceed. Nat. Acad. Sci. U S A* **2002,** *107,* 16060–16065.

Ebler, J.; Haukness, M.; Pesout, T.; Marschall, T.; Paten, B. Haplotype-aware Genotyping From Noisy Long Reads. *Bio. Rxiv*. **2018,** 293944.

Edel, J. B.; Albrecht T. Engineered Nanopores For Bioanalytical Applications. Elsevier Science: Amsterdam, NL, 2013.

Eisenstein, M. Oxford Nanopore Announcement Sets Sequencing Sector Abuzz. *Nat. Biotechnol.* **2012,** *30*, 295–296.

Feng, Y.; Zhang, Y.; Ying, C.; Wang, D.; Du, C. Nanopore-Based Fourth-Generation DNA Sequencing Technology. *Genomics Proteomics Bioinformatics* **2015,** *13*, 4–16. https://doi.org/10.1016/j.gpb.2015.01.009.

Fologea, D.; Uplinger, J.; Thomas, B.; McNabb, D. S.; Li, J. Slowing DNA Translocation in a Solid-State Nanopore. *Nanotechnol. Lett.* **2005,** *5*, 1734–1737.

Gao, N.; Xie, C. Experimental Demonstration of Free-Space Optical Vortex Transmutation With Polygonal Lenses. *Optics. Lett.* **2012,** *37*(15), 3255–3257.

Gao, N.; Li, H.; Zhu, X.; Hua, Y.; Xie, C. Quasi-Periodic Gratings: Diffraction Orders Accelerate Along Curves. *Optics Lett.* **2013,** *38*(15), 2829–2831.

Garaj, S.; Hubbard, W.; Reina, A.; Kong, J.; Branton, D.; Golovchenko, J. A. Graphene as a Subnanometre Trans-Electrode Membrane. *Nature* **2010,** *467*, 190–3.

Garalde, D. R.; Snell, E.A.; Jachimowicz, D.; Sipos, B.; Lloyd, J. H.; Bruce, M.; Serra, S. Highly Parallel Direct RNA Sequencing on an Array of Nanopores. *Nat. Meth.* **2018,** *15*, 201–206. https://doi. org/10.1038/nmeth.4577

Goodwin, S.; Mcpherson, J. D.; Mccombie, W. R. Coming of Age: Ten Years of Next-Generation Sequencing Technologies. *Nat. Rev. Gene.* **2016,** *17*, 333–351. https://doi.org/10.1038/nrg.2016.49

Goto, Y.; Akahori, R.; Yanagi, I.; Takeda, K. Solid-State Nanopores Towards Single-Molecule DNA Sequencing. *J. Human Gene.* **2019.** https://doi.org/10.1038/s10038-019-0655-8

Goto, Y.; Haga, T.; Yanagi, I.; Yokoi, T.; Takeda, K. Deceleration of Single-Stranded DNA Passing Through A Nanopore Using a Nanometre-Sized Bead Structure. *Sci. Rep.* **2015,** *5*, 16640.

Goyal, P.; Krasteva, P. V.; Van Gerven, N.; Gubellini, F.; Van den Broeck, I.; Troupiotis-Tsaïlaki, A.; Chapman, M.R. Structural and Mechanistic Insights into the Bacterial Amyloid Secretion Channel CsgG. *Nature* **2014,** *516*, 250–253. https://doi.org/10.1038/nature13768

Guasch, A.; Pous, J.; Ibarra, B.; Gomis-Ruth, F. X.; Valpuesta, J.M.; Sousa, N. Detailed Architecture of a DNA Translocating Machine: The High-Resolution Structure Of The Bacteriophage Phi 29 Connector Particle. *J. Molecul. Biol.* **2002,** *315,* 663–76.

Haque, F.; Li, J.H.; Wu, H. C.; Liang, X. J.; Guo, P. X. Solid-State and Biological Nanopore for Real-Time Sensing of Single Chemical and Sequencing of DNA. *Nano Today* **2013,** *8*, 56–74

Heng, J. B.; Ho, C.; Kim, T.; Timp, R.; Aksimentiev, A.; Grinkova, Y.V. Sizing DNA Using a Nanometer-Diameter Pore. *Biophys. J.* **2004,** *87*, 2905–11.

Howorka, S.; Cheley, S.; Bayley, H. Sequence-Specific Detection of Individual DNA Strands Using Engineered Nanopores. *Nat. Biotechnol.* **2001,** *19*, 636–639.

Huang, S.; Romero-Ruiz, M.; Castell, O. K.; Bayley, H.; Wallace, M.I. High-Throughput Optical Sensing of Nucleic Acids in a Nanopore Array. *Nat. Nanotechnol.* **2015,** *10*, 986.

Iqbal, S. M.; Akin, D.; Bashir, R. Solid-State Nanopore Channels With DNA Selectivity. *Nat. Nanotechnol.* **2007,** *2*, 243–248.

Iqbal, S. M.; Bashir, R. Nanopores: Sensing And Fundamental Biological Interactions. Springer: Heidelberg, 2011.

Jain, M.; Koren, S.; Miga, K. H.; Quick, J.; Rand, A. C.; Sasani, T. A.; Dilthey, A. T. Nanopore Sequencing and Assembly of a Human Genome With Ultra-Long Reads. *Nat. Biotechnol.* **2018,** *36*, 338–345. https://doi.org/10.1038/nbt.4060

Jain, M.; Olsen, H. E.; Paten, B.; Akeson, M. The Oxford NanoporeMinION: Delivery of Nanopore Sequencing to the Genom- ics Community. *Genome. Biol.* **2016,** *17*, 239. https://doi.org/10.1186/ s13059-016-1103-0

Kang, X. F.; Gu, L. Q.; Cheley, S.; Bayley, H. Single Protein Pores Containing Molecular Adapters at High Temperatures. *Ang. Chem. Int. Ed.* **2005,** *44*(10), 1495–1499.

Karin, H.; Hege, V. A. DNA Extraction of Microbial DNA Directly From Infected Tissue: An Optimized Protocol for use In Nanopore Sequencing. *Sci. Rep.* **2010,** *10*, 2985.

Kathryn, D.; Maximilian, W. S.; Harmeet, S. C.; Rod, S.; Björn, U. Oxford Nanopore Sequencing: New Opportunities for Plant Genomics. *J. Exp. Bot.* **2020,** *71*(18), 5313–5322.

Kchouk, M.; Gibrat, J. F.; Elloumi, M. Generations of Sequencing Technologies: From First to Next Generation. *Biol. Med.* **2017**, *9*, 395. DOI: 10.4172/0974-8369.1000395.

Kowalczyk, S. W.; Wells, D. B.; Aksimentiev, A.; Dekker, C. Slowing Down DNA Translocation Through a Nanopore in Lithium Chloride. *Nanotechnol. Lett.* **2012,** *12*, 1038–1044.

Kwok, H.; Briggs, K.; Tabard-Cossa, V. Nanopore Fabrication by Controlled Dielectric Breakdown. *PLoS One* **2014**, *9*, e92880.

Laszlo, A. H.; Derrington, I. M.; Ross, B. C.; Brinkerhoff, H.; Adey, A.; Nova, I. C. Decoding Long Nanopore Sequencing Reads of Natural DNA. *Nat. Biotechnol.* **2014,** *32*, 829–833.

Lee, T. J.; Guo, P.X. Interaction of gp16 with pRNA and DNA for Genome Packaging by the Motor of Bacterial Virus Phi29. *J. Molecul. Biol.* **2006**, *356*, 589–599.

Li, J.; Stein, D.; McMullan, C.; Branton, D.; Aziz, M. J.; Golovchenko, J. A. Ion-Beam Sculpting at Nanometre Length Scales. *Nature* **2001**, *412*, 166–169.

Liang, F.; Zhang, P. Nanopore DNA Sequencing: Are we there yet? *Sci. Bullet.* **2015**, *60*(3), 296–303.

Lindsay, S. The Promises and Challenges of Solid-State Sequencing. *Nat. Nanotechnol.* **2016,** *11*, 109–111.

Liu, S.; Zhao, Q.; Xu, J.; Yan, K.; Peng, H. L.; Yang, F.H. Fast and Controllable Fabrication of Suspended Graphene Nanopore Devices. *Nanotechnol*ogy **2012,** *23*, 6.

Manrao, E. A.; Derrington, I. M.; Laszlo, A. H.; Langford, K. W.; Hopper, M. K.; Gillgren, N. Reading DNA at Single-Nucleotide Resolution With a Mutant MspAnanoporeand phi29 DNA Polymerase. *Nat. Biotechnol*. **2012,** *30*, 349–353.

Meller, A. Dynamics of Polynucleotide Transport Through Nanometre-Scale Pores. *J. Phy. Condensed Matter* **2003,** *15*(17), R581.

Menestrina, J.; Yang, C.; Schiel, M.; Vlassiouk, I.; Siwy, Z. S. Charged Particles Modulate Local Ionic Concentrations and Cause Formation of Positive Peaks in resistive-Pulse-Based Detection. *J. Phys. Chem.* **2014,** *118*(5), 2391–2398.

Mikheyev, A. S.; Tin, M. Y. A First Look at the Oxford NanoporeMinION Sequencer. *Mol. Ecol. Res*. **2014,** *14*, 1097–1102.

Nelson, T.; Zhang, B.; Prezhdo, O. V. Detection of Nucleic Acids With Graphene Nanopores: ab Initio Characterization Of A Novel Sequencing Device. *Nano Lett.* **2010,** *10*(9), 3237–3242.

Patel, A.; Belykh, E.; Miller, E. J.; George, L. L.; Martirosyan, N. L.; Byvaltsev, V. A.; Preul, M. C. MinION Rapid Sequencing: Review of Potential Applications in Neurosurgery. *Surg. Neurol. Int*. **2018**, *9*, 157.

Pennisi, E. Search for Pore-Fection. *Science* **2012,** *336*, 534.

Pungetmongkol, P. Speculation of Nano-gap Sensor for DNA Sequencing Technology: A Review on Synthetic Nanopores. *Eng. J.* **2018,** *22*(6), 229–250.

Quick, J.; Ashton, P.; Calus, S.; Chatt, C.; Gossain, S.; Hawker, J.; Loman, N. J. Rapid Draft Sequencing and Real-Time Nanopore Sequencing in a Hospital Outbreak of Salmonella. *Genome Biol*. **2015,** *16*(1), 1–14.

Quick, J.; Grubaugh, N. D.; Pullan, S. T.; Claro, I. M.; Smith, A. D.; Gangavarapu, K.; Loman, N. J. Multiplex PCR Method for MinION and Illumina Sequencing of Zika and Other Virus Genomes Directly From Clinical Samples. *Nat. Protocols*. **2017,** *12*(6), 1261.

Quick, J.; Loman, N. J.; Duraffour, S.; Simpson, J. T.; Severi, E.; Cowley, L.; Carroll, M. W. Real-time, Portable Genome Sequencing For Ebola Surveillance. *Nature* **2016,** *530*(7589), 228–232.

Quick, J.; Quinlan, A.; Loman, N. A reference bacterial genome dataset generated on the MinIONTM portable single-molecule nanopore sequencer. *Giga. Sci*. **2014,** *3*, 22.

Rang, F. J.; Kloosterman, W. P.; De Ridder, J. From Squiggle to Basepair: Computational Approaches for Improving Nanopore Sequencing Read Accuracy. *Genome Biol.* **2018**, *19*, 90. https://doi. org/10.1186/s13059-018-1462-9

Schadt, E. E.; Turner, S.; Kasarskis, A. A Window Into Third- Generation Sequencing. *Human Molecul. Gene*. **2010,** *19*, R227–R240.https://doi.org/10.1093/hmg/ddq416

Simpson, J. T.; Workman, R. E.; Zuzarte, P. C.; David, M.; Dursi, L. J.; Timp, W. Detecting DNA Cytosine Methylation Using Nanopore Sequencing. *Nat. Methods* **2017,** *14*, 407–410. https://doi. org/10.1038/nmeth.4184

Singh, A.; Rajput, V. D., Mehrotra, R.; Pal, N.; Singh, V. K.; Minkina, T.; Chokheli, V. A.; Singh, R. K. In *Sustainable Soil Fertility Management*; Nova. Sci. Publishers, Inc., 2020; vol *1*, pp 73–100.

Singh, A., Rajput, V., Singh, A., Sengar, R., Singh, R. and Minkina, T. Transformation Techniques and Their Role in Crop Improvements: A Global Scenario of GM Crops. *Policy Issues Genetically Modified Crops* **2021**, *1,* 515–542.

Song, L. Z.; Hobaugh, M. R.; Shustak, C.; Cheley, S.; Bayley, H.; Gouaux, J.E. Structure of Staphylococcal Alpha-Hemolysin, a Heptameric Transmembrane Pore. *Science* **1996,** *274*, 1859–1866.

Squires, A. H.; Hersey, J. S.; Grinstaff, M. W.; Meller, A. A Nanoporenanofiber Mesh Biosensor to Control DNA Translocation. *J. Am. Chem. Soc.* **2013**, *135*,16304–16307.

Storm, A. J.; Chen, J. H.; Ling, X. S.; Zandbergen, H. W.; Dekker, C. Fabrication of Solid-State Nanopores With Single–Nanometre Precision. *Nat. Mater*. **2003,** *2*, 537–540.

Tang, Z.; Liang, Z.; Lu, B.; Li, J.; Hu, R.; Zhao, Q. Gel Mesh as "Brake" to Slow Down DNA Translocation Through Solid-State Nanopores. *Nanoscale* **2015,** *7*, 13207–13214.

Traversi, F.; Raillon, C.; Benameur, S. M.; Liu, K.; Khlybov, S.; Tosun, M. Detecting the Translocation of DNA Through a Nanopore Using Graphene Nanoribbons. *Nat. Nanotechnol*. **2013**, *8*, 939–945.

Venkatesan, B. M.; Bashir, R. Nanopore sensors for nucleic acid analysis. *Nat. Nanotechnol.* **2011,** *6*, 615–624.

Venkatesan, B. M.; Dorvel, B.; Yemenicioglu, S.; Watkins, N.; Petrov, I.; Bashir, R. Highly Sensitive, Mechanically Stable Nanopore Sensors for DNA Analysis. *Adv. Mat*. **2009,** *21*, 2771.

Verschueren, D. V.; Jonsson, M. P.; Dekker, C. Temperature Dependence of DNA Translocations Through Solid-State Nanopores. *Nanotechnol*ogy **2015,** *26*, 234004.

Wang, Y.; Yang, Q.; Wang, Z. The Evolution of Nanopore Sequencing. *Front. Gene*. **2014,** *5*, 449.

Wanunu, M. Nanopores: A Journey Towards DNA Sequencing. *Phy. Life. Rev*. **2012,** *9*(2), 125–158.

Wanunu, M.; Sutin, J.; McNally, B.; Chow, A.; Meller, A. DNA Translocation Governed by Interactions With Solid-State Nanopores. *Biophys. J.* **2008,** *95*, 4716–4725.

Wendell, D.; Jing, P.; Geng, J.; Subramaniam, V.; Lee, T.J.; Montemagno, C. Translocation of Double-Stranded DNA Through Membrane-Adapted Phi29 Motor Protein Nanopores. *Nat. Nanotechnol*. **2009,** *4*, 765–772.

Yang, Y.; Liu, R.; Xie, H.; Hui, Y.; Jiao, R.; Gong, Y.; Zhang, Y. Advances in Nanopore Sequencing Technology. *J. Nanosci. Nanotechnol*. **2013,** *13*(7), 4521–4538.

Yanxiao, F.; Yuechuan, Z.; Cuifeng, Y.; Deqiang, W.; Chunlei, D. Nanopore-Based Fourth Generation DNA Sequencing Technology. *Genomics Proteomics Bioinformatics* **2015,** *13*, 4–16.

Yoshida, H.; Goto, Y.; Akahori, R.; Tada, Y.; Terada, S.; Komura, M. Slowing the Translocation of Single-stranded DNA by Using Nano-Cylindrical Passage Self-Assembled by Amphiphilic Block Copolymers. *Nanoscale* **2016,** *8*, 18270–18276.

Yusko, E. C.; Johnson, J. M.; Majd, S.; Prangkio, P.; Rollings, R. C.; Li, J. L. Controlling Protein Translocation Through Nanopores With Bio-Inspired Fluid Walls. *Nat. Nanotechnol.* **2011,** *6*, 253–260.

CHAPTER 12

Bionanotechnological Methods in Crop Production and Pest Management

ZAHRA GHORBANZADEH[1], RASMIEH HAMID[2], MOHAMMAD REZA GHAFFARI[1], BAHADOR MALEKNIA[3], RUKAM S. TOMAR[4], and FEBA JACOB THOPPURATHU[5]

[1]*Department of Systems Biology, Agricultural Biotechnology Research Institute of Iran (ABRII), Agricultural Research Education and Extension Organization (AREEO), Karaj, Iran.*

[2]*Cotton Research Institute of Iran (CRII), Agricultural Research, Education and Extension Organization (AREEO), Gorgan, Iran*

[3]*Department of Entomology, Faculty of Agriculture, Tarbiat Modares University, Tehran, Iran*

[4]*Department of Biotechnology, Junagadh Agricultural University, Junagadh, Gujarat, India*

[5]*Centre for Plant Biotechnology and Molecular Biology, Kerala Agricultural University, Thrissur, India*

ABSTRACT

In this century, most agriculture products are grown by using various pesticides and fertilizers but only a small portion of pesticides applied to crops reach the target pest and most of these chemicals enter the environment and kill non-target living things more than pests. Sustainable agriculture is an eco-friendly option and reduces the negative effects of conventional agriculture,

Nanotechnology for Sustainable Agriculture: An Innovative and Eco-Friendly Approach. Vishnu D. Rajput, Abhishek Singh, Tatiana M. Minkina, Krishan K. Verma, and Awani Kumar Singh (Eds.)

as it protects both human and the environment from these unknown risks by using sustainable methods, such as crop protection, integrated pest management (IPM), and biotechnological methods. In cases where the production of organic crops is impossible, biotechnological programs would be appropriate strategies for food safety in integrated crop management (ICM) programs. However, integrated pest management methods have been developed worldwide, recently improvement of IPM is more than before required. The use of biotechnological methods to create resistant plants can enhance integrated pest management efficiency. A major improvement will come from a new cultivar with different levels of resistance to pests. Classical breeding for increased host plant resistance needs more time and special labor-intensive technique. Recent molecular methods have opened up new opportunities for pest control by providing access to novel pathways, the accessibility for changing the quality and quantity of expression of genes, and the development of transgenic crops with pesticide genes. Then, there are several biotechnological strategies for increasing plant resistance such as enhanced direct resistance against pests, a transgenic plant that promotes natural enemies' demography parameters, incorporating transgenic plants in IPM, effects of transgenic varieties on beneficial organisms, side effects of genes expression in plant resistance, and the others. In this chapter, we will discuss about the use of biotechnological methods to develop insect-resistant transgenic plants as a novel method to integrating pest management.

12.1 INTRODUCTION

Pest management is one of the most important strategies of increasing crop production, currently an estimated 15% of total agricultural production were losses by pests (Singh et al., 2020a). Using pesticides to decrease damage of insect pest, disease, and weeds has resulted in side effects to the beneficial organisms, accumulation of pesticide residues in the tritrophic levels. In pest management, it is necessary to use target-specific compounds with the low persistence. An increase in using integrated pest management (IPM) strategies, and development of alternative technologies that will allow a logical use of pesticides for sustainable agriculture production are needed (Singh et al., 2021a). Natural enemies, biopesticides, natural plant products, pest-resistant varieties, and genetic engineering methods offer a potentially safe method of managing insect pests' damages. The application of biotechnology can develop plants that are resistant to drought, insect pest, weeds, and diseases (Rajput et al., 2021a). Crop cultivars derived through biotechnological methods can play a pivotal role in IPM in different crops and cropping systems. This transgenic plant with different insecticidal genes can be exploited for sustainable crop production in the future (Singh et al., 2021b). Manipulation of plants with resistance to insects pests and disease mixed with biocontrol agents decreased using pesticides and also reduces farmer's costs while benefiting community health programs (Rajput et al., 2021b).

Several projects have been done on genome sequencing of model organisms, such as human, yeast (Piškur and Langkjær, 2004), *Caenorhabditis elegans White* (Duggan et al., 1998), *Arabidopsis thaliana* (L.) Heynh. (Burrell et al., 2011), and rice, *Oryza sativa* L. (Palevitz, 2000) in the last decade. This provided valuable information on gene, genome organization and function pathway, and has updated our understanding of crop production, and ability to manipulate traits for increase crop productivity (Iglesias et al., 2000). Such an information can improve our understanding of plant biology, and thus our ability to extract genomic information in agriculture productions. Recombinant genetic technologies, besides generating data on gene sequences and function pathways, is helpful to identification of specific chromosomal regions carrying genes to choosing the traits of economic interest (Karp et al., 1997). The improvement of genetic maps in different numbers of species has led to the realization of positional similarity of maps across species. This information will allow advances made in one species genetic traits, easily applicable to other species genetic traits (Gale and Devos, 1998). When a high level of same sequences is detected between an expressed sequence tags and a gene of known pathway in another species, it is possible to understand the gene function in the species of interest traits. However, the emphatic explanation of gene function still needs experimental verification.

Using of biotechnology methods is an important prerequisite for sustainable and economic use of biotechnology for making new crop with traits of economic interest. Genetic resistance to insects has been inserted into main crop plants, such as maize, cotton, potato, tobacco, rice, broccoli, lettuce, walnuts, apple, alfalfa, and soybean. A number of transgenic crops have now been released for on-farm production or field testing (Cranston, 2010). The increase of conventional breeding with the use of marker-assisted selection and manipulate plants promises to comfort substantial increases in agriculture crops production. However, knowledge of the physiology and biochemistry of plants will be very important for explicating the information from molecular markers and concluded new and more effective models in plant breeding methods. The application of DNA marker methods in activating the vast and huge under-utilized pool of suitable alleles existing in the wild relatives of agriculture crops will increase a huge new resource of genetic difference to fuel the next phase of crop economic trait improvement. Specifically, enormous benefits will be taken through the transfer of economic trait genes important for crop protection and crop quality. However, rapid and cost-effective improvement, and adoption of biotechnology-derived products will depend on improving a full realizing of the interaction of trait within their genomic environment, and with the environment in which their admitted phenotype trait must interact.

12.2 MAKING RESISTANCE PLANT FOR PESTS CONTROL (MECHANISMS)

It is difficult to control several insect species through currently available insecticides because of pest resistance to insecticides, which has renewed our interests in the development of insect-resistant cultivars for pest management (Abrol and Shankar, 2015). Production of varieties that are resistant to insects has not been as rapid as for disease resistance. Most timely progress in developing resistant plant cultivars has been mainly due to the difficulties involved in ensuring adequate insect infestation for resistance screening in addition to low levels of resistance to certain insect species in the cultivated plants (Cook et al., 2007). However, even if 90% of the insects are killed when farmers use high dose of insecticides, the remaining population multiplies at a much faster rate in the absence of natural enemies, which are killed by the insecticides (Samways, 2007), which finally results in failure of control operations and environmental pollution. The most effective and safe strategy for pest management is to use selective insecticides at a low dosage in combination with plant resistance to reduce the rate of evolution of insecticide-resistant insect populations (Fathipour et al., 2020). Making a resistant plant for pest control through conventional breeding, wide hybridization, marker-assisted selection, and genetic transformation will significantly contribute to sustainable agriculture and environmental conservation. Varieties with adequate rates of resistance to insect pests will provide safer farm environments and produce sustainable pest management tool.

The plants that were resistant to insect pests survived until crop harvest where the herbivore pressure resulted in plant mortality, or their proportion decreased over time where the herbivore pressure did not result in plant mortality. This process led to natural selection of plants with resistance to several kinds of stresses prevalent in an ecosystem. Because of this unintentional but continuous selection of resistance plants to insect pests, different varieties with resistance to insects were selected by the farmers (Sharma and Ruud, 2003). Resistance of plants to insects enables a plant to avoid or inhibit host selection, oviposition, and feeding, reduce insect survival, retard development, and tolerate or recover from injury from insect populations that would otherwise cause greater damage to other genotypes of the same species under similar environmental conditions (Cockbain, 1961). The ability of plants to resist insect damage is based on morphological and biochemical characteristics of the plants, which affect the behavior and biology of insects, and thus influence the extent of damage caused by the insect pests. Wild species of crops are important sources of genes for plant resistance to biotic and abiotic constraints. Host plant resistance sources will play an important role in the application of tools of biotechnology for integrated crop management and sustainable crop production.

There are several kinds of resistances: (1) Pseudoresistance or false resistance through avoidance of insect infestation; (2) Constitutive resistance is due to physicochemical characteristics of the host plant that affect the host selection and feeding behavior, survival, development, and fecundity of insect pests; (3) Inducible resistance is due to the influence of temperature, photoperiod, plant–water potential, chemicals, and pathogen or insect damage on nutritional quality of the target plant; (4) Associate resistance can be created in the presence of resistant or non-resistance plants in the vicinity that damage resistant cultivars.

Induced resistance is a mechanism against insect pests in response to extrinsic physical or chemical stimuli. Chemically induced expression systems or "gene switches" created by temporal, spatial, and quantitative control of genes inserted into target plants or those that are already present in the host plants to impart resistance to insects. In transgenic crops, this approach has provided opportunities for the management of development of resistance in insect pest populations. In addition to insect or pathogen attack, resistance can also be developed by suboptimal concentration that caused resistance to various insect pests and pathogens. Induced resistance against *E. varivestis* lasted 3 days after damage in soybean (Dermody et al., 2008). Proteinase inhibitors and oxidative enzymes, such as polyphenol oxidase, peroxidase, and lipoxygenase persist for at least 21 days after induction in damaged tomato leaflets (Cho and Park, 2000)

12.3 ABIOTIC FACTOR AFFECTED RESISTANCE

Several climatic and edaphic factors influence the level and nature of plant resistant to insect pests, such as physiological characteristics that are influenced by the environmental factors. Moisture stress alters the plant's reaction to insect damage, leading either to an increase or decrease in susceptibility to insect pests. Reproduction rates of Aphis fabae (Scop.) populations are decreased on water-stressed plants (Rivelli et al., 2013). High levels of water stress also reduce damage by sorghum shoot fly, *A. soccata* (Chamarthi et al., 2011). High humidity increases the detection of odors, and thus, may influence host finding by the insects. In cotton, frequent irrigation increases vegetative growth and subsequent damage by *H. armigera* (Bass et al., 2015).

Nutrients play a major role in plant resistance to insect pests. In some instances, high levels of nutrients increase the level of plant resistance to insects, and in others, they may increase the susceptibility. Application of nitrogenous fertilizers decreases the damage by shoot fly, *A. soccata*, and spotted stem borer, *C. partellus*, in sorghum (Kumar et al., 1996). Application of potash decreases the incidence of the top borer *Scirpophaga excerptalis* (Walker) in sugarcane. High levels of nitrogen lead to greater damage by the cotton jassid, *A. biguttula biguttula* (Crowder et al., 2010).

Temperature-induced stress changes the levels of biochemicals, enzymes, morphological defenses, or nutritional quality of the host plant. Temperature affects not only the plant growth, but also the biology, behavior, and population dynamics of the insect pest, therefore, this abiotic factor is one of the most important factors that impressed the behavioral and physiological interactions of insects and their host plants (Davies et al., 2012). In general, low temperatures have a negative effect on plant resistance to insects (Jönsson et al., 2009). In sorghum, expression of plant resistance to sorghum midge, *S. sorghicola* is influenced by temperature and the relative humidity (Sharma et al., 2003). There is significant variation in the impact of temperature on expression of resistance to insect pests, and such interactions need to be kept in mind while identifying sources of resistance to target insects for use in crop improvement programs.

Photoperiod influences plant growth and physicochemical characteristics of crop plants which in turn effects the interaction between insects and crop plants. Failure or inability to grow certain crop plants during the off season at times is largely associated with increased susceptibility to insects and diseases. Intensity and quality of light impacts the biosynthesis of phenylpropanoids (Hahlbrock and Grisebach, 1979) and anthocyanins (Smith et al., 1981). Prolonged exposure to high-intensity light induces susceptibility in PI 227687 soybean plants to the cabbage looper, *Trichoplusia ni* (Hubner) (otherwise resistant) (Cloonan, 2013). Susceptibility in sorghum to midge, *S. sorghicola*, increases under long day length in Kenya near the Equator (Sharma and Lavanya, 2002).

12.4 EFFECTS OF HOST PLANTS RESISTANCE IN INSECT POPULATION

The insect damage is determined by economic injury levels (EILs) that is used for the determination of the levels of host plant resistance. Studies on the effect of insect-resistant cultivars on EILs will also be useful in assessing the contribution of insect-resistant germ plasm in regulating pest populations, avoiding excessive insecticide use, and in determining the levels of insect resistance needed in the newly developed cultivars, as well as the effectiveness of insect-resistant cultivars in IPM for sustainable crop production. In the case of insects, in which the damage is limited to a particular stage and a short span of time (e.g., deadheart formation due to sorghum shoot fly), a cultivar can be planted up to a period when insect density is expected to be below EIL. If EIL is based on adults, which is a nondamaging stage of the insect (e.g., sorghum midge adults or number of *Helicoverpa* or Spodoptera moths caught in pheromone or light traps), the levels of EIL are depended to increase or decrease in the levels of insect resistance. One of the first and most important adjustments to crop management recommendations, is associated with the economic thresholds or action thresholds in relation to host plant resistance (Allison and Martiny, 2008). In some cases, there are several cultivars of a crop with different levels of resistance. Experimental and empirical data should be generated to determine the level of resistance of a cultivar, which is critical in deciding the nature and timing of the intervention (insecticide application or release of natural enemies) needed to suppress an increasing pest population.

12.5 HOST PLANT RESISTANCE TO HELP IPM

Host plant resistance (HPR) as a method of insect control in the context of IPM has a greater potential than any other method of pest suppression. In general, the use of insect-resistant varieties is not subjected to the vagaries of nature, unlike chemical and biological control methods. HPR along with natural enemies and cultural practices is an important component of any pest management strategy (Olaitan et al., 2017; Sharma et al., 2009). Plant resistance as a method of pest control offers many advantages, and in other cases, it is the only practical and effective method of pest management. However, there may be some problems if we rely exclusively on plant resistance for insect control, for example, high levels of resistance may be associated with low yield potential or undesirable quality traits, and resistance may not be expressed in every environment where a variety is grown. Therefore, insect-resistant varieties need to be carefully fitted into the pest management programs in different agroecosystems. The nature of deployment, alone or together with other methods of insect control, depends on the level and mechanisms of resistance, and the cropping system (Kennedy et al., 1987).

High levels of plant resistance are available against the management of several insect species and only a few insect species can be controlled by using resistant varieties alone. Varieties with low-to-moderate levels of pest resistance or those that can avoid insect damage can be deployed for pest management in combination with other components of pest management. Deployment of insect-resistant cultivars should be aimed at conservation of the natural enemies and minimizing the number of insecticide applications. Use of insect-resistant cultivars also improves the efficiency of other pest management practices, including the synthetic insecticides.

The benefits of host plant resistance depend on the pattern of insect invasion, for example, many insects, such as aphids, whiteflies, and mites, invade the crop in low numbers, and their abundance increases over several generations before reaching the economic threshold levels. For such insects, even low levels of antixenotic and antibiotic resistance would be useful in delaying the time required to reach the damaging levels. However, for insects that invade a crop in large numbers due to immigration, such as *Heliothis/Helicoverpa* and the armyworms *M. separata*, *Spodoptera exempta* (Walker), and *S. frugiperda*, grass hoppers, and locusts, the effects of HPR in suppressing insect populations and damage may not be apparent in the first few generations. When the two types of resistances are combined, the insect would take 32 generations to overcome the antibiotic resistance, and over 100 generations to overcome the antixenotic resistance. The influence of insect-resistant varieties on insect populations can be demonstrated by making use of the simple insect models of Knipling (1979) as adopted by Adkisson and Dyck (1980).

12.6 USING BIOLOGICAL CONTROL AGENTS IN COMBINATION WITH HOST PLANT RESISTANCE

Plant resistance and biological control are the key components of integrated pest management. generally, is compatible with the natural enemies for pest management. Varieties with moderate levels of resistance that allow the insect densities to remain below economic threshold levels are best suited for use in pest management in combination with natural enemies. The natural enemies not only help in controlling the target pests, but also help in reducing the population densities of other insects within their host range (Sharma and Ortiz, 2002). Insect-resistant varieties also increase the effectiveness of the natural enemies because of a favorable ratio between the densities of the target pest and its natural enemies. Such a combination is more effective in crops with a tolerance mechanism of resistance (Després et al., 2007). The use of HPR and biological control brings together unrelated mortality factors and thus reduces the insect population's genetic response to selection pressure from either plant resistance or from the natural enemies. Acting in concert, they provide a density-independent mortality at times of low insect density, and density-dependent mortality at times of pest abundance (Bergman and Tingey, 1979). In addition to the direct and indirect effects of plant resistance on insect pests, the selection pressure imposed by natural enemies can also result in the magnification of the effects of plant resistance on insect density (van Emden, 1995). In general, the rate of insect adaptation to a resistant cultivar is lower when the suppression is achieved by the combined action of plant resistance and natural enemies than by high levels of plant resistance alone (Gould et al., 1991). Restless behavior of the insects on the resistant varieties also increases their vulnerability to the natural enemies (Johnson et al., 1997). A prolonged developmental period of the immature stages also increases the susceptibility period of the target insect species to the natural enemies or result in synchronization of the insect developmental stages with the peak activity and abundance of the natural enemies.

Biological control processes involve the tritrophic interactions between the plants, the target pests, and the natural enemies. A synergism between plant resistance and biological control is a valuable phenomenon in the development of practical insect pest management. These interactions are not limited to the impact of plant chemicals on the target insect and the effect of insect fitness on natural enemies, but also include the direct effects of plants on the natural enemies. The physical or chemical characteristics of the host plant that influences the activity and abundance of natural enemies can be used for producing resistance plants. It is very likely that plants have evolved mechanisms to attract the natural enemies to reduce the extent of insect damage, for example, the female parasitoid wasp, *Campoletis sonorensis* (Cam.) responds to the volatiles of cotton plant over a short distance, while searching for its prey, *H. virescens*. It is easier for the wasp to find the host habitat first and then the prey within the vicinity of the host plant. Tobacco, cotton, and maize plants produce distinct volatile blends in response to damage by *H. virescens* and *H. zea* (Delphia et al., 2007). Therefore, the nature of insect–host plant interactions is critical in determining the extent of parasitization by the natural enemies. This type of strategy is compatible with biological control (Rodriguez-Saona et al., 2012). In contrast, secondary compounds of plants such as tomatine in the insect host's diet may affect the parasitization. In other cases, changes in host suitability due to the insect host's diet can influence the developmental rate, size, survival, parasitization success, sex ratio, fecundity, and life span of the parasitoids (Fatouros et al., 2008).

Plant characteristics can also be manipulated to increase the effectiveness of the natural enemies. For example, the hairiness of cucumbers interferes with the biological control of the greenhouse whitefly, *Trialeurodes vaporariorum* (West.) by *Encarsia formosa* (Gahan). Movement of *E. formosa* is 30% higher on cucumber hybrids with half the number of hairs (van Lenteren et al., 1995). Similarly, development of pigeonpea lines without glandular trichomes may lead to greater parasitization of *H. armigera* eggs by *Trichogramma* spp. (Sharma, H.C., unpublished). Brussels sprouts with glossy leaves are more attractive to aphid predators than the cultivars with waxy foliage (Eigenbrode, 2004), while predation by *Hippodamia convergens* (G.M.) on adults of *Plutella xylostella* (L.) larvae is significantly greater on lines with glossy leaves compared with lines with normal wax in *Brassica oleracea* L. Therefore, new works should be made to identify insect-resistant genotypes that are compatible or hospitable to the natural enemies.

12.7 USING CULTURAL CONTROL IN COMBINATION WITH HOST PLANT RESISTANCE

Cultural practices cause specific physiological changes that reduce the suitability of host plants for phytophagous insects (Hare and Andreadis, 1983). Most of these practices are compatible with other pest control tactics, including host plant resistance, and have long been associated with subsistence farming. Insect-resistant cultivars, including those that can escape pest damage are highly useful in pest management in combination with cultural practices. This will have the same effect on the population dynamics of the pest species in question as the combined action of insecticides and insect-resistant cultivars. Cultural control by itself may not reduce the pest populations below economic threshold levels, but aids in reducing the losses through interaction with plant resistance (Tamhane et al., 2007). Plant resistance in concert with cultural control can drastically reduce the need for insecticide application. For example, late planting of sorghum cultivars M 35-1 and Phule Yashoda resistant to shoot fly, *A. soccata*, can reduce the dead heart formation substantially during the post-rainy season. Cultural control is a powerful tool to suppress insect pests in different agroecosystems. This technique involves two basic approaches: making the environment less favorable to the pest or making the environment more favorable to the pest's natural enemies.

Insect-resistant varieties in combination with early-planting, early-maturity, defoliation, destruction of stalks, and deep ploughing can be used effectively to control boll weevil, *Anthonomus grandis* (Boh.) and bollworms, *H. virescens*, and *P. gossypiella*, in cotton (Tefera et al., 2016). This not only reduces the pest damage but also decreases the over-wintering populations of these pests, and thus results in reduced crop loss in the following seasons. The nectarless cotton varieties reduce pink bollworm infestation by 50%, and this in combination with cultural practices can reduce the pink bollworm infestation by 16-fold. Through careful planning, the cropping pattern can be adjusted such that the most susceptible stage of the crop avoids the peak periods of insect activity. A combination of plant resistance and short-duration cotton varieties has been quite effective for controlling the bollworms (Slosser et al., 1989). Short-season and rapid-fruiting cotton varieties mature 2–3 weeks earlier than the long-duration varieties. Early harvest coupled with area-wide stalk destruction reduces the over-wintering population of diapausing insects by 90% (Stoner, 1996). Such a system not only suppresses the pest population but also restores the biological control and significantly reduces the need for insecticide application.

Genetic diversity, through its influence on herbivores and on the natural enemies, can play a key role in pest management. Polyculture (growing more than one crop in the same area) is one way of increasing crop diversity. Polycultures are ecologically complex because of interspecific and intraspecific competition with the insects and the natural enemies. Elimination of alternate habitats leads to decreased biological control agents populations and an increase in insect pest abundance (Southwood, 1977). Polyculture farms frequently have lower population densities of insect pests (Gadanakis et al., 2015) because of associational resistance levels or resource available, and the distinct action of natural enemies (Russell, 1989). Specialist biological agents are generally less abundant in diverse natural habitats because of low concentration of their food in the habitat and increased activity of biological control agents. Plant diversity may also provide important resources for the natural enemies, such as alternative prey, nectar, pollen, and breeding sites. In diverse plant communities, a specialist predator or parasitoid is less successful to find its host because of the mixing and confusing or confusing effects of chemical stimuli from the nonhost plants, several physical barriers to movement for searching behavior, and changes in the environmental conditions of the target insect pests. Frequently, insect survival items are lower in polyculture farms than in the monoculture farms (Hall and Matos, 2010)

Insect pests or other organisms attracted to trap crops caused decrease in pest attraction on the target crop. Preventing from pest damage is achieved either by reduction in insect pests from infesting the target crop or steering them in a certain part of the field where they can be easily feeding. The truth is similar to associational plant resistance, in which the insect pests show a frequent feeding preference for certain plant species, cultivars, or a certain part of the crop. Crop stands can be manipulated in time and space so that attractive host plants are offered to the insect pests at a critical stage of insect development. The insects attract at the special part on the trap crop, and as a result, for control, the main crop does not need to be treated with insecticides in several time and species and thus the natural control agents of insect pests remain safe and operational in most of the field. Trap cropping system is one of the most important methods in subsistence farming in the developing countries, and using this method in cotton and soybean has been very successful (Sparks and Nauen, 2015).

Crop susceptibility to different insect pests changes with the amount and type of fertilizers applied. Therefore, care should be taken to apply appropriate combinations of nutrients to minimize pest damage and realize maximum crop yield in combination with insect-resistant cultivars. Expression of plant resistance to pests changes with the availability of nutrients. Sorghum shoot fly incidence decreases with an increase in the application of nitrogenous fertilizers (Peshin and Dhawan, 2009).

12.8 LIMITATION IN MAKING PLANT RESISTANCE

Plant resistance is not an absolute solution for pest problems. Certain limitations and problems will always be associated with any insect control program, and resistance in plant is no exception. Development of plant varieties resistant to insect pests is a long-time program (Singh et al., 2020b). Developing insect-resistant crop varieties requires a great deal of expertise and resources. It is usually necessary to organize a well-planned multidisciplinary team of entomologists and plant breeders. Commitment of relatively long-term funding is a critical factor in the ultimate success of crop resistance programs. Several mechanisms of plant resistance to pests may involve the diversion of some resources by the plant to extra structures or production of defense chemicals at the expense of other physiological processes, including those contributing to yield (Li et al., 2007). One might expect a negative correlation between the potential yield of a cultivar and its level of resistance to the target pest. This is illustrated by the failure to evolve insect-resistant varieties in soybean, pigeonpea, chickpea, etc. Although host plant resistance promises to contribute a great deal to pest management in several crops, progress has been slow mainly because of low yield potential of the resistant varieties (Liebhold and Tobin, 2008). However, the fundamental objective of breeding for insect resistance in crop plants is to reduce the amount of insecticides needed to achieve acceptable control of the target pests, and an acceptable level of sustainable resistance, compatible with agriculture crop, and quality of the produce. Despite many dramatic successes in host plant resistance, there are still cases where plant resistance to one insect leads to increased susceptibility to another insect (Nauen and Denholm, 2005).

Secondary plant substances have a negative effect on the behavior of herbivores that cause decrease in the fitness of insects. However, many herbivores possess remarkable potential for utilizing or metabolizing the toxic plant chemicals, and therefore, their role as plant defense chemicals is not sacrosanct. Insects also develop into biotypes to overcome antibiosis resistance. However, partially resistant varieties would probably last longer in the field than those with high levels of pest resistance (Dunse et al., 2010). The current theories on plant defense strategies do not take into account the complexity of tritrophic interactions. Plant resistance based on antibiosis may not always be compatible with biological control (Ong et al., 2016). Elucidation of these interactions can further the understanding and provide greater potential for manipulation of these systems to specific crop species and varieties. The possibility of using compounds from plants to reduce herbivore damage and increase the effectiveness of biological control agents is quite attractive. Ideally, plant resistance should strive to reduce substances attractive to herbivores, while increasing the substances attractive to the natural enemies.

Plant resistance at times may be associated with low yield or factors resulting in poor or unacceptable produce, for example, sorghum genotypes with high tannin content are resistant to sorghum midge (*S. sorghicola*) and birds, but such a grain has poor nutritional quality. Similarly, gossypol and other terpenoids in cotton confer resistance to bollworms (Tabashnik and Carrière, 2015), but high gossypol content spoils the quality of cottonseed oil. In such situations, one has to break the linkage between the factors conferring resistance to the target insects and the low yield potential or arrive at a threshold level for the resistant traits (secondary metabolites) that result in reduced pest susceptibility, do not have any adverse effect on the quality of the produce at the same time.

12.9 CONCLUSIONS AND FUTURE

Biotechnology methods will be improved for identification of plant resistance to insects. Multilocational investigation of the identified resistance sources and breeding material need to be strengthened to detect stable and diverse sources of crop resistance or establish the presence of new resistance strategy. Resistance to insects should be mentioned as much emphasis as agriculture products in order to make new varieties and hybrids. Different effects of insect-resistant varieties with constant and cumulative effect on decreasing insect pest populations over time and space have no side effects on the environment, reduce the usage of insecticides, decrease farmer's costs, and do not require inputs and application skills by the farmers. Therefore, plant resistance to insects should increase the stability of pest management programs for integrated pest management. Biotechnology methods can be used to improve introgression pace of insect-resistant genes pathway into high-yielding cultivars. Conventional plant resistance to insects can also be deployed in combination with novel genes to make plant resistance is an effective weapon for pest management and sustainable agriculture production.

KEYWORDS

- **host plant resistance**
- **integrated pest management**
- **sustainable agriculture**
- **transgenic plant**

REFERENCES

Abbaci, A.; Azzouz, N.; Bouznit, Y. A New Copper Doped Montmorillonite Modified Carbon Paste Electrode for Propineb Detection. *Appl. Clay Sci.* **2014,** *90*, 130–134. https://doi.org/ https://doi.org/10.1016/j.clay.2013.12.036

Abrol, D. P.; Shankar, U. Integrated Pest Management. In *Breeding Oilseed Crops for Sustainable Production: Opportunities and Constraints*, 2015. https://doi.org/10.1016/ B978-0-12-801309-0.00020-3

Agrawal, D. S.; Rathore, D. P. Review Article Nanotechnology Pros and Cons to Agriculture: A Review. *Int. J. Curr. Microbiol. Appl. Sci.* (ISSN: 2319-7706) **2014,** *3*, 43–55. https:// doi.org/10.13140/2.1.3352.1283

Aguila, S. A.; Shimomoto, D.; Ipinza, F.; Bedolla-Valdez, Z. I.; Romo-Herrera, J.; Contreras, O. E.; Farías, M. H.; Alonso-Núñez, G. A Biosensor Based on Coriolopsis Gallica Laccase Immobilized on Nitrogen-Doped Multiwalled Carbon Nanotubes and Graphene Oxide for Polyphenol Detection, 2015. *Http://Www.Tandfonline.Com/Action/JournalInformation?Show=aimsScope&journalCode=tsta20#.VmBmuzZFCUk*, *16*(5). https://doi.org/10.1088/1468-6996/16/5/055004

Alahmad, S.; Dinglasan, E.; Leung, K. M.; Riaz, A.; Derbal, N.; Voss-Fels, K. P.; Able, J. A.; Bassi, F. M.; Christopher, J.; Hickey, L. T. Speed Breeding for Multiple Quantitative Traits in Durum Wheat. *Plant Meth.* **2018,** *14*(1), 1–15. https://doi.org/10.1186/s13007-018-0302-y

Alessio, P.; Martin, C. S.; De Saja, J. A.; Rodriguez-Mendez, M. L. Mimetic Biosensors Composed by Layer-by-Layer Films of Phospholipid, Phthalocyanine and Silver Nanoparticles to polyphenol Detection. *Sens. Actuators B Chem.* **2016,** *233*, 654–666. https://doi.org/10.1016/J.SNB.2016.04.139

Alfadul, S. Application of Nanotechnology in the Field of Food Production. *Academia J. Sci. Res.* **2017,** *5*, 143–154. https://doi.org/10.15413/ajsr.2017.0220

Allison, S. D.; Martiny, J. B. H. Resistance, Resilience, and Redundancy in Microbial Communities. *Proc. Natl. Acad. Sci. U S A* **2008.** https://doi.org/10.1073/pnas.0801925105

Arumai Selvan, D.; Mahendiran, D.; Senthil Kumar, R.; Kalilur Rahiman, A. Garlic, Green Tea and Turmeric Extracts-Mediated Green Synthesis of Silver Nanoparticles: Phytochemical, Antioxidant and In Vitro Cytotoxicity Studies. *J. Photochem. Photobiolo. B Biol.* **2018,** *180*, 243–252. https://doi.org/10.1016/j.jphotobiol.2018.02.014

Augustin, M. A.; Sanguansri, P. Chapter 5 Nanostructured Materials in the Food Industry. *Adv. Food Nutr. Res.* **2009,** *58*, 183–213. https://doi.org/10.1016/S1043-4526(09)58005-9

Bass, C.; Denholm, I.; Williamson, M. S.; Nauen, R. The Global Status of Insect Resistance to Neonicotinoid Insecticides. In *Pesticide Biochemistry and Physiology*; 2015. https://doi.org/10.1016/j.pestbp.2015.04.004

Bergman, J. M.; Tingey, W. M. Aspects of Interaction Between Plant Genotypes and Biological Control. *Bull. Entomol. Soc. Am.* **1979**. https://doi.org/10.1093/besa/25.4.275

Bilir, K.; Weil, M.-T.; Lochead, J.; Kök, F. N.; Werner, T. Construction of an Oxygen Detection-Based Optic Laccase Biosensor for Polyphenolic Compound Detection. *Turk. J. Biol.* **2016,** *40*(6), 1303–1310.

Boonham, N.; Glover, R.; Tomlinson, J.; Mumford, R. Exploiting Generic Platform Technologies for the Detection and Identification of Plant Pathogens. *Euro. J. Plant Pathol.* **2008,** *121*(3), 355. https://doi.org/10.1007/s10658-008-9284-3

Burrell, A. M.; Taylor, K. G.; Williams, R. J.; Cantrell, R. T.; Menz, M. A.; Pepper, A. E. A Comparative Genomic Map for Caulanthus Amplexicaulis and Related Species (Brassicaceae). *Mol. Ecol.* **2011**.https://doi.org/10.1111/j.1365-294X.2010.04981.x

Butnariu, M.; Butu, A. Plant Nanobionics: Application of Nanobiosensors in Plant Biology. *Nanotechnol. Life Sci.* **2019,** 337–376. https://doi.org/10.1007/978-3-030-16379-2_12

Cai, H.; Xu, C.; He, P., Fang, Y. Colloid Au-enhanced DNA Immobilization for the Electrochemical Detection of Sequence-Specific DNA. *J. Electroanal. Chem.* **2001,** *510*(1–2), 78–85.

Cassman, K. G. Ecological Intensification of Cereal Production Systems: Yield Potential, Soil Quality, and Precision Agriculture. *Proc. Natl. Acad. Sci.* **1999,** *96*(11), 5952–5959. https://doi.org/10.1073/pnas.96.11.5952

Chamarthi, S. K.; Sharma, H. C.; Sahrawat, K. L.; Narasu, L. M.; Dhillon, M. K. Physico-Chemical Mechanisms of Resistance to Shoot Fly, Atherigona Soccata in Sorghum, Sorghum Bicolor. *J. Appl. Entomol.* **2011**. https://doi.org/10.1111/j.1439-0418.2010.01564.x

Cho, U. H.; Park, J. O. Mercury-Induced Oxidative Stress in Tomato Seedlings. *Plant Sci.* **2000**. https://doi.org/10.1016/S0168-9452(00)00227-2

Christopher, J.; Richard, C.; Chenu, K.; Christopher, M.; Borrell, A.; Hickey, L. Integrating Rapid Phenotyping and Speed Breeding to Improve Stay-Green and Root Adaptation of Wheat in Changing, Water-Limited, Australian Environments. *Procedia Environ. Sci.* **2015,** *29*(Agri), 175–176. https://doi.org/10.1016/j.proenv.2015.07.246

Cloonan, K. R. The Navel Orangeworm, Amyelois transitella: An Examination of its Biology, Pest Ecology in Almonds, and Development of Screening Bioassays to Identify Compounds for Reducing Oviposition. In *ProQuest Dissertations and Theses,* 2013.

Cockbain, A. J. Low temperature Thresholds for Flight in Aphis Fabae Scop. *Entomol. Exp. Appl*. **1961**. https://doi.org/10.1111/j.1570-7458.1961.tb02136.x

Collard, B. C. Y.; Beredo, J. C.; Lenaerts, B.; Mendoza, R.; Santelices, R.; Lopena, V.; Verdeprado, H.; Raghavan, C.; Gregorio, G. B.; Vial, L.; Demont, M.; Biswas, P. S.; Iftekharuddaula, K. M.; Rahman, M. A.; Cobb, J. N.; Islam, M. R. Revisiting Rice Breeding Methods–Evaluating the Use of Rapid Generation Advance (RGA) for Routine Rice Breeding. *Plant Prod. Sci.* 2017, *20*(4), 337–352. https://doi.org/10.1080/1343943X.2017.1391705

Cook, S. M.; Khan, Z. R.; Pickett, J. A. The Use of Push-Pull Strategies in Integrated Pest Management. *Ann. Rev. Entomol.* **2007**. https://doi.org/10.1146/annurev.ento.52.110405.091407

Cranston, P. J. G.; P. S. The Insects: An Outline of Entomology (4th Edition). *J. Insect Conservat*. **2010**.

Crowder, D. W.; Northfield, T. D.; Strand, M. R.; Snyder, W. E. Organic Agriculture Promotes Evenness and natural Pest Control. *Nature* **2010**. https://doi.org/10.1038/nature09183

da Silva, A. C. N.; Deda, D. K.; da Róz, A. L.; Prado, R. A.; Carvalho, C. C.; Viviani, V.; Leite, F. L. Nanobiosensors Based on Chemically Modified AFM Probes: A Useful Tool for Metsulfuron-Methyl Detection. *Sensors (Basel)* **2013,** *13*(2), 1477–1489. https://doi.org/10.3390/s130201477

Davies, T. G. E.; Field, L. M.; Williamson, M. S. The Re-emergence of the Bed Bug as a Nuisance Pest: Implications of Resistance to the Pyrethroid Insecticides. *Med. Veter. Entomol*. **2012**. https://doi.org/10.1111/j.1365-2915.2011.01006.x

Davis, J. J.; Coleman, K. S.; Azamian, B. R.; Bagshaw, C. B.; Green, M. L. H. Chemical and Biochemical Sensing With Modified Single Walled Carbon Nanotubes. *Chem. Euro. J.* **2003,** *9*(16), 3732–3739. https://doi.org/10.1002/CHEM.200304872

de Silva, P. A.; Nimal Gunaratne, H. Q.; Gunnlaugsson,T.; Huxley, A. J. M; McCoy, C. P.; Rademacher, J. T.; Rice, T. E. Signaling Recognition Events with Fluorescent Sensors and Switches. *Chem. Rev.* **1997,** *97*(5), 1515–1566. https://doi.org/10.1021/CR960386P

Delphia, C. M.; Mescher, M. C.; De Moraes, C. M. Induction of Plant Volatiles by Herbivores With Different Feeding Habits and the Effects of Induced Defenses on Host-Plant Selection by Thrips. *J. Chem. Ecol*. **2007**. https://doi.org/10.1007/s10886-007-9273-6

Dermody, O.; O'Neill, B. F.; Zangerl, A. R.; Berenbaum, M. R.; DeLucia, E. H. Effects of Elevated CO2 and O3 on Leaf Damage and Insect Abundance in a Soybean Agroecosystem. *Arthropod-Plant Interact*. **2008**. https://doi.org/10.1007/s11829-008-9045-4

Després, L.; David, J. P.; Gallet, C. The Evolutionary Ecology of Insect Resistance to Plant Chemicals. *Trends Ecol. Evolut*. **2007**. https://doi.org/10.1016/j.tree.2007.02.010

Ditta, A.; Arshad, M. Applications and Perspectives of Using Nanomaterials for Sustainable Plant Nutrition. *Nanotechnol. Rev.* **2016,** *5*(2), 209–229. https://doi.org/doi:10.1515/ntrev-2015-0060

Dong, Y.; Phillips, K.; Chip, Q. Immunosensing of Staphylococcus enterotoxin B (SEB) in milk with PDMS microfluidic systems using reinforced supported bilayer membranes (r-SBMs).*Lab Chip* **2006,** *6*(5), 675–681. DOI: 10.1039/b514902a. Epub 2006 Mar 15.

Duggan, A.; Ma, C.; Chalfie, M. Regulation of Touch Receptor Differentiation by the Caenorhabditis Elegans mec-3 and unc-86 genes. *Development* 1998.

Duhan, J. S.; Kumar, R.; Kumar, N.; Kaur, P.; Nehra, K.; Duhan, S. Nanotechnology: The New Perspective in Precision Agriculture. *Biotechnol. Rep.* **2017,** *15*, 11–23. https://doi.org/https://doi.org/10.1016/j.btre.2017.03.002

Dunse, K. M.; Kaas, Q.; Guarino, R. F.; Barton, P. A.; Craik, D. J.; Anderson, M. A. Molecular Basis for the Resistance of an Insect Chymotrypsin to a Potato Type II Proteinase Inhibitor. *Proc. Natl. Acad. Sci. U S A* **2010**. https://doi.org/10.1073/pnas.1009327107

Durán, N.; Marcato, P. D. Nanobiotechnology Perspectives. Role of Nanotechnology in the Food Industry: A Review. *Int. J. Food Sci. Technol.* **2013,** *48*(6), 1127–1134. https://doi.org/10.1111/IJFS.12027

Eigenbrode, S. D. The Effects of Plant Epicuticular Waxy Blooms on Attachment and Effectiveness of Predatory Insects. *Arthropod Struct. Dev*. **2004**. https://doi.org/10.1016/j.asd.2003.11.004

Etefagh, R.; Azhir, E.; Shahtahmasebi, N. Synthesis of CuO Nanoparticles and Fabrication of Nanostructural Layer Biosensors for Detecting Aspergillus Niger Fungi. *Sci. Iran.* **2013,** *20*(3), 1055–1058. https://doi.org/https://doi.org/10.1016/j.scient.2013.05.015

FAO. *Global agriculture towards 2050,* Rome, 2009.

FAO WFP, & IFAD. The State of Food Insecurity in the World 2012. *Economic Growth Is Necessary but Not Sufficient to Accelerate Reduction of Hunger and Malnutrition.* Rome, FAO, 2012.

Fathipour, Y.; Maleknia, B.; Bagheri, A.; Soufbaf, M.; Reddy, G. V. P. Functional and Numerical Responses, Mutual Interference, and Resource Switching of Amblyseius Swirskii on Two-Spotted Spider Mite. *Biol. Control* **2020**. https://doi.org/10.1016/j.biocontrol.2020.104266

Fatouros, N. E.; Dicke, M.; Mumm, R.; Meiners, T.; Hilker, M. Foraging Behavior of Egg Parasitoids Exploiting Chemical Information. *Behav. Ecol.* 2008. https://doi.org/10.1093/beheco/arn011

Ferreira, N.; Cruz, M.; Gomes, M.; Rudnitskaya, A. undefined. (n.d.). Potentiometric chemical sensors for the detection of paralytic shellfish toxins. Talanta **2018,** *181,* 380–384.

Finger, R. Evidence of Slowing Yield Growth – The Example of Swiss Cereal Yields. *Food Policy* **2010,** *35*(2), 175–182. https://doi.org/https://doi.org/10.1016/j.foodpol.2009.11.004

Fuertes, G.; Soto, I.; Carrasco, R.; Vargas, M.; Sabattin, J.; Lagos, C. Intelligent Packaging Systems: Sensors and Nanosensors to Monitor Food Quality and Safety. *J. Sens.* **2016**. https://doi.org/10.1155/2016/4046061

Gadanakis, Y.; Bennett, R.; Park, J.; Areal, F. J. Evaluating the Sustainable Intensification of arable farms. *J. Environ. Manag.* **2015**. https://doi.org/10.1016/j.jenvman.2014.10.005

Gale, M. D.; Devos, K. M. Comparative Genetics in the Grasses. *Proc. Natl. Acad. Sci. U S A* **1998**. https://doi.org/10.1073/pnas.95.5.1971

Ghorbanpour, M.; Fahimirad, S. Plant Nanobionics a Novel Approach to Overcome the Environmental Challenges. *Med. Plants Environ. Chall.* **2017,** 247–257. https://doi.org/10.1007/978-3-319-68717-9_14

Gould, F.; Kennedy, G. G.; Johnson, M. T. Effects of Natural Enemies on the Rate of Herbivore Adaptation to Resistant Host Plants. *Entomol. Exp. Applicata* **1991**. https://doi.org/10.1111/j.1570-7458.1991.tb01445.x

Hahlbrock, K.; Grisebach, H. Enzymic Controls in the Biosynthesis of Lignin and Flavonoids. *Ann. Rev. Plant Physiol.* **1979**. https://doi.org/10.1146/annurev.pp.30.060179.000541

Hall, J.; Matos, S. Incorporating Impoverished Communities in Sustainable Supply Chains. *Int. J. Phys. Distrib. Logist. Manag.* **2010**. https://doi.org/10.1108/09600031011020368

Hao, L.; Lin, G.; Wang, H.; Wei, C.; Chen, L.; Zhou, H.; Chen, H.; Xu, H.; Zhou, X. Preparation and Characterization of Zein-Based Nanoparticles via Ring-Opening Reaction and Self-Assembly as Aqueous Nanocarriers for Pesticides. *J. Agric. Food Chem.* **2020,** *68*(36), 9624–9635. https://doi.org/10.1021/acs.jafc.0c01592

Hare, J. D.; Andreadis, T. G. Variation in the Susceptibility of Leptinotarsa Decemlineata (Coleoptera: Chrysomelidae) When Reared on Different Host Plants to the Fungal Pathogen, Beauveria Bassiana in the Field and Laboratory. *Environ. Entomol.* **1983**. https://doi.org/10.1093/ee/12.6.1892

He, X.; Aker, W. G.; Fu, P. P.; Hwang, H.-M. Toxicity of Engineered Metal Oxide Nanomaterials Mediated by nano–Bio–Eco–Interactions: A Review and Perspective. *Environ. Sci.: Nano* **2015,** *2*(6), 564–582. https://doi.org/10.1039/C5EN00094G

He, X.; Aker, W. G.; Huang, M.-J.; Watts, J. D.;Hwang, H.-M. Metal Oxide Nanomaterials in Nanomedicine: Applications in Photodynamic Therapy and Potential Toxicity. *Curr. Topics Med. Chem.* **2015,** 15(18), 1887–1900). https://doi.org/http://dx.doi.org/10.2174/1568026615666150506145251

He, X.; Aker, W. G.; Hwang, H.-M. An In Vivo Study on the Photo-Enhanced Toxicities of S-doped TiO2 Nanoparticles to Zebrafish Embryos (Danio rerio) in Terms of Malformation, Mortality, Rheotaxis Dysfunction, and DNA Damage. *Nanotoxicology* **2014,** *8*(supl), 185–195. https://doi.org/10.3109/17435390.2013.874050

He, X.; Aker, W. G.; Leszczynski, J.; Hwang, H.-M. Using a Holistic Approach to Assess the Impact of Engineered Nanomaterials Inducing Toxicity in Aquatic Systems. *J. Food Drug Anal.* **2014,** *22*(1), 128–146. https://doi.org/https://doi.org/10.1016/j.jfda.2014.01.011

He, X.;, Deng, H.; Hwang, H. The Current Application of Nanotechnology in Food and Agriculture. *J. Food Drug Anal.* **2019,** *27*(1), 1–21. https://doi.org/https://doi.org/10.1016/j.jfda.2018.12.002

He, X.; Fu, P.; Aker, W. G., Hwang, H.-M. Toxicity of Engineered Nanomaterials Mediated by Nano–Bio–Eco Interactions. *J. Environ. Sci. Health C* **2018,** *36*(1), 21–42. https://doi.org/10.1080/10590501.2017.1418793

Hickey, L. T.; Germán, S. E.; Pereyra, S. A.; Diaz, J. E.; Ziems, L. A.; Fowler, R. A.; Platz, G. J.; Franckowiak, J. D.; Dieters, M. J. Speed Breeding for Multiple Disease Resistance in Barley. *Euphytica* **2017,** *213*(3). https://doi.org/10.1007/s10681-016-1803-2

Iglesias, A.; Rosenzweig, C.; Pereira, D. Agricultural Impacts of Climate Change in Spain: Developing Tools for a Spatial Analysis. *Global Environ. Change* **2000**. https://doi.org/10.1016/S0959-3780(00)00010-8

International Food Policy Research Institute. Green Revolution: Curse or Blessing? *Int. Food Policy Res. Inst.* **2002**.

Johnson, M. T.; Gould, F.; Kennedy, G. G. Effects of Natural Enemies on Relative Fitness of Heliothis Virescens Genotypes Adapted and not Adapted to Resistant Host Plants. *Entomol. Exp. Applicata* **1997**. https://doi.org/10.1046/j.1570-7458.1997.00133.x

Jönsson, A. M.; Appelberg, G.; Harding, S.; Bärring, L. Spatio-Temporal Impact of Climate Change on the Activity and Voltinism of the Spruce Bark Beetle, Ips Typographus. *Global Change Biol.* **2009**. https://doi.org/10.1111/j.1365-2486.2008.01742.x

Song, J. M.; Kasili, P. M.; Griffin, G. D.; Vo-Dinh, T. Detection of Cytochrome c in a Single Cell Using an Optical Nanobiosensor. *Anal. Chem.* **2004,** *76*(9), 2591–2594. https://doi.org/10.1021/AC0352878

Juárez-Maldonado, A.; Ortega-Ortíz, H.; Cadenas-Pliego, G.; Valdés-Reyna, J.; Pinedo-Espinoza, J.; López-Palestina, C.; Hernández-Fuentes, A. Foliar Application of Cu Nanoparticles Modified the Content of Bioactive Compounds in Moringa oleifera Lam. *Agronomy* **2018,** *8*(9), 167. https://doi.org/10.3390/agronomy8090167

Kalita, D.; Baruah, S. The Impact of Nanotechnology on Food. In *Nanomaterials Applications for Environmental Matrices: Water, Soil and Air*; Elsevier Inc., 2019. https://doi.org/10.1016/B978-0-12-814829-7.00011-2

Karp, A.; Kresovich, S.; Bhat, K. V; Ayad, W. G.; Hodgkin, T. Molecular Tools in Plant Genetic Resources Conservation: A Guide to the Technologies. In *IPGRI Technical Bulletin,* 1997.

Kasili, P. M.; Cullum, B. M.; Griffin, G. D.; Vo-Dinh, T. Nanosensor for In Vivo Measurement of the Carcinogen Benzo[a]pyrene in a Single Cell. *J. Nanosci. Nanotechnol.* (2002). *2*(6), 653–658. https://doi.org/10.1166/JNN.2002.155

Kasili, P. M.; Song, J. M.; Vo-Dinh, T. Optical Sensor for the Detection of Caspase-9 Activity in a Single Cell. *J. Am. Chem. Soc.* **2004,** *126*(9), 2799–2806. https://doi.org/10.1021/JA037388T

Kendall, H. W.; Pimentel, D. Constraints on the Expansion of the Global Food Supply. *Ambio* **1994,** *23*(3), 198–205.

Kennedy, G. G.; Gould, F.; Deponti, O. M. B.; Stinner, R. E. Ecological, Agricultural, Genetic, and Commercial Considerations in the Deployment of Insect-resistant Germplasm. *Environ. Entomol.* **1987**. https://doi.org/10.1093/ee/16.2.327

Kissinger, P. T. Biosensors—A Perspective. *Biosens. Bioelectron.* **2005,** *20*(12), 2512–2516. https://doi.org/10.1016/J.BIOS.2004.10.004

Knipling, E. F. No TitleThe Basic Principles of Insect Population Auppression and Management. *US Department of Agriculture.*, No. 512, 1979.

Kottegoda, N.; Munaweera, I.; Adassooriya, N.; Karunaratne, V. A Green Slow-Release Fertilizer Composition Based on Urea-Modified Hydroxyapatite Nanoparticles Encapsulated Wood. *Curr. Sci.* **2011,** *101*, 73–78.

Kumar, A.; Sinha, R. P.; Häder, D. P. Effect of UV-B on Enzymes Of Nitrogen Metabolism in the Cyanobacterium Nostoc Calcicola. *J. Plant Physiol.* **1996**. https://doi.org/10.1016/S0176-1617(96)80298-7

Kwak, S. Y.; Lew, T. T. S.; Sweeney, C. J.; Koman, V. B.;Wong, M. H.; Bohmert-Tatarev, K.; Snell, K. D.; Seo, J. S.;, Chua, N. H.; Strano, M. S. Chloroplast-selective gene delivery and expression in planta using chitosan-complexed single-walled carbon nanotube carriers. *Nature Nanotechnol.* **2019,** *14*(5), 447–455. https://doi.org/10.1038/S41565-019-0375-4

Lee, C.;Itoh, T.; Sasaki, G.;Suga, T. Sol-gel Derived PZT Force Sensor for Scanning Force Microscopy. *Mat. Chem. Phys.* **1996,** *44*(1), 25–29. https://doi.org/10.1016/0254-0584(95)01647-D

Lew, T. T. S.; Koman, V. B.; Gordiichuk, P.; Park, M.; Strano, M. S. The Emergence of Plant Nanobionics and Living Plants as Technology. *Adv. Mater. Technol.* **2020,** *5*(3), 1900657. https://doi.org/10.1002/ADMT.201900657

Li, X.; Huang, Q.; Yuan, J.; Tang, Z. Fipronil Resistance Mechanisms in the Rice Stem Borer, Chilo Suppressalis Walker. *Pestic. Biochem. Physiol.* **2007**. https://doi.org/10.1016/j.pestbp.2007.06.002

Li, Y.; Yang, D.; Cui, J. Graphene Oxide Loaded With Copper Oxide Nanoparticles As an Antibacterial Agent Against Pseudomonas Syringae pv. tomato. *RSC Adv.* **2017,** *7*(62), 38853–38860. https://doi.org/10.1039/C7RA05520J

Liebhold, A. M.; Tobin, P. C. Population Ecology of Insect Invasions and Their Management. *Ann. Rev. Entomol.* **2008**. https://doi.org/10.1146/annurev.ento.52.110405.091401

Lim, T. C.; Ramakrishna, S. A Conceptual Review of Nanosensors. *Zeitschrift Fur Naturforschung A J. Phys. Sci.* **2006,** *61*(7–8), 402–412. https://doi.org/10.1515/zna-2006-7-815

Liu, T.; Tang, J.; Jiang, L. The Enhancement Effect of Gold Nanoparticles as a Surface Modifier on DNA Sensor Sensitivity.Biochem. Biophys. Res. Commun. **2004,** *313*(1), 3–7. DOI: 10.1016/j.bbrc.2003.11.098. Lu, J.; Bowls, M. How will nanotechnology affect agricultural supply chains? Int. Food Agribus.Manag. Assoc. **2013,** *16*(2), 21–42.

Malik, P.; Katyal, V.; Malik, V.; Asatkar, A.; Inwati, G.; Mukherjee, T. K. Nanobiosensors: Concepts and Variations. *ISRN Nanomater.* **2013,** *2013*, 1–9. https://doi.org/10.1155/2013/327435

Marchiol, L.; Filippi, A.; Adamiano, A.; Degli Esposti, L.; Iafisco, M.; Mattiello, A.; Petrussa, E.; Braidot, E. Influence of Hydroxyapatite Nanoparticles on Germination and Plant Metabolism of Tomato (Solanum lycopersicum L.): Preliminary Evidence. *Agronomy* **2019,** *9*(4), 161. https://doi.org/10.3390/agronomy9040161

McClements, D. J. Nanotechnology Approaches for Improving the Healthiness and Sustainability of the Modern Food Supply. *ACS Omega* **2020,** *5*(46), 29623–29630. https://doi.org/10.1021/acsomega.0c04050

Mello, L.; Kubota, L. T. Review of the Use of Biosensors as Analytical Tools in the Food and Drink Industries. *Food Chem.* **2002,** 77(2), 237–256.

Mobini, S. H.; Lulsdorf, M.; Warkentin, T. D.;Vandenberg, A. Plant Growth Regulators Improve In Vitro Flowering and Rapid Generation Advancement in Lentil and Faba Bean. *In Vitro Cell. Dev. Biol. Plant* **2015,** *51*(1), 71–79. https://doi.org/10.1007/s11627-014-9647-8

Mousavi, S. R.; Rezaei, M. Nanotechnology in Agriculture and Food Production. *J. Appl. Environ. Biol. Sci.* 2011, *1*, 414–419.

Narang, J.; Chauhan, N.; Rani, P.; Pundir, C. S. Construction of an Amperometric TG Biosensor Based on AuPPy Nanocomposite and Poly (indole-5-carboxylic acid) modified Au Electrode. *Bioproc. Biosyst. Eng.* **2012,** *36*(4), 425–432. https://doi.org/10.1007/S00449-012-0799-9

Nauen, R.; Denholm, I. Resistance of Insect Pests to Neonicotinoid Insecticides: Current Status and Future Prospects. *Arch. Insect Biochem. Physiol.* **2005**. https://doi.org/10.1002/arch.20043

Neethirajan, S.; Freund, M. S.; Jayas, D. S.; Shafai, C.; Thomson, D. J.; White, N. D. G. Development of Carbon Dioxide (CO2) Sensor for Grain Quality Monitoring. *Biosyst. Eng.* 2005, *106*(4), 395–404. https://doi.org/https://doi.org/10.1016/j.biosystemseng.2010.05.002

Niraimathi, K. L.; Sudha, V.; Lavanya, R.; Brindha, P. Biosynthesis of Silver Nanoparticles Using Alternanthera Sessilis (Linn.) Extract and Their Antimicrobial, Antioxidant Activities. *Coll. Surf B Biointerf.* **2013,** *102*, 288–291. https://doi.org/10.1016/j.colsurfb.2012.08.041

Lucchinger, N. A.; Loher, S.; Athanassiou, E. K.; Grass, R. N.; Stark, W. J. Highly Sensitive Optical Detection of Humidity on Polymer/Metal Nanoparticle Hybrid Films. *Langmuir* 2007, *23*(6), 3473–3477. https://doi.org/10.1021/LA062424Y

Nuruzzaman, M.; Rahman, M. M.; Liu, Y.; Naidu, R. Nanoencapsulation, Nano-guard for Pesticides: A New Window for Safe Application. *J. Agric. Food Chem.* 2016, *64*(7), 1447–1483. https://doi.org/10.1021/acs.jafc.5b05214

O'Connor, D. J.; Wright, G. C.; Dieters, M. J.; George, D. L.; Hunter, M. N.; Tatnell, J. R.; Fleischfresser, D. B. Development and Application of Speed Breeding Technologies in a Commercial Peanut Breeding Program. *Peanut Sci.* **2013,** *40*(2), 107–114. https://doi.org/10.3146/ps12-12.1

Ochatt, S. J.; Sangwan, R. S.; Marget, P.; Assoumou Ndong, Y.; Rancillac, M.; Perney, P. New Approaches Towards the Shortening of Generation Cycles for Faster Breeding of Protein Legumes. *Plant Breed.* **2002,** *121*(5), 436–440. https://doi.org/10.1046/j.1439-0523.2002.746803.x

OECD, Food, & of the United Nations, A. O. (2012). *OECD-FAO Agricultural Outlook 2012*. https://doi.org/https://doi.org/https://doi.org/10.1787/agr_outlook-2012-en

Olaitan, A. F.; Abiodun, T. A.; Foluke, A. O.; Oluwaseyi, O. E. Comparative Assessment of Insect Pests Population Densities of Three Selected Cucurbit Crops. *Acta Fytotech. Zootech.* **2017**. https://doi.org/10.15414/afz.2017.20.04.78-83

Omanović-Mikličanin, E.; Maksimović, M. *Nanosensors applications in agriculture and food industry,* (n.d.).

Omanović, E.; Maksimović, M. Nanosensors Applications in Agriculture and Food Industry. *Bull. Chem. Technol. Bosnia Herzegovina* **2016,** *47*, 59–70.

Ong, B.; Lin-Heng, L.; Chun, J. Biological Diversity Conservation Laws in South East Asia and Singapore: A Regional Approach in Pursuit of the United Nations' Sustainable Development Goals? In *Asia Pacific Journal of Environmental Law*, 2016. https://doi.org/10.4337/apjel.2016.01.05

Palevitz, B. A. Genetic Parasites and a Whole Lot More. *Scientist* **2000**.

Parisi, C.; Vigani, M.; Rodríguez-Cerezo, E. Agricultural Nanotechnologies: What are the Current Possibilities? *Nano Today* **2015,** *10*(2), 124–127. https://doi.org/https://doi.org/10.1016/j.nantod.2014.09.009

Peltonen-Sainio, P.; Jauhiainen, L.; Laurila, I. P. Cereal Yield Trends in Northern European Conditions: Changes in Yield Potential and its Realisation. *Field Crops Res.* **2009,** *110*(1), 85–90. https://doi.org/https://doi.org/10.1016/j.fcr.2008.07.007

Peshin, R.; Dhawan, A. K. Integrated Pest Management. In *Integrated Pest Management,* 2009. https://doi.org/10.1007/978-1-4020-8992-3

Pingali, P. L. Green Revolution: Impacts, limits, and the Path Ahead. *Proc. Natl. Acad. Sci.* **2012,** *109*(31), 12302 LP – 12308. https://doi.org/10.1073/pnas.0912953109

Piškur, J.; Langkjær, R. B. Yeast Genome Sequencing: The Power of Comparative Genomics. *Mol. Microbiol.* **2004**. https://doi.org/10.1111/j.1365-2958.2004.04182.x

Prasad, R.; Bhattacharyya, A; Nguyen, Q. D. Nanotechnology in Sustainable Agriculture: Recent Developments, Challenges, and Perspectives. *Front. Microbiol.* **2017,** *8*, 1014. https://doi.org/10.3389/fmicb.2017.01014

Prasad, S. Nanobiosensors: The Future for Diagnosis of Disease? *Nanobiosens. Dis. Diag.* **2014,** *3*, 1–10. https://doi.org/10.2147/NDD.S39421

Rajput, V.; Singh, A.; Minkina, T.; Shende, S.; Kumar, P.; Verma, K.; Bauer, T.; Gorobtsova, O.; Deneva, S.; Sindireva, A. Potential Applications of Nanobiotechnology in Plant Nutrition and Protection for Sustainable Agriculture. In *Nanotechnology in Plant Growth Promotion and Protection*, 2021a, pp 79–92.

Rajput, V.; Minkina, T.; Feizi, M.; Kumari, A.; Khan, M.; Mandzhieva, S.; Sushkova, S.; El-Ramady, H.; Verma, K.; Singh, A.; Hullebusch, E.; Singh, R.; Jatav, H.; Choudhary, R. Effects of Silicon and Silicon-Based Nanoparticles on Rhizosphere Microbiome, Plant Stress and Growth. *Biology* **2021b**, *10*(8), 7–9.

Rana, M. M.; Takamatsu, T.; Baslam, M.; Kaneko, K.; Itoh, K.; Harada, N.; Sugiyama, T.; Ohnishi, T.; Kinoshita, T.; Takagi, H.;Mitsui, T. Salt Tolerance Improvement in Rice Through Efficient SNP Marker-Assisted Selection Coupled With Speed-Breeding. *Int. J. Mol. Sci.* **2019,** *20*(10). https://doi.org/10.3390/ijms20102585

Rashidi, L.; Khosravi-Darani, K. The Applications of Nanotechnology in Food Industry, 2011. *Http://Dx.Doi.Org/10.1080/10408391003785417*, *51*(8), 723–730. https://doi.org/10.1080/10408391003785417

rathbun 2013 nanosensors - Google विद्वान. (n.d.).

Ray, D. K.; Mueller, N. D.; West, P. C.; Foley, J. A. Yield Trends Are Insufficient to Double Global Crop Production by 2050. *PLoS One* 2013, *8*(6), e66428.

Ray, D. K.; Ramankutty, N.; Mueller, N. D.; West, P. C.; Foley, J. A. Recent Patterns of Crop Yield Growth and Stagnation. *Nat. Commun.* **2012,** *3*(1), 1293. https://doi.org/10.1038/ncomms2296

Richardson, J.; Hawkins, P.; Bioelectronics, R. L.-B. and, & 2001, undefined. (n.d.). The use of coated paramagnetic particles as a physical label in a magneto-immunoassay. *Elsevier*.

Rivelli, A. R.; Trotta, V.; Toma, I.; Fanti, P.; Battaglia, D. Relation Between Plant Water Status and Macrosiphum Euphorbiae (hemiptera: Aphididae) population dynamics on three cultivars of tomato. *Eur. J. Entomol.* https://doi.org/10.14411/eje.2013.084

Rizal, G.; Karki, S.; Alcasid, M.; Montecillo, F.; Acebron, K.; Larazo, N.; Garcia, R.; Slamet-Loedin, I. H.; Quick, W. P. Shortening the breeding cycle of sorghum, a model crop for research. *Crop Sci.* **2014,** *54*(2), 520–529. https://doi.org/10.2135/cropsci2013.07.0471

Rodriguez-Saona, C., R., B., Isaacs, R. Manipulation of Natural Enemies in Agroecosystems: Habitat and Semiochemicals for Sustainable Insect Pest Control. In *Integrated Pest Management and Pest Control - Current and Future Tactics*. 2012. https://doi.org/10.5772/30375

Russell, E. P. Enemies Hypothesis: A Review of the Effect of Vegetational Diversity on Predatory Insects and Parasitoids. *Environ. Entomol.* **1989**. https://doi.org/10.1093/ee/18.4.590

Samways, M. J. Insect conservation: A synthetic management approach. In *Annual Review of Entomology*, 2007. https://doi.org/10.1146/annurev.ento.52.110405.091317

Shang, Y.; Hasan, M. K.; Ahammed, G. J.; Li, M.; Yin, H.; Zhou, J. Applications of Nanotechnology in Plant Growth and Crop Protection: A Review. *Molecules (Basel)* **2019,** *24*(14). https://doi.org/10.3390/molecules24142558

Sharma, K.; Lavanya, M. Recent developments in transgenics for abiotic stress in legumes of the semi-arid tropics. *JIRCAS Working Report No. 23,* 2002.

Sharma, H. C.; Ortiz, R. Host plant resistance to insects: An eco-friendly approach for pest management and environment conservation. *J. Environ. Biol.* 2002.

Sharma, S.; Ruud, A. On the path to sustainability: integrating social dimensions into the research and practice of environmental management. *Business Strategy and the Environment,* 2003. https://doi.org/10.1002/bse.366

Sharma, H. C.; Sujana, G.; Manohar Rao, D. Morphological and chemical components of resistance to pod borer, Helicoverpa armigera in wild relatives of pigeonpea. *Arthropod-Plant Interact.* **2009**. https://doi.org/10.1007/s11829-009-9068-5

Sharma, H. C.; Venkateswarulu, G.; Sharma, A. Environmental factors influence the expression of resistance to sorghummidge, Stenodiplosis sorghicola. *Euphytica* **2003**. https://doi.org/10.1023/A:1023041713713

Shivakumar, M.; Nataraj, V.; Kumawat, G.; Rajesh, V.; Chandra, S.; Gupta, S.; Bhatia, V. S. Speed breeding for Indian agriculture: A rapid method for development of new crop varieties. *Curr. Sci.* **2018,** *115*(7), 1241. https://doi.org/10.18520/cs/v115/i7/1241-1241

Shivendu, R.; Dasgupta, N.; Lichtfouse, E. *Nanoscience in Food and Agriculture 2*. Springer International Publishing, 2016.

Shivendu, R.; Dasgupta, N.; Lichtfouse, E. *Nanoscience in Food and Agriculture 5*. Springer International Publishing, 2017.

Singh, S. P.; Rahman, M. F.; Murty, U. S. N.; Mahboob, M.; Grover, P. Comparative study of genotoxicity and tissue distribution of nano and micron sized iron oxide in rats after acute oral treatment. *Toxicol. Appl. Pharmacol.* **2013,** *266*(1), 56–66. https://doi.org/https://doi.org/10.1016/j.taap.2012.10.016

Singh, A.; Rajput, V. D., Mehrotra, R.; Pal, N.; Singh, V. K.; Minkina, T.; Chokheli, V. A.; Singh, R. K. In *Sustainable Soil Fertility Management*; Nova. Sci. Publishers, Inc., 2020a, vol *1*, pp 73–100.

Singh, A., Rajput, V., Rawat, S., Kumar Singh, A., Bind, A., Kumar Singh, A., Chernikova, N.; Voloshina, M.; Lobzenko, I. Monitoring Soil Salinity and Recent Advances in Mechanism of Salinity Tolerance in Plants. *Biogeosyst. Tech.* **2020b**, *7*(2). https://doi.org/10.13187/bgt.2020.2.66

Singh, A.; Rajput, V. D., Rawat, S.; Sharma, R.; Singh, A. K.; Singh, A. K.; Tomar, R. S. In *Emerging Tools for Sustainable Agriculture and Food Security*; Deepika Book Agency: New Delhi, Delhi, 2021a, vol 1, pp 1–15.

Singh, A., Rajput, V., Singh, A., Sengar, R., Singh, R. and Minkina, T. Transformation Techniques and Their Role in Crop Improvements: A Global Scenario of GM Crops. *Policy Issues Genetically Modified Crops*, **2021b**, *1,* 515–542.

Slosser, J. E.; Price, J. R.; Puterka, G. J. Evaluation of Furrow Diking and Early-Season Insecticide Applications on Boll Weevils (Coleoptera: Curculionidae), Bollworms (Lepidoptera: Noctuidae), and Cotton Yield in the Texas Rolling Plains. *J. Eco. Entomol.* **1989**. https://doi.org/10.1093/jee/82.2.599

Smith, S. L.; Slywka, G. W.; Krueger, R. J. Anthocyanins of strobilanthes dyeriana and their production in callus culture. *J. Natl. Prod.* **1981**. https://doi.org/10.1021/np50017a020

Sotiropoulou, S.; Gavalas, V., … V. V.-B. and, & 2003, undefined. (n.d.). Novel carbon materials in biosensor systems. *Elsevier*.

Southwood, T. R. E.. Habitat, the Templet for Ecological Strategies? *J. Animal Ecol.* **1977**. https://doi.org/10.2307/3817

Sparks, T. C.; Nauen, R. IRAC: Mode of action classification and insecticide resistance management. *Pestic. Biochem. Physiol*, **2015**. https://doi.org/10.1016/j.pestbp.2014.11.014

Stoner, K. A. Plant Resistance to Insects: A Resource Available for Sustainable Agriculture. *Biol. Agric. Hortic*. **1996**. https://doi.org/10.1080/01448765.1996.9754764

Su, X., Chew, F., sciences, S. L.-A., & 2000, undefined. (n.d.). Design and application of piezoelectric quartz crystal-based immunoassay. *Jstage.Jst.Go.Jp*.

Sumbal, Nadeem, A.; Naz, S.; Ali, J. S.; Mannan, A.; Zia, M. Synthesis, characterization and biological activities of monometallic and bimetallic nanoparticles using Mirabilis jalapa leaf extract. *Biotechnol. Rep. (Amsterdam)* **2019,** *22*, e00338. https://doi.org/10.1016/j.btre.2019.e00338

Sunding, D., & Zilberman, D. B. T.-H. of A. E. (2001). Chapter 4 The agricultural innovation process: Research and technology adoption in a changing agricultural sector. In *Agricultural Production* (Vol. 1, pp. 207–261). Elsevier. https://doi.org/https://doi.org/10.1016/S1574-0072(01)10007-1

Swift, T. A.; Oliver, T. A. A.; Galan, M. C.; Whitney, H. M. Functional nanomaterials to augment photosynthesis: evidence and considerations for their responsible use in agricultural applications. *Interface Focus* **2019,** *9*(1), 20180048. https://doi.org/10.1098/rsfs.2018.0048

Tabashnik, B. E.; Carrière, Y. Successes and failures of transgenic bt crops: Global patterns of field-evolved resistance. In *Bt Resistance: Characterization and Strategies for GM Crops ProducingBacillusthuringiensisToxins,* 2015. https://doi.org/10.1079/9781780644370.0001

Tamhane, V. A.; Giri, A. P.; Sainani, M. N.; Gupta, V. S. Diverse forms of Pin-II family proteinase inhibitors from Capsicum annuum adversely affect the growth and development of Helicoverpa armigera. *Gene* **2007**. https://doi.org/10.1016/j.gene.2007.07.024

Tefera, T.; Mugo, S.; Beyene, Y. Developing and deploying insect resistant maize varieties to reduce pre-and post-harvest food losses in Africa. *Food Sec*. **2016**. https://doi.org/10.1007/s12571-015-0537-7

Tilman, D.; Balzer, C.; Hill, J.; Befort, B. L. Global food demand and the sustainable intensification of agriculture. *Proc. Natl. Acad. Sci.* **2011,** *108*(50), 20260 LP–20264. https://doi.org/10.1073/pnas.1116437108

United Nations, D. of E. and S. A. (UN-D. (2017). *Population Division, 2017. World Population Prospects: The 2017 Revision, Key Findings and Advance Tables. Working Paper No. ESA/P/WP/248.*

van Emden, H. F. Host plant-Aphidophaga interactions. *Agric. Ecosyst. Environ.* **1995**. https://doi.org/10.1016/0167-8809(94)09001-N

van Lenteren, J. C.; Hua, L. Z.; Kamerman, J. W.; Rumei, X. The parasite-host relationship between Encarsia formosa (Hym., Aphelinidae) and Trialeurodes vaporariorum (Hom., Aleyrodidae) XXVI. Leaf hairs reduce the capacity of Encarsia to control greenhouse whitefly on cucumber. *J. Appl. Entomol.* **1995**. https://doi.org/10.1111/j.1439-0418.1995.tb01335.x

Vo-Dinh, T.; Cullum, B. Biosensors and biochips: advances in biological and medical diagnostics. *Fresenius' J. Anal. Chem.* **2000,** *366*(6), 540–551. https://doi.org/10.1007/S002160051549

Warheit, D. B.; Brown, S. C.; Donner, E. M. Acute and subchronic oral toxicity studies in rats with nanoscale and pigment grade titanium dioxide particles. *Food Chem. Toxicol.* **2015,** *84*, 208–224. https://doi.org/https://doi.org/10.1016/j.fct.2015.08.026

Wassmann, R.; Jagadish, S. V. K.; Heuer, S.; Ismail, A.; Redona, E.; Serraj, R.; Singh, R. K.; Howell, G.; Pathak, H.; Sumfleth, K. Chapter 2 Climate Change Affecting Rice Production. The Physiological and Agronomic Basis for Possible Adaptation Strategies. In *Advances in Agronomy;* 1st ed.; Elsevier Inc., 2009; vol. 101, Issue January 2009). https://doi.org/10.1016/S0065-2113(08)00802-X

Zeigler, R. S.; Mohanty, S. Support for international agricultural research: current status and future challenges. *New Biotechnol.* **2010,** *27*(5), 565–572. https://doi.org/10.1016/j.nbt.2010.08.003

Zhang, Y.; Arugula, M. A.; Wales, M.; Wild, J.; Simonian, A. L. A novel layer-by-layer assembled multi-enzyme/CNT biosensor for discriminative detection between organophosphorus and non-organophosphorus pesticides. *Biosens. Bioelectron.* **2015,** *67*, 287–295. https://doi.org/10.1016/j.bios.2014.08.036

Zhang, Y.; Liu, B.; Huang, K.; Wang, S.; Quirino, R. L.; Zhang, Z.; Zhang, C. Eco-Friendly Castor Oil-Based Delivery System with Sustained Pesticide Release and Enhanced Retention. *ACS Appl. Mater. Interfaces* **2020,** *12*(33), 37607–37618. https://doi.org/10.1021/acsami.0c10620

CHAPTER 13

Nanobiotechnological Approaches for Improved Plant Breeding

SAPNA RAWAT[1], ABHISHEK SINGH[2], VISHNU D RAJPUT[3], RAGINI SHARMA,[4] SAGLARA MANDZHIEVA[3], and AWANI KUMAR SINGH[5]

[1]*Department of Botany, University of Delhi, Delhi, India*

[2]*Department of Agricultural Biotechnology, College of Agriculture, Sardar Vallabhbhai Patel University of Agriculture and Technology, Meerut, UP, India*

[3]*Academy of Biology and Biotechnology, Southern Federal University, Rostov-on-Don, Russia*

[4]*Department of Zoology, Punjab Agriculture University, Ludhiana, India*

[5]*Centre for Protected Cultivation, ICAR-Indian Agricultural Research Institute, New Delhi, India*

ABSTRACT

Nanobiotechnological approaches can make a huge advancement in crop improvement through molecular-marker-assisted breeding. It is a cost-effective, rapid, and accurate platform for identification of genetic polymorphism with high throughput and automated detection. Thus, nanogenomics is facilitating plant breeders to attain their target of precision breeding as well as providing a new prospect for selection and gene transfer. In this chapter,

Nanotechnology for Sustainable Agriculture: An Innovative and Eco-Friendly Approach. Vishnu D. Rajput, Abhishek Singh, Tatiana M. Minkina, Krishan K. Verma, and Awani Kumar Singh (Eds.)

emphasized nanobiotechnological advances and their application in plant breeding and focused on single nucleotide polymorphism detection.

13.1 INTRODUCTION

Nanobiotechnology is a combination of molecular biology and engineering of atomic and molecular size with the aim of developing structures and devices (Jain, 2003). These nanodevices have application in various biological fields such as diagnostics, drug delivery, and therapeutics. Nanobiotechnology is growing enormously in the field of pharmaceuticals (Seetharam, 2006) and is targeted to drug delivery and noninvasive medical procedures as used in cancer (Shelley, 2006). It utilizes materials with particle size of 1–100 nm. They have unique properties in various terms such as chemistry, morphology, reactivity, and magnetic characters. Nanotechnology uses these nanoparticles as advanced biotechnological tool as well as for better understanding and development of living beings and biomolecules.

Plant breeding can produce genome-wide molecular markers quickly and simply, which can lead to a substantial improvement, QTL analysis, MAS, and varieties identification (Kanazin et al., 2002). Molecular marker method has not been implemented properly (Gupta et al., 2001) due to various factors such as cost of technology, time consuming, cost of measuring traits, and large segregating populations are not easy to manage (Lamkey and Lee, 1993) thus restricting broader execution of plant breeding based on molecular markers (Gupta et al., 2001). Eventually, identifying an allele, trait, or individual using molecular marker technology remains a major challenge. The nanogenomics method is helping plant breeders to achieve their target of precision breeding as well as providing a new opportunity for selection and gene transfer. This can also allow the gene transfer from distant plant. DNA sequencing in a nanofabricated gel-free system that can allow more rapid DNA sequencing. This sequencing data can be used to know a lot of information about the relationship between molecular markers and economically important features of crop germplasm that help in production. Thus, nanobiotechnology can improve the progress of breeding for crop improvement assisted by molecular markers. Thus, nanobiotechnology can improve the progress of Molecular-marker Assisted Breeding for the purpose of development of crops. In near future, advances in the field of nanotechnology can facilitate genotyping by generating materials by microfabrication, decreasing reaction volume, increasing volume rates, automated sample handling, and evolving single-platform genotyping, data analysis as well as instrumentation for storage. In this chapter, we highlighted nanobiotechnological advances and its application in plant breeding and focused on single nucleotide polymorphism (SNP) detection.

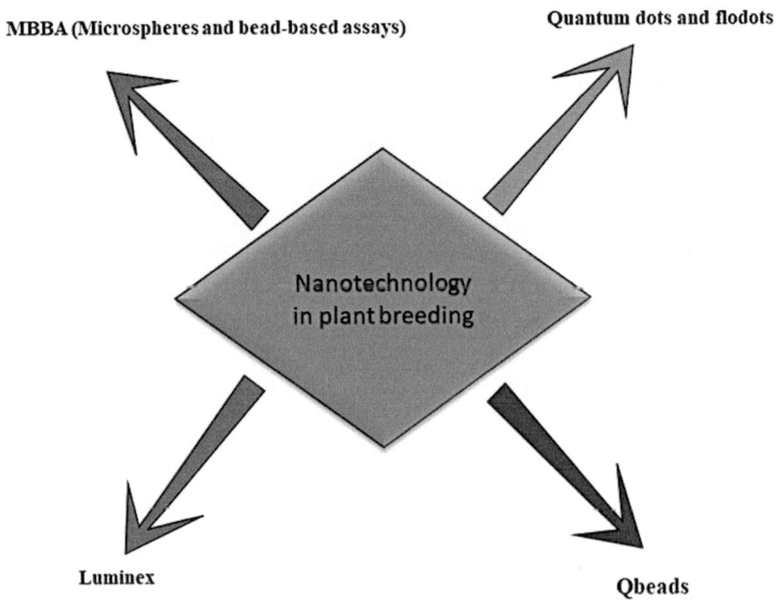

FIGURE 13.1 Diagrammatic representation of nanotechnology products in plant breeding for crop improvement.

13.2 QUANTUM DOTS AND FLODOTS

Quantum dots are semiconductor nanocrystals made up of CdSe core, various times capped with ZnS to enhance the quantum yield of light which is the ratio of absorbed light and emitted light (Chan et al., 2002). Semiconductors are made through the process of doping, adding conductive metal atoms onto the insulator surface that changes the available electrons. Electrons move freely into vacant orbitals upon excitation, thus semiconductor QD can absorb and emit photons. The emission properties depend on size and composition. A single wavelength of light can be used to stimulate emitted wavelengths ranging from blue to near-infrared (Chan et al., 2002) that is detected through instrumentation. Thus, QDs can be great alternative to traditional fluorescent dyes. QDs can be linked to biomolecules themself.

Using various combinations of QDs generating light of six distinct hues at varied intensities, they can generate 40,000 unique optical codes (Han et al., 2001; Ng and Liu, 2006). This unique optical address with specific QD composition can be used to create large screening libraries by conjugation of allele with particular oligonucleotide probes to microspheres (Han et al., 2001). This optical signal can be used to detect the target, however, the complex of labeled target/probe specifies the existence or nonexistence as well as richness of the target (Han et al., 2001). Standard microscopes can image and examine this. Flow cytometers, on the other hand, may be used for high-throughput detection and screening (Gao and Nie, 2004).

FloDots is a silica matrix with thousands of luminous dye molecules incorporated in it. When stimulated, it produces a powerful emission signal. These silica surfaces, which may be changed by adding functional groups, are utilized as a substrate for the immobilization of biomolecules. It offers a number of benefits over QD, including water dispersion, the ability to overcome limited solubility, and agglutination (Yao et al., 2006).

13.3 MICROSPHERES AND BEAD-BASED ASSAYS

Microspheres (diameter of 2–500) permit the attachment of numerous compounds on each and every bead thus can provide solid platforms for attaching biological recognition compounds for bead-based assays (Trau and Battersby, 2001). These spherical microsphere beads are made up of polystyrene, latex, and silica thus allow spatial homogeneity of probe attachment and consequent target hybridization (Spiro et al., 2000). Assays based on microspheres can create large number of combinational libraries with varying number and intensities of attached dyes (Battersby et al., 2001).

13.4 LUMINEX

Luminex is a fluorescent microspore technology, which is made up of polystyrene beads with red and orange fluorophores that are impregnated in various ratios to produce a library that can be differentiated by their relative intensities of red and orange fluorescence. Probes that are complimentary to the target of interest can be attached directly to the microsphere with its unique code (Fulton et al., 1997). Detection and analysis of fluorescent wavelength and intensity were done using flow cytometers. Thus, the identification with quantification of prokaryotic DNA sequences in heterogeneous

DNA samples and can allow the quantification of target (Spiro et al., 2000). The microspheres form aggregate when targeted DNA was found to be complementary to oligos trapped on the surface of the microsphere, producing a FRET-induced fluorescence (Ihara et al., 2004). That can be measured by fluorescent microscopy. This technique requires improvement in color differentiation for high-throughput applications (Ihara et al., 2004).

13.5 QBEADS

Latex microspheres embedded with QDs are Qbeads. Qbeads of various colors and intensities are used to color label Qbeads. This assay is based on dyes used and allele-specific target hybridization. They are conjugated with allele-specific oligos get hybridized to biotinylated targets. TaqMan PCR is used to confirm the results of Qbead test and can identify homozygous and heterozygous SNP alleles with more precision than sequencing (Xu et al., 2003). Multiplexed amplification step was made to save genomic template, time, and expense associated with PCR (Xu et al., 2003).

13.6 SILICA MICROPHORES

Silica microspores are porous and can facilitate attachment of various biomolecules with covalently attached dyes that are optically encoded (Johnston et al., 2005). These dyes can be used in various combinations thus generating optical signals with a wide range (Johnston et al., 2005). Microsphere surface produce fluorescence when oligonucleotides attached with microspheres get hybridized to the target site, while mismatch show less fluorescence significantly (Johnston et al., 2005; Corrie et al., 2006). The microsphere interior has free either thiol or amines groups uniformly that form covalent bonds with dyes (Miller et al., 2005). This fluorescence is quantified by flow cytometer (Wedemeyer and Potter, 2001). Thus, high-throughput detection and quantification of various molecular interactions can be done via flow cytometer (Nolan and Sklar, 1998).

13.7 NANOPORE

It is a nanometer pore sized, made of pore-forming protein or a hole of synthetic compounds such as silicon or grapheme. It is used as single-molecule detector. Nanopore-based DNA sequencing can electrically detect DNA sequence with low cost, high speed, and less sample preparation. It is a rapid and lost cost alternative of next-generation DNA sequencing (Khiyami et al., 2014).

13.8 GOLD NPs

Gold NPs can act as attractive labels for biosensors that can be perceived through various analytical techniques such as electrical conductivity, fluorescence, and optical absorption (Jain, 2003). Storhoff et al. (2004) conducted an experiment that used gold nanoparticle labeled with oligonucleotide probe (GNP-DNA) for detection of bacterial DNA that was unamplified. When probe became hybridized to the target and was identified on a glass slide, scattered yellowish orange light when illuminated with white light. Simultaneously, nonhybridized DNA scattered green light. Though, it is rapid and inexpensive technique that uses visual readings. Microarray bound allele-specific probes modified with gold nanoparticle can be used to detect the target genomic DNA site. Scattered light captured on photo sensor.

13.9 NANOBARCODES

Nanobarcodes are microscopic metal wires that are sequentially barcoded by electroplating of metal ions of diverse reflectivity into fine channels by means of lithography (Nicewarner-Pena et al., 2001; Sha et al., 2006). Varying sequence and kind of deposited metal are used to increase barcode complexity. Each Nanowire with unique coding is fixed with different fluorescently oligonucleotide probe added into PCR reaction. When they perfectly hybridized with target, nanowire imaged and analyzed for SNP detection (Sha et al., 2006).

13.10 NANOCHIPS AND NANO ARRAYS

It is an ultra-miniature form of microarray for ultramicro and nanoscale fluid delivery. Large number of samples such as nucleic acids, antibodies, and nanomaterials can be analyzed using thousands of nanospots (1–20 mm). They are patterned directly by the nanoarrays through microcantilever print head (Jain, 2003).

13.11 NANOPRODUCT USE IN PLANT BREEDING

Nanoparticles are easier for a plant to absorb due to their small size and that means plant spend less energy in doing so. This energy conservation means enhanced growth and development for crop. Plant have to work less in taking up nutrient more it can mobilize their energy in promoting healthy, vigorous growth, faster germination, increase initial growth, and powerful root development. For plant breeding 24 nanoproducts are available in the commercial market in six countries developed by six companies (Table 13.1).

13.11.1 NANOTECH T5 REFLECTORS

As the light travel through target area, plenty of available light is lost as that was absorbed by ballast. Recovery of that lost light is our foremost objective. Once it is achieved can enhance the performance of previously outstanding grow lamp. Each and every T5 lighting fixtures had similar designs. They collect 100% of lost light and reflect 99% of it back to the target area as diffused light of improved quality. This diffused light penetrates deeper into the canopy, resulting in better beginnings and quicker development cycles. As a result, Nanotech's superior crystal reflecting layer surface improved overall performance while also providing a number of additional benefits. Improvement in light quality enhanced light effectiveness, availability of lumen, improved coverage, and light penetration of the canopy.

13.11.2 SUNBLASTER NANODOME

SunBlaster 7″ NanoDome can speedup germination, improve seedling growth, and maintaining high humidity with controlled temperature with adjustable vents that controls air circulation in all 1020 seedling trays or flats. It has a H-shaped groove that gives the dome superior strength. It permits both our T5 NanoTech lights and LED Strip Lights to fitted in the dome, ensuring centered lights, and plants can receive even light coverage. The dome is fit with Quad Thick and Double thick trays. The light channels integrated into these domes (a patented feature) retain lights in place during the propagation stage and easily detached for crop inspection or misting.

TABLE 13.1 Number of Nanoproducts, Companies and the Countries That Involved in Plant Breeding Related Nano Products Development (*Source:* Statnano, 2021).

Plant-breeding nanoproducts	Company	Country	Properties	Applications
NanoTech-T5 reflectors	SunBlaster Holdings ULC	Canada	Seed germination promotion, hydroponic, increased lighting, increased lumen availability	Use in plant breeding
SunBlasterNanoDome	SunBlaster Holdings ULC	Canada	High humidity management, temperature control, seed germination promotion, air circulation control	Use in plant breeding
Primo MAXX growth regulator	Syngenta	Switzerland	Regulation of plant growth, acceleration of plant development, plant resistance, stress withstand, and drought tolerance.	Use in plant breeding
Plant growth promoter HTNY-03	Nanjing Hgh Technology Nano Material	China	Anti-bacterial activity, nontoxic, inorganic	Use in plant breeding
Anabolic plant roids	Plant vitality Ltd	United Kingdom	Plant growth regulation, plant growth acceleration	Use in plant breeding
PK boost	Plant vitality Ltd	United Kingdom	Plant growth regulation, plant growth acceleration	Use in plant breeding
Cal mag nitrate	Plant vitality Ltd	United Kingdom	Plant growth regulation, plant growth acceleration	Use in plant breeding
Nano silicate+	Plant vitality Ltd	United Kingdom	Plant growth regulation, plant growth acceleration, seed germination promotion	Use in plant breeding
Hydro: grow	Plant vitality Ltd	United Kingdom	Plant growth regulation, root activity improvement, seed germination promotion	Use in plant breeding
Root zone mass	Plant vitality Ltd	United Kingdom	Plant growth regulation, root activity improvement, seed germination promotion	Use in plant breeding
Seed speed	Plant vitality Ltd	United Kingdom	Plant growth regulation, root activity improvement, seed germination promotion	Use in plant breeding
Nano-boron	Kanak Biotech	India	Plant growth regulation	Use in plant breeding
Nano-molybdenum	Kanak Biotech	India	Plant growth regulator	Use in plant breeding

TABLE 13.1 *(Continued)*

Plant-breeding nanoproducts	Company	Country	Properties	Applications
Nanosilver	Kanak Biotech	India	Plant growth regulation, nutritional, plant growth acceleration	Use in plant breeding
Nano-magnesium	Kanak Biotech	India	High specific surface area, plant growth regulation, nutritional, plant growth acceleration	Use in plant breeding
Nano-iron	Kanak Biotech	India	High specific surface area, plant growth regulation, nutritional, plant growth acceleration	Use in plant breeding
Nano-nitrogen	Kanak Biotech	India	High specific surface area, plant growth regulation, nutritional, plant growth acceleration	Use in plant breeding
Nano-sulfur	Kanak Biotech	India	High specific surface area, plant growth regulation, nutritional, plant growth acceleration	Use in plant breeding
Nano-potassium	Kanak Biotech	India	High specific surface area, plant growth regulation, nutritional, plant growth acceleration	Use in plant breeding
Nano-phosphorous	Kanak Biotech	India	High specific surface area, plant growth regulation, nutritional, plant growth acceleration	Use in plant breeding
Nano-Zinc	Kanak Biotech	India	High specific surface area, plant growth regulation, nutritional, plant growth acceleration	Use in plant breeding
NeuSelen-X®	Neufarm GmbH	Germany	Crop yield enhancement	Use in plant breeding
NeuCytokin®	Neufarm GmbH	Germany	Plant growth regulator with nano technology (PGR), crop yield enhancement	Use in plant breeding
Nano silicate+	Plant vitality Ltd	United Kingdom	Plant growth regulation, root activity Improvement, seed germination promotion	Use in plant breeding

13.11.3 PRIMO MAXX GROWTH REGULATOR

Primo Maxx® plant growth regulator can promote denser and healthier turfgrass that can endure a various stresses comprising heat, drought, and diseases. It stimulates gibberellic acid production, which promotes elongation of cell. Shoot growth is slowed vertically, whereas lateral and below-ground growth of rhizomes, stolons, tillers, and roots is boosted. It has the potential to reduce water use and improve drought resistance. To inhibit basal rot anthracnose, this innovative formulation combines water and may be sprayed at rates of 0.1–0.2 fl. oz. per 1000 sq. ft. every 7–14 days.

13.11.4 PLANT GROWTH PROMOTER HTNY-03

These products are made from a variety of nano oxides, with various characteristics such as nontoxic and harmless to human, purified water, support the plant growth, advance pest, and disease resistance. Product benefits contain improvement in agricultural production, easy to use, low quantity to be used with high yield, nontoxicity, harmless, and safety.

13.11.5 ANABOLIC PLANT ROIDS

It contains natural steroids found in plants. It can be absorbed through roots enhancing metabolic and anabolic rate, enhances bud development, increase yield, and flowers aided through larger rooting systems.

13.11.6 PK BOOST

Pk boost 9/18+ is a flowering cycle upregulation enhancer that aims to improve the size and weight of vegetables, flowers, and fruits.

13.11.7 CAL MAG NITRATE

It is formed to promote fast growth of plants by avoiding nutrient deficiencies, enhanced nutrient uptake, and enhance plant growth and development. Help in uptake of vital elements and providing a critical surge of trace elements to the plants.

13.11.8 NANO SILICATE+

Nanoparticle product can faster the plant growth, larger flower development, reduction in harvesting time, resistance to drought and diseases, and are easy to use.

13.11.9 HYDRO: GROW

A **hydroponic** system can **produce** plants and vegetables at faster rate than growing them in outdoors in soil. Plants **grown hydroponically** often more yield, need less space, and water than with conventional gardening.

13.11.10 ROOT ZONE MASS (RZM)

RZM is a root growth stimulator that contains a combination of NPK, Humic Acid, Fulvic Acid, chosen plant vitamins, Lecithin, and Protein molecules to stimulate multiroot development and allow plants to carry nutrients to their maximum potential. It may be found in all types of substrates, as well as hydroponics and airoponics.

13.11.11 SEED SPEED

It is a new development of unlocking dormant seeds and encourages quicker emerging of seeds. Soak for 12–24 h after blending speed seed with warm water as it is safest for seed soaking. Some seeds can tolerate boiling water that can differ significantly from species to species.

13.11.12 NANO-BORON

Boron is a crucial nutrient for the growth and development of healthy plants. It is also used as a herbicide, algaecides, and as a pest controller. Vital for upholding a balance between sugar and starch and aids in the translocation of sugar & carbohydrates crops are sensitive to boron deficiency. Boron deficiency can lead to impaired cell expansion in fast-growing organs (leaves, roots, pollen tube), malformation of roots and shoots, male and female flower sterility, and reduced seed set due to inhibition of pollen growth.

13.11.13 NANO-MOLYBDENUM

Molybdenum is crucial to plant growth as a constituent of the enzymes nitrate reductase and nitrogenase. Legumes need more molybdenum than any other crops, in the root nodules of legumes want molybdenum for nitrogen fixation. Almost all crops are sensitive to molybdenum deficiency. An overall light green tint, stunted development with tiny leaves, and reddening of veins on immature leaves are all indicators of N deficiency.

13.11.14 NANOSILVER

Eco-friendly and organic-certified SILVER (Ag) nanoparticles consider as an effective antimicrobial mechanism because silver ions are toxic to 650 microorganisms (bacteria, yeast, fungi, molds, viruses, etc.). Nanosilver can absorb the oxygen atom on its surface area, will combine with the cell walls of microorganism thus have ability to interact and oxidize with the microbe's surface protein (-SH functional group), eventually destroying the bacteria.

Nano Silver is a highly effective, eco-friendly sanitizing solution. It indicates advanced approach of clean, harmless, environmentally friendly biocides for the effective control of micro-pathogens on hard surfaces and in water systems.

13.11.15 NANO-MAGNESIUM

The only mineral constituent of the chlorophyll molecule, and a component of chlorophyll on which photosynthesis depends. It has an essential role in the translocation of phosphorus. Deficiency can lead to yellowing between the leaf veins or reddish-brown tints Leaves become brittle and necrotic and may drop prematurely.

13.11.16 NANO-IRON

The most important mineral for respiration and photosynthesis is iron. All crops are sensitive to Iron deficiency. Deficiency can lead to inhibited growth and reduced fruit development, decreased leaf size, and discolored premature leaf fall. It can also cause tissue between the veins to become yellow, even when the veins say green leading to necrosis due to chlorosis (leaf mottling).

13.11.17 NANO-NITROGEN

Nitrogen is the most crucial nutrient for plants. It is an important component of chlorophyll and while leaves encompass adequate nitrogen, photosynthesis occurs at high rates. Its deficiency can lead to light green or yellow and some may develop red or orange. Stunted growth and decrease the protein levels in pasture and grain.

13.11.18 NANO-SULFUR

Sulfur is an essential plant nutrient for nutrient and protein synthesis, as well as for enzyme and vitamin function. It acts as supplement that provides dynamic plant growth. Deficiency can lead to uniform pale green to yellow leaf, poor low-yielding, low protein, and pale green and yellow leaves in wheat.

13.11.19 NANO-POTASSIUM

Potassium is required by the plant for protein synthesis as well as the opening and shutting of stomata. It is linked with water movement, nutrients, and carbohydrates in plant tissues. All crops are especially all fruits and vegetables. Deficiency includes brown scorching and curling of leaf tips as well chlorosis reduces in plant growth, root development, and fruits formation.

13.11.20 NANO-PHOSPHORUS

It is another nutrient that plants require for their development and reproduction. It is reflected as one of the three major nutrients due to the relatively large amounts utilized by plants. All crops are sensitive to phosphorous deficiency. Reduced growths lead to stunted and shortened inter.

13.11.21 NANO-ZINC

Zinc is an important nutrient that promotes crop growth and reproduction. It functions as a structural component or regulatory co-factor in a variety of enzymes. Its deficiency can result in stunted growth and reduce development.

13.11.22 NEUSELEN-X®

It is an innovative formulation of a chemical enzymatic synthesis through which the division of a tripeptide (Glutapione) as valuable to retain the selenium inside potato. For potatoes a tripeptide that permits an improved absorption of elementary nutrients, the amino acids content advances the growth, speckling, endogenous defense, and the production. Infect, from field trials resulted in the production is improved of about 7–8%.

13.11.23 NEUCYTOKIN®

NeuCytokin® is a liquid plant hormone produce comprising cytokinins from natural seaweed source and water soluble humic/fulvicacid. NeuCytokin® is used in foliar and in the soil as a supplement to the natural hormone levein plants to encourage and withstand flowering, flower set, fruit retention, and fruit development to proliferate production and yields. For those conventional and organic cultivators needing the profits of foliar fertilizers, NeuCytokin® is a foliar fertilizer made from natural potassium and seaweed that promotes growth over that experienced with fertilizers alone.

13.12 CONCLUSIONS

Miniaturization has reduced cost of plant genotyping as well as increasing throughput. Development of cost-effective, rapid, and accurate platform needed to identify genetic polymorphism with multiplex and high-throughput automated detection (Shi, 2001). The main barrier is the cost of developing and validating innovative high-throughput tests. PCR amplification should be eliminated to remove additional steps for the generation of target amplicons, eliminate chances of cross-contamination and variation in the efficacy of target amplification (Griffin and Smith, 2000; Galvin, 2002). Unamplified targets can be detected through gold nanoparticles, due to their sensitivity

(Storhoff et al., 2004; Bao et al., 2005). Bead-based assays have capability for SNP genotyping at low price, fast fluid-phase reaction rates, small reaction volume, and automated sample processing. This one platform of genotyping with fast data analysis and storage in flow cytometer can provide cost effective genotyping in plant sciences.

ACKNOWLEDGMENTS

The research was financially supported by the Ministry of Science and Higher Education of the Russian Federation (no. FENW-2023-0008) and the Strategic Academic Leadership Program of the Southern Federal University ("Priority 2030").

KEYWORDS

- **automation**
- **high throughput**
- **nanogenomics**
- **SNP**

REFERENCES

Battersby, B. J.; Lawrie, G. A.; Trau, M. Optical Encoding of Microbeads for Gene Screening: Alternatives to Microarrays. *Drug Discov. Today* **2001,** *6,* S19–S26.

Chan, W. C. W.; Maxwell, D.; Gao, X.; Bailey, R.E.; Han, M.; Nie, S. Luminescent Quantum Dots for Multiplexed Biological Detection and Imaging. *Curr. Opin. Biotechnol.* **2002,** *13,* 40–46.

Corrie, S. R.; Lawrie, G. A.; Trau, M. Quantitative Analysis and Characterization of Biofunctionalized Fluorescent Silica Particles. *Langmuir* **2006,** *22,* 2731–2737.

Fulton, R. J.; McDade, R. L.; Smith, P. L; Kienker, L. J.; Kettman Jr.; J. R. Advanced Multiplexed Analysis With the Flow Metrix (TM) System. *Clin. Chem.* **1997,** *43,* 1749–1756.

Galvin, P. A nanobiotechnology roadmap for high-throughput single nucleotide polymorphism analysis. *Psychiatric Genetics*, 2002, 12, 75–82.

Gao, X.; and Nie, S. Quantum Dot-Encoded Mesoporous Beads With High Brightness and Uniformity: Rapid Readout Using Flow Cytometry. *Anal. Chem.* **2004,** *76,* 2406–2410.

Griffin, T. J.; Smith, L. M. Single-Nucleotide Polymorphism Analysis by MALDI-TOF Mass Spectrometry. *Trends Biotechnol.* **2000,** *18,* 77–84.

Gupta, P. K.; Roy, J. K; Prasad, M. Single Nucleotide Polymorphisms: A New Paradigm for Molecular Marker Technology and DNA Polymorphism Detection With Emphasis on Their use in Plants. *Curr. Sci.* **2001,** *80,* 524–535.

Han, M.; Gao, X.; Su, J. Z; Nie, S. Quantum-Dot-Tagged Microbeads for Multiplexed Optical Coding of Biomolecules. *Nat. Biotechnol.* **2001,** *19,* 631–635.

Ihara, T.; Tanaka, S.; Chikaura, Y; Jyo, A. Preparation of DNA-Modified Nanoparticles and Preliminary Study for Colorimetric SNP Analysis Using Their Selective Aggregations. *Nucleic Acids Res.* **2004,** *32,* e105.

Jain, K. K. Nanodiagnostics: Application of Nanotechnology in Molecular Diagnostics. *Expert Rev. Mol. Diagn.* **2003,** *3,* 153–161.

Johnston, A. P. R.; Battersby, B. J.; Lawrie, G. A.; Trau, M. Porous Functionalized Silica Particles: A Potential Platform for Bimolecular Screening. *Chem. Commun.* **2005,** *7,* 848–850.

Kanazin, V.; Talbert, H.; See, D.; DeCamp, P.; Nevo, E; Blake, T. Discovery and Assay of Single-Nucleotide Polymorphisms in Barley (*Hordeum vulgare*). *Plant Mol. Biol.* **2002,** *48,* 529–537.

Khiyami, M. A.;Almoamma, H.; Awad, Y. M.; Alghuthaymi, M. A.; Abd-Elsalam, K. A. Plant Pathogen Nanodiagnostic Techniques: Forthcoming Changes? *Biotechnol. Biotechnol. Equip.* **2014,** *28,* 5, 775–785. DOI: 10.1080/13102818.2014.960739

Lamkey, K. R; Lee, M. In *Quantitative Genetics, Molecular Markers, and Plant Improvement*, 10th Australian Plant Breeding Conference, Focused Plant Improvement: Towards Responsible and Sustainable Agriculture; Imrie, B. C., Hacker, J. B., Eds.; Gold Coast, Australia, 1993; pp 104–115.

Miller, C. R.; Vogel, R.; Surawski, P. P. T.; Corrie, S. R.; Ruhmann, A; Trau, M. Biomolecular Screening With Novel Organosilica Microspheres. *Chem. Commun.* **2005,** 4783–4785.

Nicewarner-Pena, S. R.; Freeman, R. G.; Reiss, G. D.; He, L.; Pena, D. J.; Walton, I. D.; Cromer, R.; Keating, C. D; Natan, M. J. Submicrometer Metallic Barcodes. *Science* **2001,** *294,* 137–141.

Nolan, J. P; Sklar, L. A. The Emergence of Flow Cytometry for Sensitive, Real-Time Measurements of Molecular Interactions. *Nat. Biotechnol.* **1998,** *16,* 633–638.

Seetharam, R. N. Nanomedicine – Emerging Area of Nanobiotechnology Research. *Curr. Sci.* **2006,** *91,* 260.

Sha, M. Y.; Walton, I. D.; Norton, S. M.; Taylor, M.; Yamanaka, M.; Natan, M. J.; Xu, C.; Drmanac, S.; Huang, S.; Bordherding, A.; Drmanac, R; and Penn, S. G. Multiplexed SNP Genotyping Using Nanobarcode Particle Technology. *Anal. Bioanal. Chem.* **2006,** *384,* 658–666.

Shelley, S. A. Nanobiotechnology: Cancer's Newest Deadly Foe. *Chem. Eng. Prog.* **2006,** *102,* 43–47.

Shi, M. M. Enabling Large-Scale Pharmacogenetic Studies by High-Throughput Mutation Detection and Genotyping Technologies. *Clin. Chem.* **2001,** *47,* 164–172.

Spiro, A.; Lowe, M; Brown, D. A Bead-Based Method for Multiplexed Identification and Quantitation of DNA Sequences Using Flow Cytometry. *Appl. Environ. Microbiol.* **2000,** *66,* 4258–4265.

Statnano, [Online] 2021. https://product.statnano.com/industry/agriculture

Storhoff, J. J.; Lucas, A. D.; Garimella, V.; Bao, Y. P; Muller, U. R. Homogeneous Detection of Unamplified Genomic DNA Sequences Based on Colorimetric Scatter of Gold Nanoparticle Probes. *Nat. Biotechnol.* **2004,** *22,* 883–887.

Trau, M.; Battersby, B. J. Novel Colloidal Materials for High-Throughput Screening Applications in Drug Discovery and Genomics. *Adv. Mater.* **2001,** *13,* 975–979.

Wedemeyer, N.; Potter, T. Flow Cytometry: An 'Old' Tool for Novel Applications in Medical Genetics. *Clin. Genet.* **2001**, *60,* 1–8.

Xu, H.; Sha, M. Y.; Wong, E. Y.; Uphoff, J.; Xu, Y.; Treadway, J. A.; Truong, A.; O'Brien, E.; Asquith, S.; Stubbins, M.; Spurr, N. K.; Lai, E. H.; Mahoney, W. Multiplexed SNP Genotyping Using the Qbead (TM) System: A Quantum Dot-Encoded Microsphere-Based Assay. *Nucleic Acids Res.* **2003,** *31,* e43.

Yao, G.; Wang, L.; Wu, Y.; Smith, J.; Xu, J.; Zhao, W.; Lee, E.; Tan, W. FloDots: Luminescent Nanoparticles. *Anal. Bioanal. Chem.* **2006,** *385,* 518–524

CHAPTER 14

Nanobionics Aid in Agriculture

ABHISHEK SINGH[1], PRIYADARSHANI RAJPUT[2], SAPNA RAWAT[3], SVETLANA SUSHKOVA[2], and VISHNU D. RAJPUT[2]

[1]*Department of Agricultural Biotechnology, College of Agriculture, Sardar Vallabhbhai Patel University of Agriculture and Technology, Meerut, UP, India*

[2]*Academy of Biology and Biotechnology, Southern Federal University, Rostov-on-Don, Russia*

[3]*Department of Botany, University of Delhi, Delhi, India*

ABSTRACT

Plant nanobionics provides an interface between nanotechnology and plant biology. It is an area of plant science that uses nanoparticles and their interaction with plant system that comes up with a novel function. The application of nanoparticles in plants system provides them with novel functions is termed as plant nanobionics. Plant nanobionics has various applications in the field of agriculture, biosensors, defense, etc. Thus, can help in crop management.

14.1 INTRODUCTION

Nanobionics is the combination of two words, nano + bionics, so nanobionics can be defined as the study of electronic interactions at nanoscale in the biological systems. The broader vision is to generate an extensive array of wild-type plants that are trained in imaging objects in their environment, infrared communication devices, working as self-powered light sources,

Nanotechnology for Sustainable Agriculture: An Innovative and Eco-Friendly Approach. Vishnu D. Rajput, Abhishek Singh, Tatiana M. Minkina, Krishan K. Verma, and Awani Kumar Singh (Eds.)

and also running as self-powered groundwater sensors (Lew et al., 2020). Plants are exclusively fit to execute these roles due to their ability to generate energy from sunlight through photosynthesis. In the field of nanobiotechnology, researchers target to develop bionic plants that display the highest possible photosynthesis efficiency and biochemical sensing (Butnariu and Butu, 2019).

14.2 PLANT NANOBIONICS

Nanotechnology has been enhanced to a greater degree, where nanoscale materials are designed for drugs delivery onto particular cells or tissues. Nanoparticles can expand the stability of the drug, the pharmacokinetics, and the creation of therapeutics into nanoparticles that can lessen toxicity (Ghorbanpour and Fahimirad, 2017). The nanoparticle development for drug delivery systems and production of these drugs on larger scale takes a lengthy time, signifying major obstacles to its success. Plant viruses are an evolving class of biologics used in drug delivery systems (Czapar and Steinmetz, 2017). The main improvement of using these plant sources to produce biologics is that the plant cell wall defends expressed PDs from acids and enzymes in the stomach via bioencapsulation. Though, gut microbes have progressed to break down every cell wall component; thus, when whole plant cells comprising PDs reach the gut, commensal microbes digest the cell walls and discharge the proteins. When they are fused to PDs, tags (receptor-binding proteins) proficiently cross the intestinal epithelium (the largest mucosal area in the body, measuring 1.8–2.7 m^2) and are carried to the circulatory system or the immune system (Kwon and Daniell, 2015). Plant cell walls are consisted of various polysaccharides. Thus, the cell wall acts as an obstacle for avoiding entry of foreign particles and nanoparticles into the cell, where they could cause damage to the internal cellular part. Layered double hydroxides (LDHs) are a novel class of nanoparticles and have been described to efficiently deliver biomolecules into plant cells. The main benefits of using LDH nanoparticles are that RNA activity is secure and their penetration through cell walls into intact cells suffers only minimal damage. In general, LDHs are usually represented as [M2+ 1−xM3+x(OH)2] [An−]x/n·zH2O, where M2+ and M3+ are divalent and trivalent metal ions, respectively, and a interlayer anion. The positively charged ion sheets are intercalated with charge-balancing anions to establish the LDH complex, which is tunable, and the ion sheets can be delaminated into ultrathin nanostructures. LDHs hold very promising properties such as good compatibility, simply biodegradable, low cytotoxicity, and precise release of drugs (Bao et al., 2017). ENMs have wonderful physiochemical skins appropriate for various applications in medicine, biotechnology, energy, pharmaceuticals, cosmetics, and electronics due to their ultrafine size and high surface reactivity. Engineered nanoparticles have made a solid impact over the previous few decades, thus they tend to possess properties such as a huge capability to absorb target molecules, the capacity to release molecules more efficiently within target cells or to precise organelles, and the capacity to penetrate into cell walls without any damage. ENMs play a part by arriving plant cells and interacting with metabolic pathways and intracellular structures, which may cause toxicity or promote plant growth and development by varied mechanisms.

14.2.1 APPLICATION OF PLANT NANOBIONICS

(A) *Crop production*

(1) Improve photosynthesis efficiency

In most varieties of plants, thylakoid membranes in chloroplasts are chief site of the photosynthetic machinery. Chloroplast has ability to absorb visible range of the light spectrum, which encompass of 50% of the incident solar energy radiation (SY et al., 2019). Besides, Plants normally use only around 10% of the sunlight accessible to them (Zhu et al., 2010). Hence, researchers have tried to improve photosynthetic efficiency by extending the range of solar light absorption (Blankenship et al., 2011). Nanomaterials using seamless chemical and physical traits in photocatalytic complexes based on chloroplast improved novel functional properties (Giraldo et al., 2014). Single-walled carbon nanotubes (SWNTs) have ability to capture visible and near-infrared spectra of light wavelengths, while absorption rate of chloroplast antenna pigments is limited in this case. Giraldo et al. (2014) effectively designed highly charged SWNTs coated with DNA and chitosan (a biomolecule derivative from shrimp and other crustacean shells) which have ability to spontaneously enter into chloroplasts. This new lipid exchange envelope penetration procedure for integrating the nanostructures comprises wrapping carbon nanotubes (CNTs) or nanoparticles with highly charged DNA or polymer molecules, permitting them to penetrate into the fatty, hydrophobic membranes that surround chloroplasts. SWNTs embedded within chloroplasts has the potential to improve the photosynthetic light reactions of with their typical optical properties. SWNTs are capable to capture visible and near-infrared spectra of light wavelengths though chloroplast antenna pigments absorption rates are inadequate in this case. SWNTs alter this absorbed solar energy into excitations, which transfer electrons to the photosynthetic machinery (Han et al., 2010). Integration of CNTs into chloroplasts extracted from plants improved chloroplast's photosynthetic activity by 49% as compared with the control. When these nanocomposites were incorporated into living leaf chloroplasts, the electron flow related with photosynthesis was enhanced by 30%. SWNT real-time sensing of NO in extracted chloroplasts and leaves could also be stretched to identify anextensive range of plant signaling molecules and exogenous compounds such as pesticides, herbicides, and environmental pollutants.

(2) Plant nanobionic with higher plant ROS-savaging ability

Remarkably, SWNT-based nanosensors have ability to display single-molecule free radical dynamics within the chloroplasts for optimizing photosynthetic environmental conditions (light and CO_2) (Zhang et al., 2010). The main limitation in the usage of extracted chloroplasts for solar energy applications is that they easily breakdown because of light and oxygen-induced damage to the photosynthetic proteins. Giraldo et al. (2014) demonstrated that cerium oxide nanoparticles (nanoceria) were united with a highly charged polymer (polyacrylic acid) pass through the outer membranes of the chloroplast and localize in the stroma, and remarkably inhibit damage to the photosystems (PS) by quenching reactive oxygen species that are extensively distributed throughout the chloroplast and permit real-time monitoring of free-radical species and environmental pollutants using in vivo and ex vivo embedded nanosensors (Siddiqui et al., 2015). Additionally, solar energy is caught by chlorophylls in the two types of pigment-protein complexes (PSs I and II, designated PSI and PSII, respectively) and is converted into electrochemical energy to yield ATP and NADPH that are used for CO_2 fixation. PSII accomplishes the light-induced oxygen evolution reaction and transfers electrons from water to plastoquinone in the membrane and PSI yields strong reducing power using electrons provided by PSII and decreases ferredoxin and NADPH. Noji et al. (2011) showed that nanomesoporous silica compound (SBA) conjugated with PSII retained the high and stable oxygen-evolving capacity of *Thermosynechococcus vulcanus* PSII even inside silica nanopores. The movement lasted for more than 3 h under the adequate illumination/dark cycles. Combination of PSII-SBA conjugates with the mediator recycling systems can eliminate the detrimental effects of electron acceptors and light-induced radicals and have properties to develop for photosensors and artificial photosynthetic system.

(B) *The aspect of plant nanobionic in environment*

(1) Nanobionic plants as nitroaromatics

It may appear exquisite but imitate from simple plant transpiration process. Plants pull up water and other analytics from the ground and can gather even trace levels of analytics within tissues. Knowing this rule, Wong et al. (2016) made a nanobionic plant that can notice explosives in groundwater and aware a user to their occurrence in the area. IR-fluorescent CNTs-based sensors that selectively answer to nitro aromatics that were inoculated into a spinach plant's leaves. These nanotubes quench in fluorescence intensity in the occurrence of nitro-aromatics. Then a reference sensor that is invariant in signal intensity was designed. The plant pulls the nitro-aromatics or the common explosives component picric acid (2,4,6-trinitrophenol) up via the roots into its leaves, where the repressed IR signal is imaged with a night-vision camera and directed to a smartphone through a Wi-Fi signal. With the reference sensor implanted in the leaf as well, the technique creates high-contrast images.

(2) Nanobionic plant as temperature detector

Cyberwood was deliberated by using a new synthetic CNT with mechanical and structural properties like wood embedded and intricately sensitive to temperature changes into a matrix of plant cells from the tobacco. Conserving plant cells' natural capacity to sense temperature differences even after their death cause electrical conductivity of this kind of CNTs change with temperature. The incidence of multiwalled carbon nanotubes (MWCNTs) deliberates structural stability and a great electrical conductivity, which can be exploited to attach the samples to an external circuit (Di Giacomo et al. 2015). In fact, pectin and charged atoms (ions) perform a crucial role in the temperature sensitivity of both living plant cells and the dry cyber wood. Pectinisa sugar molecules found in plant cell walls that can be cross-linked, dependent on temperature, to form a gel. Calcium and magnesium ions both exist in this gel. With rise in the temperature, the contacts of the pectin break apart, the gel becomes softer, and the ions can travel more freely. In the result, the material conducts electricity better with increase in temperature. For the synthesis of cyberwood, several conventional nano-synthesis methodologies were pooled. At first phase, undifferentiated tobacco BY-2 cells derived from the callus seedlings of *Nicotianatabacum* were cultured in a growth medium that contained MWCNTs. Spontaneous accumulation of cells was detected with tobacco cells combined with MWCNTs. A gel-like material formed after 24 h, was collected and dried at 47 °C for 15 days. The material designed had a complex, hierarchical structure, similar to wood. This construct may discover applications in thermal sensors, for example, for thermal cameras, or in distance sensors for consumer products and security systems. Because of its exquisite temperature sensitivity, the cyberwood sensor can recognize warm bodies even at distance; for example, a hand impending the sensor from a distance of a few dozen centimeters. The sensor's conductivity depends straight on the hand's distance from the sensor. The very high responsively to temperature changes of cyberwood proposes that it can be used as a temperature distance sensor. The distance of a warm body from the sensor can be inferred from temperature measurements achieved at constant environmental conditions (Di Giacomo et al., 2015).

14.3 CONCLUSIONS

Plants are the lifeguards on earth with huge influence in our life due to their valuable products. Now with the help of biotechnology, novel features have been introduced into plants as well as in animals. Nanobionics is a mixture of nanotechnology and biological processes. Nanoparticles are introduced into plant cells and organelles like chloroplasts, imparting alteration in plant trait. The major idea behind it is to empower plants with novel features in their vicinity. Agricultural production can be enhanced with nanoformulations such as agrochemicals in plants, nanosensor for plant protection, and nanodevices for genetic manipulations of plant. Thus it can help in better crop management.

ACKNOWLEDGMENTS

The research was financially supported by the Ministry of Science and Higher Education of the Russian Federation (no. FENW-2023-0008) and the Strategic Academic Leadership Program of the Southern Federal University ("Priority 2030")

KEYWORDS

- **agriculture**
- **biosensor**
- **defense**
- **nanobionic**
- **nanoparticles**

REFERENCES

Bao, W.; Wan, Y.; Baluška, F. Nanosheets for Delivery of Biomolecules Into Plant Cells. *Trends Plant Sci.* **2017,** *22*(6), 445–447. https://doi.org/10.1016/j.tplants.2017.03.014

Blankenship, R. E.; et al. Comparing Photosynthetic and Photovoltaic Efficiencies and Recognizing the Potential for Improvement. *Science* **2011,** *332,* 805–809.

Butnariu, M.; Butu, A. Plant Nanobionics: Application of Nanobiosensors in Plant Biology. Nanotechnol. Life Sci. **2019,** 337–376. https://doi.org/10.1007/978-3-030-16379-2_12

Czapar, A. E.; Steinmetz, N. F. Plant Viruses and Bacteriophages for Delivery in Medicine and Biotechnology. *Curr. Opin. Chem. Biol.* **2017,** *38,* 108–116. https://doi.org/10.1016/j.cbpa.2017.03.013

Di Giacomo, R.; Daraio, C.; Maresca.; B. Plant Nanobionic Materials With a Giant Temperature Response Mediated by Pectin-Ca2+. *Proc. Natl. Acad. Sci. U S A* **2015,** *112*(15), 4541–4545. DOI: 10.1073/pnas

Giraldo, J. P.; Landry, M. P.; Faltermeier, S. M.; McNicholas, T. P.; Iverson, N. M.; Boghossian, A. A.; Reuel, N. F.; Hilmer, A. J.; Sen, F.; Brew, J. A.; Strano, M. S. Plant Nanobionics Approach to Augment Photosynthesis and Biochemical Sensing. *Nat. Mater*. **2014**. DOI: 10.1038/nmat3890

Han, J. H.; et al. Excitation Antennas and Concentrators From Core-Shell and Corrugated Carbon Nanotube Filaments of Homogeneous Composition. *Nat. Mater*. **2010,** 9, 833–839.

Kwak, S. Y.; Lew, T. T. S.; Sweeney, C. J.; Koman, V. B.; Wong, M. H.; Bohmert-Tatarev, K.; Snell, K. D.; Seo, J. S.; Chua, N. H; Strano, M. S. Chloroplast-Selective Gene Delivery and Expression in Planta Using Chitosan-Complexed Single-Walled Carbon Nanotube Carriers. Nat. Nanotechnol. **2019,** *14*, 447–455. https://doi.org/10.1038/s41565-019-0375-4

Kwon, K. C.; Daniell, H. Low-Cost Oral Delivery of Protein Drugs Bioencapsulated in Plant Cells. *Plant Biotechnol. J.* 2015, *13*(8), 1017–1022. https://doi.org/10.1111/pbi.12462

Noji, T.; Kamidaki, C.; Kawakami, K.; Shen, J. R.; Kajino, T.; Fukushima, Y.; Sekitoh, T, Itoh, S. Photosynthetic Oxygen Evolution in Mesoporous Silica Material: Adsorption of Photosystem IIreaction Center Complex into 23 nm Nanopores in SBA. *Langmuir* **2011,** *27*(2), 705–713.

Siddiqui, M. H.; Al-Whaibi, M. H.; Mohammad, F. Nanotechnology and Plant Sciences: Nanoparticles and Their Impact on Plants. *Nanotechnol. Plant Sci. Nanopart. Impact Plants* **2015,** 1–303.

Wong, M. H.; Giraldo, J. P.; Kwak, S. Y.; Koman, V. B.; Sinclair, R.; Lew, T. T.; Bisker, G.; Liu, P.; Strano, M. S. Nitroaromatic Detection and Infrared Communication From Wild-Type Plants Using PlantNanobionics. *Nat. Mater.* **2016,** *16*(2), 264–272. doi:10.1038/nmat4771

Zhang, J.; Boghossian, A. A.; Barone, P. W.; Rwei, A.; Kim, J. H.; Lin, D.; Heller, D. A., et al. Single Molecule Detection of Nitric Oxide Enabled by d(AT)(15) DNA Adsorbed to Near Infrared Fluorescent Single-Walled Carbon Nanotubes. *J. Am. Chem. Soc.* **2010**, *20*, 567–581.

Zhu, X. G.; Long, S. P.; Ort, D. R. Improving Photosynthetic Efficiency for Greater Yield. *Annu. Rev. Plant Biol.* **2010,** *61*, 235–261

CHAPTER 15

Nanosensors and Nanobiosensors: Nanoparticles as Sensing Materials

SHIVANGI MISHRA, RAKHI SINGH, and SUKH VEER SINGH

Department of Food Science and Technology, National Institute of Food Technology Entrepreneurship and Management, Kundli, Sonepat, Haryana, India.

ABSTRACT

Food safety and quality is the foremost duty of food-business operator to make hazard-free human life. Therefore, it is a primary concern of human well-being. To make proper food safety and quality system, it is necessary to monitor every step of food processing from farm to fork, including food product storage. However, monitoring all steps is not an easy task. It necessitates developing rapid, sensitive, safe, and efficient methods to monitor quality changes in biological, chemical, and physical hazards to meet the demand for eco-friendly processing. Nanotechnology plays a tremendous role in developing such kinds of detectors, nanosensors and biosensors. nanosensors and biosensors can be defined as a type of analytical device biologically derived material that help in the detection of any undesirable hazard or chemical changes in the food system. It is a compassionate monitoring strategy to detect the shelf-life of food products and even slight changes in packaging material responsible for food safety and quality. Use of nanosensors are varied from product to product, but the overall role of these sensors is to ensure food quality. This chapter mainly focuses on different nanosensors, their roles, and their impact on food quality monitoring, such as pesticides, contaminants, and foodborne pathogens.

Nanotechnology for Sustainable Agriculture: An Innovative and Eco-Friendly Approach. Vishnu D. Rajput, Abhishek Singh, Tatiana M. Minkina, Krishan K. Verma, and Awani Kumar Singh (Eds.)

15.1 INTRODUCTION

Nowadays, nanosensors are a very emerging and promising tool for various agriculture and food production sectors (Rajput et al., 2021a). Nanosensors are advanced and give better results than traditional physicochemical, microbial, and biological analysis methods (Singh et al., 2020a; Singh et al., 2021a). We can utilize nanosensors in various agriculture and food processing fields to detect microbial infection and microbial toxins, pollutants, and toxicants. Nanosensors assume a vital part in checking the period of usability of the item (Joyner and kumar, 2015). In food manufacturing, nanosensors play a significant role in food packaging and food transport. Nanosensors can further improve food packaging through their chemical and electro-optical properties (Omanović-Mikličanin and Maksimović, 2016). Nanosensors are like any other sensors, yet their creation is at the nanoscale. In this way, nanosensors can be characterized as tiny/little gadgets that can undoubtedly tie to what is needed to be distinguished and convey back a message in the form of signals. However, all these nanosensors have an excellent capability to detect and give a response in the form of signals that can be used by humans (Lu and Bowle, 2013).

Impact of nanosensors on food production and agriculture-related field is increasing day by day because nanosensors provide a better and easy way to detect something like inactive packaging various sensing agents are used and they easily detect the problem and give the response in various ways like by changing the color of packaging material (Rajput et al., 2021b). It is also used in shelf-healing composite. Many techniques are used for the development of nanosensors:

1. Top-down lithography
2. Self-assembly of molecule
3. Assembly of bottom-up.

According to Omanović and Maksimović (2016), nanosensors gadget can be classified into the following five categories:

1. Nanostructured materials, for example, porous silicon
2. Sensors made with nanoparticles
3. Nano probes
4. Nanowire nanosensors
5. Nano system: cantilevers, nanoelectromechanical system

For food analysis, nanosensors can be divided into

1. Nanoparticle-based nanosensors
2. Electrochemical nanosensors
3. Optical nanosensors

There are various nanosensors used to monitor the food quality like moisture sensors for monitoring the moisture level, oxygen sensors for monitoring oxygen level, compact nano sensing devices to identify synthetic substances that are not good for consumption, microorganism, and poisonous matter in food. Nanosensors are helpful to improve the quality of packaging material because they have specific unique physical and chemical properties. The use of nanosensors in food packaging ensures that the food that comes in contact with consumers is safe and free from any foodborne diseases, in this way, it ensures food safety. Significant advantages of nanosensors are in analyzing and sensing real-time information regarding the product from production to the end consumers (Omanović and Maksimović, 2016).

15.2 CONCEPTUAL IDEA ABOUT NANOSENSORS

Nanosensor can be defined as a sensing device that has a sensing dimension not more than 100 nm. Nanosensors perform various roles and are used as instruments for: (1) identifying the biochemical level in cellular organelles; (2) measuring the nanoscopic material which has small particles in the food processing and other industry and ecosystem; (3) use as the sensors for packaging materials (Kissinger, 2005).

The nanosensors are grouped into the following four major categories:

1. Optical nonosensors
2. Mechanical and vibrational nanosensors
3. Electromagnetic nanosensors
4. Nanobiosensors

15.2.1 OPTICAL NANOSENSORS

Optical nanosensors that were invented for the first time were based on fluorescein, and this fluorescein caught inside a polyacrylamide nanoparticle, and these nanosensors were utilized for the pH estimations (Lee et al., 1996). The most basic concept about fluorescent chemosensors is that molecules of these chemosensors have at least one substrate binding unit and one photoactive component (de Silva et al., 1997). Molecule base color or dye probe inside a cell is the most basic optical nanosensors, which is essential for a direct cell loading of fluorescent dyes. Many other types of optical nanosensors are given in the following subsections.

15.2.1.1 FIBER OPTIC NANOSENSORS

These nanosensors have the remarkable capacity to measure crucial cellular processes in vivo. Basic conceptual understanding and monitoring of biological and cellular processes at the molecular cell level is essential to understand dynamic cell capacities further. When the target molecule A and receptor R interface, they produce a physicochemical perturbation that can change into electrical signals or other quantifiable signals. Then, they are picked by the optical probe and afterward sent into the information base (Song et al., 2004; Vo-Dinh and Cullum, 2000).

The fiber optical nanosensors have been successful in various applications namely:

1. In the measurement of benzopyrene tetrol and benzopyrene inside a single cell (Vo-Dinh and Cullum, 2000; Kasili et al., 2004)
2. For the monitoring of apoptosis (Kasili et al., 2004)
3. In the measurement of cytochrome (Song et al., 2004)
4. Optical nanosensors are helpful for the detection of cellular pH value as well as some ions such as K^+ and Ca^{2+}, NO, NO_2^-, Cl^-, Na^+ (Lim and Ramakrishna, 2006)

15.2.1.2 PHOTONIC EXPLORER FOR BIOANALYSIS WITH BIOLOGICALLY LOCALIZED EMBEDDING SENSORS (PEBBLES)

There are some problems associated with the free moving fluorescent dye procedure and optical fiber methods. The invention of photonic explorer was done to deal with these issues, which has the capabilities of bioanalysis with a biologically embedding system. A nanoscale sensor encapsulates analyze specific dye and a reference dye inside a bio-based inert matrix. The short probe is less disruptive to the cellular environment and the inert matrix to protect the sensing phase from the cell environment, and this system prevents chemical interference (Lim and Ramakrishna, 2006). PEBBLEs are classified into the following four categories:

1. Polyacrylamide
2. Polydecylmethacrylate

3. Sol–gel
4. Organically modified silicates

15.2.2 ELECTROMAGNETIC NANOSENSORS

Base on their physical mechanism, electromagnetic nanosensors are divided into two broad categories.

1. Measurement system based on electrical current measurement.
2. Analysis by magnetic measuring system.

The electrical current measuring system is basically for two cases: analysis by current reduction and analysis by current enhancement. In the category of current reduction, Geng et al. (2005) studied the hydrogen sulfide and gold nanoparticles and interaction of these two components, and he found that when the adsorption of hydrogen sulfide molecule occurs on the nanoparticles. It changes the hopping behavior of electrons by its particles and this is called suppressed hopping phenomenon and these electrons are measured by recording the current across chromium and gold electrodes. When no exposure of hydrogen sulfide increases in current with applied voltage, but when exposure with hydrogen sulfide loss is observed. In the category of current advancement as nanosensors, carbon nanotubes are used, and these tubes are helpful in the monitoring of oxides such as glucose; apart from glucose, lactate oxidase is also detected by these tubes, and it also plays a crucial role in the detection of enzymes like dehydrogenase, peroxidase hydrogen peroxides, etc.

15.2.3 MECHANICAL NANOSENSORS

The first mechanical nanosensors (Binh et al., 1994) were proposed to measure the vibrational and elastic characteristics of a nanosphere attached to a tapered cantilever. This work is vital for application in nanodevice components and nanoscale subassemblies in microelectronic devices (Lim and Ramakrishna, 2006).

15.2.4 BIOSENSORS

This is also a type of sensing device or a measurement system designed specifically for the estimation of material by using the biological interaction then convert these accessing into the readable form with the help of electromechanical transduction interpretation. The prominent functional role of biosensors is to sense a biologically specific material. It can be antibodies, amino acids, enzymes, immunological molecules, and so on. In the making of nanobiosensors, various nanomaterials are used, which improve the quality of biosensors (Kissinger, 2005). The transduction mechanism is a crucial component of biosensors; this mechanism is responsible for converting responses of bioanalytical interaction in an identifiable and reproducible manner (Malik et al., 2013). Some nanomaterials are listed in Table 15.1, which are used for preparing biosensors.

TABLE 15.1 Nanomaterials Used for the Designing of Biosensors and Their Key Benefits.

Nanomaterials used in biosensors	Benefits of nanomaterials	References
Carbon nanotubes	Improved enzyme loading, provide better electrical communication	(Malik et al., 2013)
Nanoparticles	It helps in possessing good catalytic properties, provides better loading of analyte	(Malik et al., 2013)
Quantum dots	Brilliant fluoresce, quantum restriction of charge carriers,	(Malik et al. 2013)
Nanowires	Have outstanding electrical and detecting properties for bio and chemical substance	(Malik et al., 2013)

A broad range of nanosensors are available, and their application and working have been discussed by various scientists. Some of the nanosensors are helpful in the detection of chemicals which is present inside a single cell, and some of them are useful in analysis of physicochemical properties of the product. In a short duration of their development, achievements of nanosensors are remarkable and in upcoming days it is going to be more advance and compatible (Lim and Ramakrishna, 2006). Malik et al. (2013) reported that nanobiosensors can be classified into three major categories based on their function and materials to prepare sensors.

15.2.4.1 NANOPARTICLE-BASED SENSORS

It is also further classified in a different category. Acoustic-wave biosensors are created to intensify the sensing reactions to exact the restriction limit of bio-detection. In these sensors, various stimulus-based effects are present (Su et al., 2000; Liu et al., 2004).

15.2.4.1.1 Magnetic Biosensors

These kinds of sensors utilize specially designed nanoparticles. The ferrite-based material may be used as single or in combination—the prominent use of these sensors in the biomedical fields. The material with magnetic properties empowers much variety for the few logical applications that the constituent includes in screening have an iron with other charge material, and all have the various properties. Furthermore, with the consolidation of attractive nanoparticles, bio-detection have further become more sensitive and powerful (Richardson et al., 2001).

15.2.4.1.2 Electrochemical Biosensors

Electrochemical biosensors are used to investigate chemical and biological reactions with enhanced electrical means. Moreover, these types of instruments are made up of metallic nanoparticles. Metallic nanoparticles help in a chemical reaction between biomolecules which help in accomplished to immobilize one of the reactants (Cai et al., 2001).

15.2.4.2 NANOTUBE-BASED SENSORS

In the realm of material science and optoelectronic application nowadays, carbon nanotubes are perhaps the most mainstream nanomaterials. These nanotubes have electronic conductivity, adoptable actual calculation highlights, and ever dynamic physical–mechanical properties. Moreover, due to these properties, both single-walled and multiwall nanotubes are utilized to design biosensors that provide better performance (Davis et al., 2003; Sotiropoulou et al., 2003).

15.2.4.3 NANOWIRE-BASED NANOSENSORS

Nanowires are the arrangement of one-dimensional nanostructure which have good electron transport properties. Usually, nanowire-based sensors are few. However, some research and literature report some great utilization of nanowires altogether which enhance the performance of these sensors for the detection of biological materials. Cullum et al. (2000) reported that gold-electrode-coated ZnO nanotubes are utilizing for the detection of hydrazine by using amperometric response. Nanowires are versatile and much better than nanotubes. Nanowires provide various modifications in their working and arrangement profile by optimized procedural parameters during synthesis. Some scholars reported the striking highlights of nanowires and clarified their utilization in excellent conduction and detection of biological stimulus (Malik et al., 2013). By exploring various literatures, we can say that nanomaterials are profoundly prosperous for lighting up the sensing and detecting innovation and have worked on the diagnostic, analytical, and detection procedures in a considerable amount at a time. Some nano-based materials like quantum dots have been added as tags with the dyes, and these kinds of sensors yielded thermochromic, photochromic, and electrochromic material shows extreme sensitive detection that can be easily monitored. In this way, various nanosensors positively impact food quality because due to these easy and fast monitoring patterns, food quality analysis becomes precise, and convenient day by day.

15.3 APPLICATIONS OF NANOSENSORS IN THE FOOD INDUSTRY AND FOOD QUALITY MONITORING

Nowadays, various nanosensors are developed for use in different applications in the agriculture and food industry. For example, for the detection of growth of microorganism, nanosensors used in food packaging other than this nanosensors use in supply-line process control and for keeping an eye on the storage environment which is very helpful for fend off food poisoning (Augustin and Sanguansri, 2009). For example, a team of scientists developed a nanoparticle made of gold and coated these nanoparticles with that molecule, which can bind to substances like pesticides. By using these nanoparticles, farmers can detect the chemical like pesticides (Debnath and Das, 2013).

One more literature reported that nanosensors are ideally suited for food forensic like investigation of food origin, adulteration, and contamination. Nanosensors are very convenient in detecting and monitoring instantaneous information related to the product. Nanosensors can get information quickly and analyze the data and generate the result in a concise duration. Because of this, nanosensors are very helpful in monitoring the critical control point in the food supply chain from the point where food is processed or manufactured until is consumed. Nanosensors can provide quality assurance as they help track microbes, toxins, pathogens, pesticides, metals throughout the food processing chain by using data for automated control function and documentation (Prasad, 2014). List of commercially available intelligent packaging is given in Table 15.2. A redox dye has been developed for in-package oxygen detection, based on nanosized TiO_2 or SuO_2 particles (Mills, 2005). Nanoparticle-based nanosensors have also been developed to detect the moisture content within the food (Luechinger et al., 2007).

TABLE 15.2 Commercially Available Intelligent Packaging and Their Application.

Application	Trade name	Company name
Time and temperature indicators	Cook-check	Pymah corp.
	Time strip	Time strip plc
	Thermax	Thermographic measurement limited
		Vista
	Checkpoint	
Integrity indicators	Novas	Insignia Technologies Ltd.
	Time strip	Time strip Ltd.
	Ageless eye	Mistubishi Gas Chemical Inc.
Radiofrequency identification	Tempt rip	Tempt rip LLC.

Source: Fuertes et al. (2016)

In recent years, nanotechnology market proliferates; according to a survey study, the leading manufacturer of nanofood (in which nanotechnology is used) is in the United States, followed by Japan and China. Around 400 plus companies are currently producing the nanofood. Nonchocolate are very much famous nowadays in delivering the nutrient to cell without affecting the color and taste of food products (Kalita and Baruah, 2019).

Various nanosensors based on nanoparticles are developed to analyze the moisture content inside a food packaging (Luechinger et al., 2007). There is a vast application of nanosensors in food and food-related industries. Nanosensors are used as a flavor enhancer, which is nano encapsulated with some flavoring compound and delivers the flavor in a particular food. Some nanotubes and nanoparticles act as a scoffing agent. Figure 15.1 shows the various applications of nanosensors in various fields of food processing. The development of DNA biochips helps to detect pathogens and toxicants caused by bacteria, yeast, and molds. Some nanosensors act as electronic tongues as they detect different flavors when incorporated into food material. It also helps in the detection of some gases like ethylene. In simple words, we can say that nanosensors play a significant role in every field of food processing and agriculture processing (Omanović-Mikličanin and Maksimović, 2016).

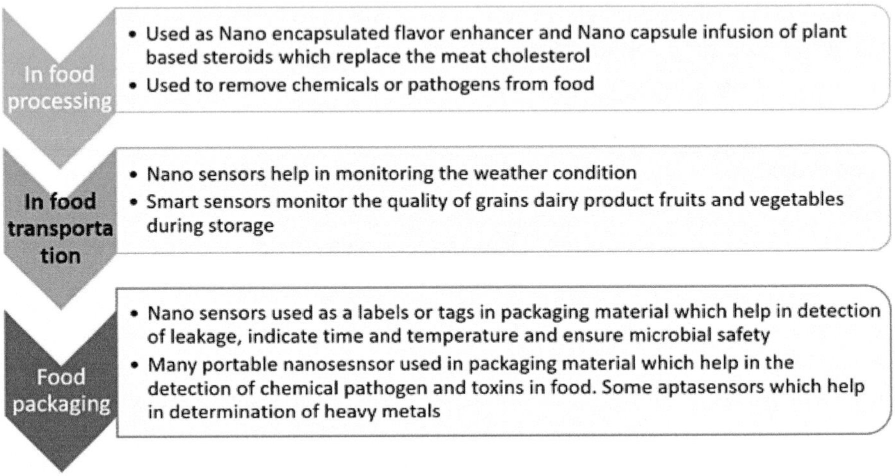

FIGURE 15.1 Application of nanosensors in various fields of food technology.

15.4 ROLE OF NANOSENSORS IN FOOD QUALITY

The role of nanosensors in food quality is a multifactorial and multilevel concern because nanosensors play a crucial role in food quality monitoring, and due to the development of nanosensors, many of the critical and arduous tasks and monitoring procedures related to food quality become easy. Therefore, crucial role of the nanosensors in the food quality are given below.

15.4.1 NANOSENSORS IN THE DEVELOPMENT OF INTELLIGENT FOOD PACKAGING

The significant impact of nanosensors in food quality monitoring is basically on the development of intelligent packaging material. We all know that safe and wholesome food products are one of the main objectives of food law. Packaging materials play a critical role in food safety because packages provide protection, better conversation mediums with consumers to convey food-related information, ergonomics, and marketing. The primary function of packages is containment means ensure the right amount of product to delay or keep away from any spill or leakage. Moreover, it is observed that the physical properties of packages and improvement in the safety aspect of food packaging are the expectation that affects consumer acceptability and sale of products. In this context, intelligent packaging is a better option. Intelligent packaging is any food container or packaging box that provides a particular property beyond the essential functions of packaging material. Intelligent packaging is a packaging technology that monitors the interaction of food, packaging material, and the environment through internal and external indicators, these packaging materials analyze the system, collect information, and present it without affecting food quality.

Nowadays, when consumers are very conscious about what they eat, and due to this, they are very curious about the ingredient of product and how the product store, use and discarded after use, intelligent packaging technology provides packaging material with intelligent tags and stickers, which can have a conversation with the consumer at first hand with the help of packing film instruments that provide visual information (Singh et al., 2020b, 2021b). The primary application of these packaging materials is time and temperature, integrity, freshness, and radio-frequency identification of food products. This effective use of nanosensors for time and temperature indicators is due to their intelligibility, economical, affordable, and efficient work system. This indicator is majorly utilizing to observe and interpret to the consumer about standard characteristics of food products. In the form of tiny chips or stickers, time, and temperature indicator (TTI) is attached with the containers of food product. TTI gives a particular indication by chemical changes when the food is exposed to some different environment which is not suitable for their quality. Time and temperature indicators are specifically essential for monitoring chilled or frozen food's quality characteristics and safety.

After TTI, most used sensors are freshness indicator and integrity indicator, freshness and novelty indicator directly indicate the characteristics of a food product either it is edible or not. It is also present in the form of labels on the packaging material. A significant role of the freshness indicator is to monitor the very first kind of change and detect these changes. Usually, a color reaction is generated that can be easily analyzed and matched up with the freshness of food. So, biosensors are used in this type of indicator. Radiofrequency identification detector (RFID) tags are more advanced kind of indicators that support specific statistical information that can find any food product with specific marked tag/labels that emit radio waves. These kinds of tags categorizes into four classes:

1. Active
2. Passive
3. Semiactive
4. Semipassive

Depending on their function these tags are attached in the same way as the other tags are attached with packaging material, but these tags can be easily accessible from several meters away beyond the limit of sight. Nevertheless, RFID is very much expensive than other indicators that are significantly less expensive. These current advancements of nanosensors have high potential benefits for the food industry, and intelligent and active packaging systems focus on food security and safety, and the demand for this technology increasing day by day the eventual fate of food security is majorly dependent on the innovation head way of nanosensors. These new bundling frameworks can aid in the location tracking, and convey throughout the store duration (Fuertes et al., 2016).

15.4.2 NANOSENSORS IN FOOD PROCESSING AND SAFETY

During food processing, nanoparticle nanomaterials are added to food to improve its nutritional quality. Sometimes, these particles give some specific taste, color, and texture to food and increase the product's shelf-life. Recent advancements in the field of nanotechnology help to develop nanosensors for the detection of foodborne pathogens or toxins (Rashidi and Khosravi-Darani, 2011). For example, immunosensing of *Staphylococcus* sp. Enterotoxin B using polydimethylsiloxane chips with reinforced fluid bilayer membrane and specific antibodies to the toxins (Dong et al., 2006). Development of G-liposomal nanovesicles-based immune magnetic bead sandwich assay for detecting *Escherichia coli* O157:H7, salmonella sp., and listeria monocytes in food. (Rashidi and Khosravi-Darani, 2011). In the agriculture fields, nanoparticles-based innovative nanosensors are used for rapid warning of changing conditions that can respond to different conditions. Several other bio-based nanosensors are developed to detect the most common food pathogens and mycotoxin in food (Durán and Marcato, 2013). Nanobiosensors are utilized as taking care of supplement media and substrate combination into the bioreactors. Thus, much large-scale commercial preparation and separation can be ameliorated with these sensors. In addition, nanobiosensors are used for filtration purposes at the industrial level. According to the working condition and composition, various nanobiosensors are available as thermal biosensors that are used to analyze samples based on the heat over time (Mello and kubota, 2002; Fuertes et al., 2016).

Nanobiosensors play a significant role in detecting foodborne diseases whereas traditional methods like ELISA and HPLC are expensive and time-consuming. However, nanobiosensors may not just assist in pathogenic food identification in a limited time, and it is effectively versatile and gives a speedy, explicit, and designated approach. Biosensors play a significant role in detecting foodborne toxins and help detect many pesticides that come in food from agriculture form, and these pesticides affect the quality of food and affect the quality of water and the environment. Acetylcholinesterase-based electrochemical and optical biosensors alongside nanomaterials like carbon nanotube, quantum dab, graphene, and nanocomposite are utilized to identify various pesticides like carbonyl and monocrotophos, DDVP (insecticide), and phoxim in the test sample within 10 min (Chauhan et al., 2019).

A wide range of toxins present in food and the environment and their detection on time is a big challenge, but these problems are somewhat overcome by using the third generation of nanobiosensors. For example, graphene-based fluorescence resonance energy transfer biosensors were developed similarly to other graphene-based immunosensors to detect fumonisin B_1, aflatoxin, diarrheic shellfish poisoning toxin, and sax toxin in food samples (Ferreira et al., 2018). Various other studies also concentrated on polyphenol measurement in food samples. Phospholipid, phthalocyanine, and AgNP-based biosensor and optic laccase biosensor were mapped out to quantify the level catechol in coffee samples (Aguila et al., 2015). Alessio et al. (2016) and Bilir et al. (2016) designed nitrogen-doped MWCNT (Lac/CNx-MWCNT) and on graphene oxide (Lac/GO)-based biosensor on sensing both important compounds one is catechol and the second one is catechin in wine sample which have a detection limit of 10−8 mol/L. An amperometric biosensor was also developed to analyze polyphenol in various food products such as tea leaves, alcoholic beverages, and water (Narang et al., 2013). In this way, food quality and safety are majorly affected by the nanosensors. In future, the use of these nanosensors increases with some advancement in the technology of nanosensors. Quality monitoring and analysis of food quality become straightforward with the nanosensors.

15.4.3 NANOSENSORS IN FOOD STORAGE

Various nanosensors are developed, which help to detect bacteria and several other contaminants, such as *Salmonella* sp., in food packaging plants. These sensors are beneficial in regulating the environment of foodstuff. In addition, this system allows monitoring to the self-system and reduce the cost of sending the sample to the laboratory for testing. More than 80% of the market with bakery and meat products utilizes nanosensors-based packaging material correctly monitor the storage and quality of products. In this field, massive demand of nanosensor is found as oxygen scavenger, moisture observer, and barrier packing material. Some examples are gold nanoparticle incorporated enzymes used for microbial detection and gas sensing related to the environmental condition in which food is placed. ZnO and TiO_2 nanocomposite are used for the volatile organic compound and nano barcodes are widely used for the tagging and security. In this way, the use of nanosensors or intelligent sensors for the storage, packaging, and different parts of the food processing system is beneficial for the consumer by which they know very well about the quality of food. Moreover, this technology is beneficial for the producer itself because they can rapidly distribute the authentic product (Kalita and Baruah, 2019).

15.5 CONCLUSIONS

The nanosensors and nanotechnology give many benefits and positively affect the quality of food. The application of nanosensors and nanotechnology increase in a swift manner. Due to its fast and accurate result, it is beneficial for the food producer. It reduces wastage and the extra cost for keeping and storage facility. It helps in enhancing the production, increase the shelf-life, and monitor food quality throughout storage without extra effort. In the future, the use of nanosensors would be increased for monitoring and analysis of food quality which would be straightforward. Consequently, the use of nanosensors or intelligent sensors in the storage, packaging, and different part of the food processing system is beneficial for the consumers by which they know very well about the quality of food.

ACKNOWLEDGMENTS

This book chapter was supported by the National Institute of Food Technology Entrepreneurship and Management (NIFTEM), set up by Ministry of Food Processing Industries (MOFPI), Government of India, Kundli, Sonepat district, Haryana, under Delhi NCR.

AUTHOR DISCLOSURES:

The authors declare no conflict of interest, financial, or otherwise.

KEYWORDS

- **biosensors**
- **monitoring**
- **nanosensors**
- **quality**
- **safety**

REFERENCES

Aguila, S. A.; Shimomoto, D.; Ipinza, F.; Bedolla-Valdez, Z. I.; Romo-Herrera, J.; Contreras, O. E.; Alonso-Núñez, G. A Biosensor Based on Coriolopsis Gallica Laccase Immobilized on Nitrogen-Doped Multiwalled Carbon Nanotubes and Graphene Oxide for Polyphenol Detection. *Sci. Technol. Adv. Mater*. **2015.**

Alessio, P.; Martin, C. S.; De Saja, J. A.; Rodriguez-Mendez, M. L. Mimetic Biosensors Composed by Layer-by-Layer Films of Phospholipid, Phthalocyanine and Silver Nanoparticles to Polyphenol Detection. *Sens. Actuators B Chem*. **2016,** *233*, 654–666.

Augustin, M. A.; Sanguansri, P. Nanostructured Materials in the Food Industry. *Adv. Food Nutr. Res*. **2009,** *58*, 183–213.

Bilir, K.; Weil, M. T.; Lochead, J.; KÖK, F. N.; Werner, T. Construction of an Oxygen Detection-Based Optic Laccase Biosensor for Polyphenolic Compound Detection. *Turk. J. Biol*. **2016,** *40*(6), 1303–1310.

Binh, V. T; Garcia, N.; Levanuyk, A. L. A Mechanical nanosensor in the Gigahertz Range: Where Mechanics Meets Electronics. *Surf. Sci*. **1994,** *301*(1--3), L224–L228.

Cai, H.; Xu, C.; He, P.; Fang, Y. Colloid Au-enhanced DNA Immobilization for the Electrochemical Detection of Sequence-Specific DNA. *J. Electroanal. Chem*. **2001,** *510*(1–2), 78--85.

Chauhan, N.; Jain, U.; Soni, S. Sensors for Food Quality Monitoring. In *Nanoscience for Sustainable Agriculture*; Springer: Cham, **2019;** pp. 601–626.

Cullum, B. M.; Griffin, G. D.; Miller, G. H.; Vo-Dinh, T. Intracellular Measurements in Mammary Carcinoma Cells Using Fiber-Optic nanosensors. *Anal. Biochem*. **2000,** *277*(1), 25–32.

Davis, J. J.; Coleman, K. S.; Azamian, B. R.; Bagshaw, C. B.; Green, M. L. Chemical and Biochemical Sensing With Modified Single Walled Carbon Nanotubes. *Chem. A Euro. J*. **2003,** *9*(16), 3732–3739.

De Silva, A. P.; Gunaratne, H. N.; Gunnlaugsson, T.; Huxley, A. J.; McCoy, C. P.; Rademacher, J. T.; Rice, T. E. Signaling Recognition Events With Fluorescent Sensors and Switches. *Chem. Rev.* **1997,** *97*(5), 1515–1566.

Debnath, N.; Das, S. Nanobiosensor: Current Trends and Applications. In *NanoBioMedicine*; Springer: Singapore, **2020;** pp 389--409.

Dong, Y.; Phillips, K. S.; Cheng, Q. Immunosensing of Staphylococcus Enterotoxin B (SEB) in Milk With PDMS Microfluidic Systems Using Reinforced Supported Bilayer Membranes (r-SBMs). *Lab Chip* **2006,** *6*(5), 675–681.

Durán, N.; Marcato, P. D. Nanobiotechnology Perspectives. Role of Nanotechnology in the Food Industry: A Review. John Wiley & Sons, Ltd; *Int. J. Food Sci. Technol.* **2013,** 48(6), 1127–34. DOI: 10.1111/IJFS.12027.

Ferreira, N. S.; Cruz, M. G.; Gomes, M. T. S.; Rudnitskaya, A. Potentiometric Chemical Sensors for the Detection of Paralytic Shellfish Toxins. *Talanta* **2018**, *181*, 380–384.

Fuertes, G.; Soto, I.; Carrasco, R.; Vargas, M.; Sabattin, J.; Lagos, C. Intelligent Packaging Systems: Sensors and nanosensors to Monitor Food Quality and Safety. *J. Sens.* **2016.**

Geng, J.; Thomas, M. D.; Shephard, D. S.; Johnson, B. F. Suppressed Electron Hopping in a Au Nanoparticle/H 2 S System: Development Towards a H 2 S nanosensor. *Chem. Commun.* **2005**, (14), 1895–1897.

Joyner, J. J.; Kumar, D. V. nanosensors and Their Applications in Food Analysis: A Review. *Int. J. Sci. Technol.* **2015**, *3*(4), 80.

Kalita, D.; Baruah, S. The Impact of Nanotechnology on Food. In *Nanomaterials Applications for Environmental Matrices*; Elsevier, **2019;** pp 369–379.

Kasili, P. M.; Song, J. M.; Vo-Dinh, T. Optical Sensor for the Detection of Caspase-9 Activity in a Single Cell. *J. Am. Chem. Soc.* **2004**, *126*(9), 2799–2806.

Kissinger, P. T. Biosensors—a perspective. *Biosens. Bioelectron.* **2005,** *20*(12), 2512–2516.

Lee, C.; Itoh, T.; Sasaki, G.; Suga, T. Sol-gel Derived PZT Force Sensor for Scanning Force Microscopy. *Mater. Chem. Phys.* **1996,** *44*(1), 25–29.

Lim, T. C.; Ramakrishna, S. A Conceptual Review of nanosensors. *Zeitschrift für Naturforschung A*. **2006,** *61*(7–8), 402–412.

Liu, T.; Tang, J. A.; Jiang, L. The Enhancement Effect of Gold Nanoparticles as a Surface Modifier on DNA Sensor Sensitivity. *Biochem. Biophys. Res. Commun.* **2004,** *313*(1), 3–7.

Lu, J.; Bowles, M. How Will Nanotechnology Affect Agricultural Supply Chains? *Int. Food Agribus. Manag. Rev.* **2013,** *16*(1030-2016-82815), 21–42.

Luechinger, N. A.; Loher, S.; Athanassiou, E. K.; Grass, R. N.; Stark, W. J. Highly Sensitive Optical Detection of Humidity on Polymer/Metal Nanoparticle Hybrid Films. *Langmuir* **2007,** *23*(6), 3473–3477.

Malik, P.; Katyal, V.; Malik, V.; Asatkar, A.; Inwati, G.; Mukherjee, T. K. Nanobiosensors: Concepts and Variations. *Int. Sch. Res. Notices* **2013**.

Mello, L. D.; Kubota, L. T. Review of the Use of Biosensors as Analytical Tools in the Food and Drink Industries. *Food Chem.* **2002,** *77*(2), 237–256.

Mills, A. Oxygen Indicators and Intelligent Inks for Packaging Food. *Chem. Soc. Rev.* **2005,** *34*(12), 1003–1011.

Narang, J.; Chauhan, N.; Rani, P.; Pundir, C. S. Construction of an Amperometric TG Biosensor Based on AuPPy Nanocomposite and Poly (indole-5-carboxylic acid) Modified Au Electrode. *Bioproc. Biosyst. Eng.* **2013**, *36*(4), 425–432.

Omanović-Mikličanina, E.; Maksimović, M. nanosensors Applications in Agriculture and Food Industry. *Bull. Chem. Technol. Bosnia. Herzegovina*. **2016**, *47*, 59–70.

Prasad, S. Nanobiosensors: Ihe Future for Diagnosis of Disease? *Nanobiosens. Dis. Diagn.* **2014**, *3*, 1–10.

Rajput, V.; Singh, A.; Minkina, T.; Shende, S.; Kumar, P.; Verma, K.; Bauer, T.; Gorobtsova, O.; Deneva, S.; Sindireva, A. Potential Applications of Nanobiotechnology in Plant Nutrition and Protection for Sustainable Agriculture. *Nanotechnol. Plant Growth Promot. Protect* **2021a**, 79–92

Rajput, V.; Minkina, T.; Feizi, M.; Kumari, A.; Khan, M.; Mandzhieva, S.; Sushkova, S.; El-Ramady, H.; Verma, K.; Singh, A.; Hullebusch, E.; Singh, R.; Jatav, H.; Choudhary, R. Effects of Silicon and Silicon-Based Nanoparticles on Rhizosphere Microbiome, Plant Stress and Growth. *Biology* **2021b**, *10*(8), 7–9.

Rashidi, L.; Khosravi-Darani, K. The Applications of Nanotechnology in Food Industry. *Crit. Rev. Food Sci. Nutr*. **2011,** *51*(8), 723–730.

Richardson, J.; Hawkins, P.; Luxton, R. The Use of Coated Paramagnetic Particles as a Physical Label in a Magneto-Immunoassay. *Biosens. Bioelectron*. **2001,** *16*(9–12), 989–993.

Singh, A.; Rajput, V. D.; Mehrotra, R.; Pal, N.; Singh, V. K.; Minkina, T.; Chokheli, V. A.; Singh, R. K. In *Sustainable Soil Fertility Management*; Nova. Sci. Publishers, Inc., **2020a;** vol 1, pp 73–100.

Singh, A.; Rajput, V.; Rawat, S.; Kumar Singh, A.; Bind, A.; Kumar Singh, A.; Chernikova, N.; Voloshina, M., Lobzenko, I. Monitoring Soil Salinity and Recent Advances in Mechanism of Salinity Tolerance in Plants. *Biogeosyst. Techniq*. **2020b**, *7*(2). https://doi.org/10.13187/bgt.2020.2.66

Singh, A.; Rajput, V. D., Rawat, S.; Sharma, R.; Singh, A. K.; Singh, A. K.; Tomar, R. S. In: *Emerging Tools for SustainableAgriculture and Food Security*; Rajput, D.; Book Agency: New Delhi, Delhi, **2021a;** vol 1, pp 1–15.

Singh, A.; Rajput, V.; Singh, A.; Sengar, R.; Singh, R.; Minkina, T. Transformation Techniques and Their Role in Crop Improvements: A Global Scenario of GM Crops. *Policy Issues Genetically Modified Crops* **2021b**, *1,* 515–542.

Song, J. M.; Kasili, P. M.; Griffin, G. D.; Vo-Dinh, T. Detection of Cytochrome C in a Single Cell Using an Optical Nanobiosensor. *Anal. Chem*. **2004,** *76*(9), 2591--2594.

Sotiropoulou, S.; Gavalas, V.; Vamvakaki, V.; Chaniotakis, N. A. Novel Carbon Materials in Biosensor Systems. *Biosens. Bioelectron*. **2003,** *18*(2--3), 211--215.

Su, X.; Chew, F. T.; Li, S. F. Design and Application of Piezoelectric Quartz Crystal-Based Immunoassay. *Anal. Sci*. **2000,** *16*(2), 107–114.

Vo-Dinh, T.; Cullum, B. Biosensors and Biochips: Advances in Biological and Medical Diagnostics. *Fresenius' J Anal.* **2000**.

Index

O

P

T

W

Z